Low Power Memory/Memristor Devices and Systems

Low Power Memory/Memristor Devices and Systems

Editors

Alex Serb
Adnan Mehonic

MDPI • Basel • Beijing • Wuhan • Barcelona • Belgrade • Manchester • Tokyo • Cluj • Tianjin

Editors

Alex Serb
University of Southampton
UK

Adnan Mehonic
University College London
UK

Editorial Office
MDPI
St. Alban-Anlage 66
4052 Basel, Switzerland

This is a reprint of articles from the Special Issue published online in the open access journal *Journal of Low Power Electronics and Applications* (ISSN 2079-9268) (available at: https://www.mdpi.com/journal/jlpea/special_issues/Low_Power_Memory).

For citation purposes, cite each article independently as indicated on the article page online and as indicated below:

LastName, A.A.; LastName, B.B.; LastName, C.C. Article Title. *Journal Name* **Year**, *Volume Number*, Page Range.

ISBN 978-3-0365-6185-1 (Hbk)
ISBN 978-3-0365-6186-8 (PDF)

Cover image courtesy of Themis Prodromakis.

Contents

Preface to "Low Power Memory/Memristor Devices and Systems"

Memristive technologies are slowly but steadily progressing towards maturity, driven by their promise of cheap and extremely scalable memory. Indeed, over the years, a large body of research has carefully investigated how the presence of memristors can lead to improvements beyond the Pareto surface normally achievable using standard CMOS technology. In this booklet, we have collected research that illustrates both the multifaceted nature of memristors and the extremely diverse areas that are being reshaped by these emerging devices. We begin with a couple of perspectives on how in-memory computing, a powerful design paradigm, is quickly becoming a competitive approach as a result of the high-performance memory achievable only via memristive devices. We then exhibit a series of articles spanning from how memristor manufacturing technology is overcoming technological challenges, the construction of memristor-based neural networks, and memristor-enhanced logic circuits, to random number generators relying on the inherent stochasticity that some flavors of memristors exhibit. Simultaneously, a selection of articles illustrating work in more conventional approaches (which may be seen as the main competitors of memristor-based design) are also included. For example, an article on how floating-gate-based analogue memories can be programmed exemplifies the challenges and solutions found as the field of floating-gate electronics progresses. Finally, a tutorial paper showing how a non-trivial mathematical problem can be resolved using memristive circuits closes the booklet. We hope that this collection provides readers with an interesting overview of this field, some understanding of its place within the wider context of electronics, and some inspiration on how it may be developed further in the most diverse of application areas.

Alex Serb and Adnan Mehonic

Editors

Journal of
Low Power Electronics and Applications

MDPI

Article

A Novel Inductorless Design Technique for Linear Equalization in Optical Receivers

Diaaeldin Abdelrahman [1,2,*], Christopher Williams [3], Odile Liboiron-Ladouceur [4] and Glenn E. R. Cowan [2]

1 Electrical Engineering Department, Faculty of Engineering, Assiut University, Assiut 71543, Egypt
2 Department of Electrical and Computer Engineering, Concordia University, Montreal, QC H3G 1M8, Canada; gcowan@ece.concordia.ca
3 Nokia, New York, NY 10016, USA; cwilliams.eng@gmail.com
4 Department of Electrical and Computer Engineering, McGill University, Montreal, QC H3A 0G4, Canada; odile.liboiron-ladouceur@mcgill.ca
* Correspondence: diaaeldin@aun.edu.eg

Abstract: To mitigate the trade-off between gain and bandwidth of CMOS multistage amplifiers, a receiver front-end (FE) that employs a high-gain narrowband transimpedance amplifier (TIA) followed by an equalizing main amplifier (EMA) is proposed. The EMA provides a high-frequency peaking to extend the FE's bandwidth from 25% to 60% of the targeted data rate (f_{bit}). The peaking is realized by adding a pole in the feedback paths of an active feedback-based wideband amplifier. By embedding the peaking in the main amplifier (MA), the front-end meets the sensitivity and gain of conventional equalizer-based receivers with better energy efficiency by eliminating the equalizer stages. Simulated in TSMC 65 nm CMOS technology, the proposed front-end achieves 7.4 dB and 6 dB higher gain at 10 Gb/s and 20 Gb/s, respectively, compared to a conventional front-end that is designed for equal bandwidth and dissipates the same power. The higher gain demonstrates the capability of the proposed technique in breaking the gain-bandwidth trade-off. The higher gain also reduces the power penalty incurred by the decision circuit and improves the sensitivity by 1.5 dB and 2.24 dB at 10 Gb/s and 20 Gb/s, respectively. Simulations also confirm that the proposed FE exhibits a robust performance against process and temperature variations and can support large input currents.

Keywords: low-bandwidth TIA; equalizer; multi-stage main amplifier; amplitude response; group delay variation

Citation: Abdelrahman, D.; Williams, C.; Liboiron-Ladouceur, O.; Cowan, G.E.R. A Novel Inductorless Design Technique for Linear Equalization in Optical Receivers. *J. Low Power Electron. Appl.* **2022**, *12*, 19. https://doi.org/10.3390/jlpea12020019

Academic Editor: Andrea Acquaviva

Received: 27 December 2021
Accepted: 4 February 2022
Published: 1 April 2022

1. Introduction

Traffic in data centers has grown rapidly over the past decade due to the rapid growth of cloud computing and big data applications. This in turn drives the development of robust, high-speed, and energy-efficient interconnects to transfer the data around the data center. Several 100+ Gb/s optical links have recently been reported to satisfy the bandwidth and reach requirements [1–3]. However, the associated cost and power dissipation prevent their widespread adoption within the data center. In short-reach photonic links, the transmitted optical modulation amplitude (OMA) must be sufficiently large that, in spite of coupling and fiber losses, the received optical power exceeds the receiver's sensitivity limit. This sensitivity is a function of the input-referred noise current of the receiver as well as the voltage amplitude requirements of the decision circuit driven by the receiver front-end [4]. Therefore, an energy-efficient link design requires a low-noise as well as a high-gain receiver.

As data rates increase, traditional approaches to receiver design dictate that the bandwidth of the receiver also increases, which limits the maximum achievable gain [5]. This trade-off is less pronounced in SiGe BiCMOS technologies where the transistor has higher intrinsic gain and transit frequency. Therefore, a reasonable gain is still achievable in wideband deigns. However, in CMOS, the trade-off limits the per-stage gain which necessitates cascading several gain stages to achieve the targeted output voltage amplitude. With increased number of stages, both noise and power dissipation increase.

This paper presents a novel inductorless design technique for high-gain optical receiver front-ends. Figure 1 illustrates the operation of the proposed front-end (FE) in contrast to the traditional wideband FE. Conventionally, the transimpedance amplifier (TIA) and the follow-on main amplifier (MA) are designed to have bandwidths in the order of $0.6f_{bit}$ and f_{bit}, respectively, to achieve an overall bandwidth of approximately $0.5f_{bit}$ [4]. In the proposed receiver, first, the TIA's bandwidth is reduced to approximately 25 % of the targeted data rate. The reduced TIA bandwidth allows for higher gain, lower input-referred noise, and fewer follow-on gain stages. The reduction in bandwidth also introduces inter-symbol interference (ISI) to the extent that the TIA's output eye diagram is fully closed. Unlike a bandlimited electrical channel which can introduce more than 30 dB of channel loss at the Nyquist frequency ($f_N = 0.5f_{bit}$), the low-bandwidth TIA introduces a moderate frequency-dependent attenuation. Consequently, a few dBs of amplitude peaking at f_N is sufficient to restore the required bandwidth (for example, the equalizer in [6] introduces only 7 dB of peaking). Therefore, in the second step of the proposed design technique, a high-frequency peaking is intentionally introduced in the main amplifier's amplitude response without impairing its low-frequency gain. Various possible designs of active feedback-based MA architectures [7–10] can introduce the required peaking by adding a pole in their feedback loops. The amplitude peaking in the equalizing main amplifier (EMA) is then used to compensate for the TIA's limited bandwidth to restore an overall bandwidth of approximately $0.5f_{bit}$. Although Figure 1 shows only the magnitude response of the TIA and EMA, group-delay variation must also be considered.

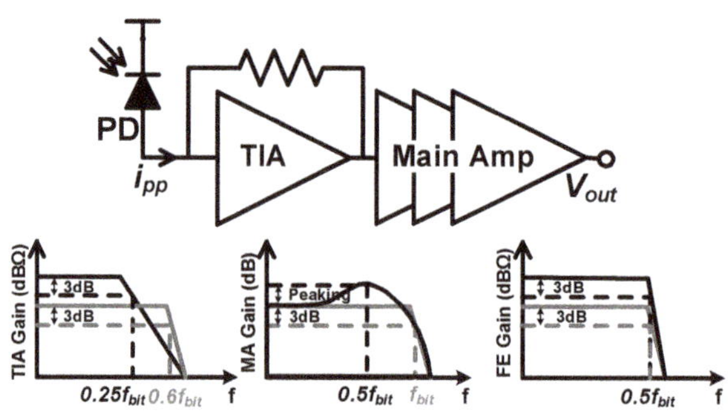

Figure 1. The proposed and the conventional receivers are represented by the same block diagram (**top**). The bottom graph illustrates the operation of the proposed receiver (black) in contrast to that of the conventional receiver (gray).

In contrast to traditional continuous-time linear equalizer (CTLE)-based designs [6,11], the proposed front-end attains the improved sensitivity and high-gain of these designs, while achieving better energy efficiency due to the elimination of the standalone equalizer stage(s). Traditional approaches to CTLE design, based on RC degenerated common-source amplifiers, suffer from a limited bandwidth and consequently insufficient peaking at high frequencies. When CTLEs are cascaded the reduction in overall bandwidth due to repeated real poles follows the same trend as that of 1st-order gain cells. Further, a 1st-order CTLE stage has a limited capability in equalizing a second-order TIA [6,12] which necessitates cascading several equalizer stages, further increasing power and area overheads. On the other hand, various inductorless feedback techniques can be used to design main amplifiers with gain-bandwidth products (GBW) far superior to a cascade of first-order stages [7–10]. The improvement is the result of poles moving away from the negative real axis. A combination of poles with high- and low-quality factors gives better GBW for the same pole magnitude. The proposed approach to design an EMA improves the overall receiver performance by increasing the gain of the TIA and improving noise performance as argued [6], but with the wideband performance of state-of-the-art MA designs.

The proposed design technique requires co-designing the TIA and the subsequent equalizing amplifier. Therefore, both stages receive equal attention in the analysis. Section 2 in this paper provides a detailed analysis of the TIA, highlighting the trade-off between its gain and bandwidth. Section 3 introduces the concept and the block diagram of the proposed EMA. The performance of the overall FE (TIA/EMA) is studied in Section 4. Section 5 shows the circuitry and simulation results of the proposed FE in comparison to the conventional full bandwidth design. Finally, Section 6 concludes the work.

2. Transimpedance Amplifier

2.1. Small-Signal Model and Freuqency Response

The inverter-based TIA (Inv-TIA) is used in this work due to its superior noise performance over its common-gate (CG) counterpart. Further, unlike the CG-TIA, the Inv-TIA is a self-biased topology which decouples the gain from the transconductance of the input device and allows for performance optimization without being limited by DC biasing constraints. The circuitry and the small-signal model of the Inv-TIA are shown in Figure 2. The CMOS inverter is modeled by its total transconductance g_m and output resistance R_A. Therefore, the core voltage amplifier has an open-loop transfer function of $A_{core}(s) = -A_0/(1+sT_A)$, where $A_0 = g_m R_A$ is the DC voltage gain and T_A is the time-constant formed by the output resistance R_A and the total output capacitance C_L. In the model, C_T is the total input capacitance, including both the photodiode capacitance C_D and the circuit's input capacitance C_I. Considering this model, the Inv-TIA exhibits a second-order transfer function characterized by a natural oscillation frequency ω_0, a pole quality factor Q_0, and a midband transimpedance gain of $Z_{TIA,0} \cong R_F$ [13]. The natural oscillation frequency ω_0 is converted to the corresponding TIA's 3dB-bandwdith (f_{TIA}) through a coefficient ρ that depends on the shape of the TIA's amplitude response (i.e., ρ is a function of Q_0). Detailed expressions for ω_0, Q_0, and ρ can be found in [14].

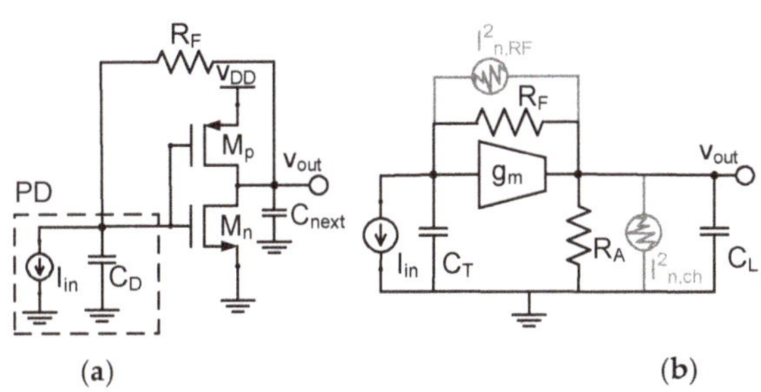

Figure 2. Inverter-based TIA (**a**) circuit and (**b**) small-signal model with noise sources indicated.

In the Inv-TIA, A_0 is constant for a given biasing condition, fixed ratio of W_p/W_n, and technology node. For example, an inverter with $V_{DD} = 1$ V, $W_n = W_p$, and simulated in TSMC 65 nm CMOS technology achieves A_0 of 6 V/V. Further, the gain-bandwidth product of the core amplifier ($GBW_A = A_0/(2\pi T_A)$) is also constant. The circuit's input capacitance (C_I) is determined by the total transistor width and is usually chosen as a fraction of the photodiode capacitance based on the noise and power constraints [14]. Therefore, for a given C_D, once C_I is fixed, the TIA's performance is controllable only through the feedback resistor. In this work, unless mentioned otherwise, A_0, GBW_A, and C_T are set to 6 V/V, 75 GHz, and 200 fF, respectively.

Figure 3a shows that both the 3 dB bandwidth and the pole quality factor Q_0 decrease with larger feedback resistor R_F. The bandwidth degrades almost linearly with the feedback resistor. In fact, the bandwidth does not follow the square-law relation

$\left(R_F \propto f_{TIA}^{-2}\right)$ predicted by the Transimpedance Limit [13]. This discrepancy can be explained as follows: Unlike [13], the model in this work allows Q_0 to change with R_F $\left(Q_0 = \sqrt{(A_0+1)R_F C_T T_A}/(R_F C_T + T_A)\right)$. For sufficiently large R_F that makes $R_F C_T \gg T_A$, Q_0 is proportional to $R_F^{-0.5}$. Consequently, it is reasonable to assume that ρ is also proportional to $R_F^{-0.5}$ with a percentage error of less than $\pm 8\%$, as shown in Figure 3b. Using this relation to rearrange the transimpedance limit from [13]:

$$f_{TIA}^2 = \frac{(A_0+1)}{A_0}\frac{GBW_A\,\rho^2}{2\pi R_F C_T} \tag{1a}$$

which implies:

$$f_{TIA}^2 \sim \frac{1}{R_F^2} \rightarrow f_{TIA} \sim \frac{1}{R_F} \tag{1b}$$

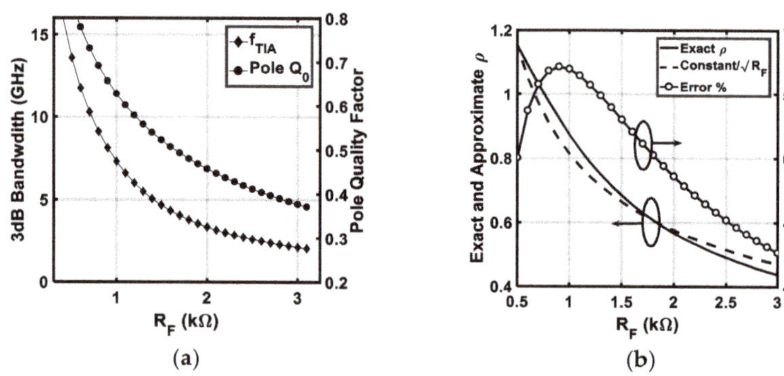

Figure 3. (**a**) TIA's 3dB bandwidth and pole Q_0 as a function of the feedback resistor. (**b**) The exact and the approximate calculations of ρ as a function of the feedback resistor along with the difference between the two as a percentage error.

This means that changing R_F changes both the pole magnitude (ω_0) and the pole quality factor (Q_0) which modifies the bandwidth dependency on the feedback resistor from that given in [13] where Q_0 is assumed to be constant. Assuming a constant Q_0 when R_F is increased by a factor of r requires both A_0 and T_A to scale up by a factor of $\sqrt{r}$. Practically, this approach is not feasible since the voltage gain of a single-stage CMOS inverter is constant for a given biasing and its maximum value is limited by the technology node.

2.2. Effective Gain

When f_{TIA} is reduced far below f_{bit}, severe ISI is introduced to the extent that the output eye diagram is fully closed. Therefore, the DC value of $Z_{TIA}(s)$ becomes a deceptive measure of the gain. The effective gain must be calculated from the transient response; more precisely, from the pulse response [15]. The TIA's pulse response is the response to an isolated binary one transmitted in a long sequence of binary zeros. Assuming a linear time-invariant (LTI) operation, if the TIA's response to a step input with a peak-to-peak value of i_{pp} is defined as $x(t)$, then its pulse response is calculated as $y(t) = x(t) - x(t - UI)$, where UI is the unit interval. The output pulse response of the Inv-TIA is plotted in Figure 4a for a data rate of 10 Gb/s with $i_{pp} = 10\ \mu A_{pp}$ and a bandwidth ranging from $0.2 f_{bit}$ to $0.6 f_{bit}$. To quantify the ISI, $y(t)$ is sampled at the symbol rate relative to its peak (as shown by the marker points in Figure 4a), resulting in a discrete-time sequence $V_{h,n}$ given by:

$$V_{h,n} = y(nT_b) \qquad -\infty < n < \infty \tag{2}$$

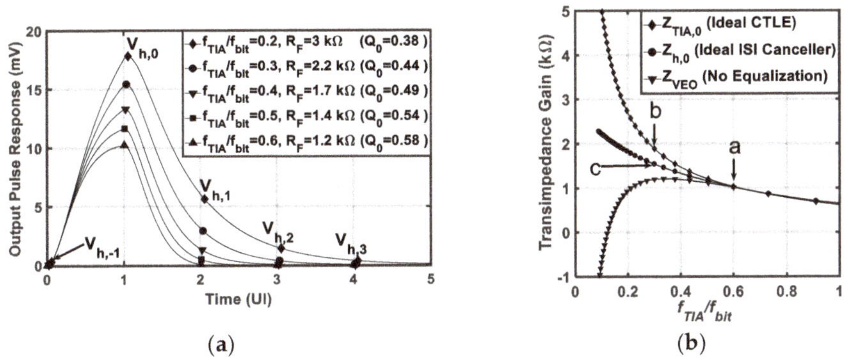

Figure 4. (**a**) Output pulse response for various values of f_{TIA}/f_{bit}. The input current pulse has a peak-to-peak value of 10 μA$_{pp}$ and a unit interval of 100 ps. (**b**) Different gains as a function of f_{TIA}/f_{bit}. f_{bit} are fixed at 10 Gb/s while f_{TIA} is swept by varying R_F. The labeled points in (**b**) illustrate that linear equalization is favorable for applications that require high gain in the receiver FE.

The sample at the peak of the pulse is denoted as the main-cursor sample ($V_{h,0}$). An effective gain of $Z_{h,0}$ can be interpreted as $V_{h,0}/i_{pp}$ if all ISI is cancelled. In the absence of equalization, the ISI samples ($V_{h,n\neq 0}$) can be subtracted from $V_{h,0}$, closing the vertical eye opening (VEO) to:

$$VEO = V_{h,0} - \sum_{\substack{n=-\infty \\ n\neq 0}}^{\infty} |V_{h,n}| \tag{3}$$

The effective gain is calculated based on the receiver architecture as follows: 1. The VEO can be used to determine an effective gain of $Z_{VEO} = VEO/i_{pp}$ for the case in which the ISI is not removed or is only partially removed. The midband gain $Z_{TIA,0}$ can also be interpreted as an effective gain if an ideal unity-gain continuous-time linear equalizer (CTLE) is employed. The CTLE compensates for the bandwidth limitation of the TIA and restores an overall bandwidth on the order of $0.6f_{bit}$ without impairing the low-frequency gain. Therefore, the TIA's midband gain $Z_{TIA,0}$ at the low bandwidth point can be used as the effective gain for the combined (TIA/CTLE).

Figure 4b shows that linear equalization improves the effective gain over both full-bandwidth and ISI canceller-based designs. For example, if the TIA's bandwidth is reduced from $0.6f_{bit}$ (point a) to $0.3f_{bit}$ and an ideal CTLE is employed (point b), the effective gain improves by a factor of 1.86× compared to point a. The gain at $0.3f_{bit}$ (point b) is also 1.23× larger than that where an ideal ISI-canceller is employed (point c). That is, ISI cancellers have no bearing on the TIA's bandwidth which means that the output pulse of a limited-bandwidth TIA does not have enough time to settle at the voltage value ($i_{pp}Z_{TIA,0}$). Further, ideal cancelers that remove all pre- and post-cursor ISI are not implementable. For example, decision feedback equalizers (DFEs) [12] cancel only the post-cursor ISI. DFEs also suffer from a tight timing constraint where the feedback signal from the previously decided bit must arrive within one unit interval (UI) to resolve the current bit. These limitations make linear equalization a more attractive choice for applications that require high gain in the receiver FE. DFEs, on the other hand, are favorable over CTLEs from the noise point of view. That is, CTLEs extend the noise bandwidth to be a function of the bandwidth of the combined TIA/CTLE instead of being a function of the bandwidth of the low-bandwidth TIA as in the DFE-based receivers [15]. Therefore, a fair comparison between different receiver architectures should consider more complete figures of merit, such as OMA sensitivity and energy efficiency. The noise performance of the presented front-end is carefully examined in Section 4.3 in the presence of the equalizing main amplifier.

3. Equalizing Main Amplifier

In addition to high-gain and broadband operation, adjustable high-frequency peaking (HFP) is a desirable feature in MA design. The amplitude peaking at the Nyquist frequency can mitigate the bandwidth limitation introduced by other components in the optical link. For example, in [16], shunt and series passive inductors are employed between cascaded stages of a programmable gain amplifier to realize an HFP. The HFP is then used to partially compensate for the varying performance of the multi-mode fiber. In this work, passive inductors are avoided because they consume significant silicon area and potentially increase substrate coupling. The HFP is realized by introducing a pole in the feedback loop of an active feedback-based MA architecture and used to compensate for the TIA's limited bandwidth.

Equalizing MA Based on a Third-Order Gain Stage

The block diagrams of the conventional and proposed gain stages are shown in Figure 5a,b, respectively. The conventional architecture is presented in [17], where a third-order nested feedback technique achieves high-speed operation while maintaining robust stability compared to the traditional third-order gain stage. In the block diagram in Figure 5a, the first-order gain cell, $A(s)$, is modeled by the transconductance of the input device g_{m1}, load resistance R_1, and load capacitance C_1. The adjustable active feedback $\beta_{con}(s)$ cell is modeled by the transconductance $-g_{mf}$. Therefore, the transfer functions of the first-order gain and feedback cells are given by:

$$A(s) = \frac{A_1}{\frac{s}{\omega_1}+1}, \qquad \beta_{con}(s) = \frac{\beta_1}{\frac{s}{\omega_1}+1}, \tag{4}$$

where $A_1 = g_{m1}R_1$ and $\omega_1 = (R_1C_1)^{-1}$ are the DC gain and cut-off frequency of the first-order gain cell, respectively. $\beta_1 = g_{mf}R_1$ is the DC feedback gain. The transfer function of the overall architecture in Figure 5a is given by:

$$H_{MA}(s) = \frac{A^3(s)}{A^2(s)\beta_{con}(s) + A(s)\beta_{con}(s) + 1} \tag{5}$$

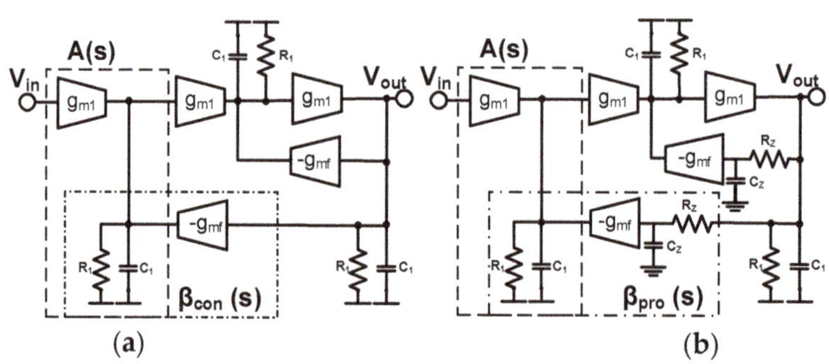

Figure 5. Block diagram of (**a**) the third-order gain stage in [17] (**b**) the proposed EMA with a LPF inserted in each feedback path.

In this work, two poles are introduced in the feedback loops to create an adjustable HFP without impairing the low-frequency gain. The transfer function of the proposed EMA is calculated using (5) by replacing $\beta_{con}(s)$ by β_{Pro} given in (6):

$$\beta_{pro}(s) = \frac{\beta_1}{\left(\frac{s}{\omega_1}+1\right)\left(\frac{s}{\omega_Z}+1\right)} \tag{6}$$

where $\omega_Z = (R_Z C_Z)^{-1}$ is the cut-off frequency of the introduced low-pass filter which is assumed to have negligible loading on the output node. Therefore, the transfer function of the EMA in Figure 5b is given by:

$$H_{EMA}(s) = \frac{A_1^3\left(\frac{s}{\omega_Z}+1\right)}{\left(\frac{s}{\omega_1}+1\right)^3\left(\frac{s}{\omega_Z}+1\right)+A_1\beta_1\left(\frac{s}{\omega_1}+1\right)+A_1^2\beta_1} \tag{7}$$

The pole-zero locations of (7) are plotted in Figure 6a in comparison with that of (5) for β_1, A_1, and ω_1, fixed at 0.25, 2.5, and $2\pi \times 30$ GHz, respectively. The poles of the conventional architecture are indicated by black x-markers. For the proposed EMA, ω_Z is swept from $0.5\omega_1$ to $5\omega_1$. The insertion of the LPF in the feedback loops of the proposed EMA creates a real zero at ω_Z (shown in blue). It also increases the order of the denominators of $\beta_{pro}(s)$ and $H_{EMA}(s)$ compared to their conventional counterparts. As a result, for low values of ω_Z, the proposed EMA has two sets of complex-conjugate poles (P_A and P_B) (shown in red). As ω_Z increases, P_A travels toward the complex poles of (5) while the damping factor of P_B increases until the two poles become real and start traveling in opposite directions. At sufficiently high ω_Z, P_{B2} and the real zero cancel each other, P_{B1} reaches the real pole of (5) and the overall architecture degenerates to the third-order gain stage in [17].

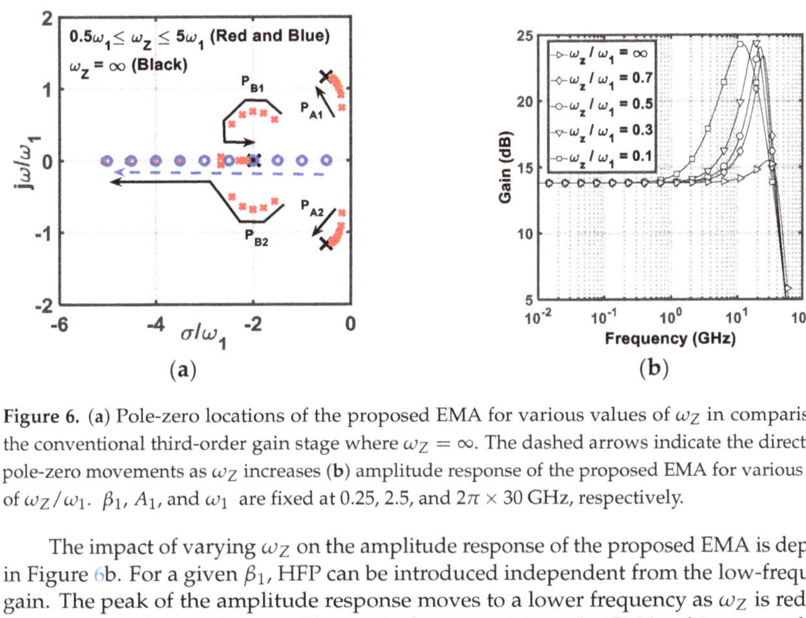

Figure 6. (**a**) Pole-zero locations of the proposed EMA for various values of ω_Z in comparison to the conventional third-order gain stage where $\omega_Z = \infty$. The dashed arrows indicate the direction of pole-zero movements as ω_Z increases (**b**) amplitude response of the proposed EMA for various ratios of ω_Z/ω_1. β_1, A_1, and ω_1 are fixed at 0.25, 2.5, and $2\pi \times 30$ GHz, respectively.

The impact of varying ω_Z on the amplitude response of the proposed EMA is depicted in Figure 6b. For a given β_1, HFP can be introduced independent from the low-frequency gain. The peak of the amplitude response moves to a lower frequency as ω_Z is reduced. As a numerical example from Figure 6b, for $\omega_Z = 0.1\omega_1$, the EMA achieves amplitude peaking of 6 dB at 5 GHz and increases to 10.5 dB at 11 GHz. In the presence of such high amplitude peaking, it is not instructive to explore the bandwidth of the EMA. Instead, the bandwidth extension ratio and the signal integrity are inspected in the following section for the overall front-end which includes the limited-bandwidth TIA and the EMA.

4. Front-End Performance Analysis

4.1. Performance Requirements for the EMA

The noise-limited input signal produces a peak-to-peak voltage of $V_O^{PP} = SNR\ i_{n,in}^{rms}\ Z_{TIA,0}\ A_{EMA,0}$ at the output of the front-end, assuming that the EMA restores a wide overall bandwidth, where SNR is the required signal-to-noise ratio and is equal to 14.07 for a BER of 10^{-12} [4], $i_{n,in}^{rms}$ is the input-referred noise current, and $A_{EMA,0}$ is the DC gain of the EMA. V_O^{PP} is sufficiently large to drive an ideal clock-and-data recovery (CDR) circuit to achieve the desired BER. However, the decision circuit in a realistic CDR has finite sensitivity and requires a minimum input voltage amplitude (V_{CDR}^{PP}). Therefore, the FE's output voltage needs to be increased by V_{CDR}^{PP} to attain the same BER as an ideal CDR. The finite sensitivity of the CDR incurs a power penalty (PP) of:

$$PP = \frac{V_O^{PP} + V_{CDR}^{PP}}{V_O^{PP}} = 1 + \frac{V_{CDR}^{PP}}{SNR\ i_{n,in}^{rms}\ Z_{TIA,0}\ A_{EMA,0}} \tag{8}$$

The optical power incident to the photodiode, the electrical current generated from the photodiode, and the voltage produced at the output of the main amplifier are linearly proportional. As a result, the amount of additional optical power required to overcome any nonideality (power penalty) can be expressed in voltage terms as shown in (8). The equation reveals that higher transimpedance gain relaxes the gain requirements for the EMA for a given PP. Figure 4b shown earlier indicates that reducing the ratio f_{TIA}/f_{bit} is beneficial for the gain as long as the equalizer is able to recover an overall bandwidth of approximately 50% to 60% of the targeted data rate. Therefore, the equalizer's capability in restoring the bandwidth determines how far the TIA's bandwidth can be reduced below the data rate. That is, excessive reduction of the TIA's bandwidth would require the equalizer to introduce a large amount of amplitude peaking which translates into large group delay variation (GDV). The latter causes horizontal and vertical eye closure which reduces the gain and noise improvements gained from equalization. In [6], it is concluded that the equalizer can restore the bandwidth by a factor of approximately 2× while simultaneously maintaining a good noise performance and a good quality of the equalized eye diagram.

For the conventional wideband TIA, a feedback resistor of 1.25 kΩ is chosen to achieve a bandwidth of $0.57 f_{bit}$, sufficiently large to introduce no ISI. The TIA's bandwidth drops almost linearly with R_F as observed in Figure 3a. Therefore, in the proposed design, the value of the feedback resistor is doubled, leading to a bandwidth of $0.26 f_{bit}$. At this bandwidth, the TIA achieves a $Z_{TIA,0}$ of 66.6 dBΩ (2143 Ω) while introducing an attenuation of 7.2 dB at the Nyquist frequency ($f_N = 0.5 f_{bit} = 5$ GHz). The EMA is now required to recover the bandwidth by a factor ranging from 1.9× to 2.3× to achieve an overall bandwidth on the order of 50% to 60% of f_{bit}. For example, using the gain of the low-bandwidth TIA while assuming V_{CDR}^{PP}, SNR, and $i_{n,in}^{rms}$ of 50 mV_{pp}, 14.07, and 1 μA_{rms}, respectively, the PP defined in (8) can be used to calculate the required gain of the EMA. In addition to recovering the bandwidth, the EMA is required to amplify the TIA's output by a low-frequency gain of approximately 20 dB to reduce the PP to less than 0.67 dB (1.17). Practically, the EMA's gain is determined to reduce the PP to a pre-determined value obtained from link budget analysis.

4.2. Bandwidth Extension and Signal Integrity

Figure 7 shows the block diagram of the proposed front-end where the limited-bandwidth TIA is followed by a two-stage EMA. The EMA's second stage is added to relax the gain requirements. The two-stage EMA is modified based on the two-stage MA presented in [10] by inserting low pass filters in the feedback loops of the second stage. Therefore, the transfer function of the overall front-end (FE) is given by $Z_{FE}(s) = Z_{TIA}(s)H_{2-EMA}(s)$, where $H_{2-EMA}(s)$ is the transfer function of the two-stage EMA and given by:

$$H_{2-EMA}(s) = \frac{A^5(s)}{Den(s)} \tag{9}$$

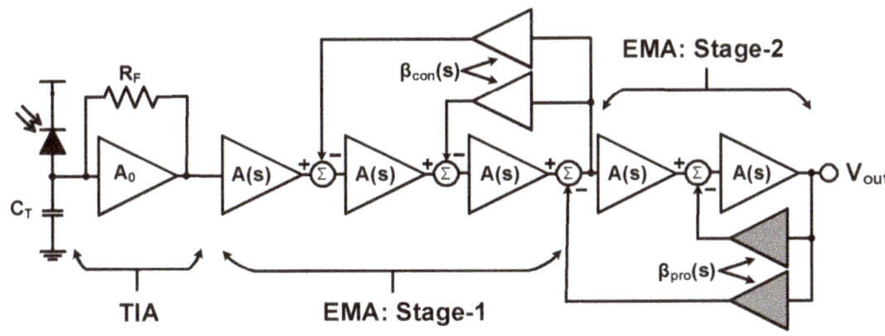

Figure 7. Block diagram of the proposed front-end. The two-stage EMA is modified based on the two-stage MA in [10]. The grayed feedback cells indicate the locations of the inserted poles.

The denominator $Den(s)$ is expressed as:

$$\begin{aligned} Den(s) = 1 + A(s)\left[\beta_{con}(s) + \beta_{pro}(s)\right] + A^2(s)\left[\beta_{con}(s) + \beta_{pro}(s) + \beta_{con}(s)\beta_{pro}(s)\right] \\ + A^3(s)\beta_{con}(s)\beta_{pro}(s) \end{aligned} \quad (10)$$

Once the TIA's feedback resistor is fixed, the full design space is reduced to only two variables: ω_Z and β_1. These two variables are swept, and the following equations are solved numerically to calculate the bandwidth (f_{FE}), the low-frequency gain $(Z_{FE,0})$, and the peaking (M_p) of the overall FE:

$$|Z_{FE}(2\pi f_{FE})| = \frac{1}{\sqrt{2}}|Z_{FE}(j\omega)|_{\omega=0} \quad (11a)$$

$$Z_{FE,0} = 20\log_{10}|Z_{FE}(j\omega)|_{\omega=0} \quad (11b)$$

$$M_P = 20\log_{10}\frac{\max(|Z_{FE}(j\omega)|)}{|Z_{FE}(j\omega)|_{\omega=0}} \quad (11c)$$

Several combinations of β_1 and ω_z can achieve the required bandwidth extension but with different noise performance. The noise analysis is presented in the following section. The feedback gain β_1 directly impacts the low-frequency gain of the EMA and is chosen to satisfy the power penalty condition indicated earlier. Then, ω_z is swept to achieve the required bandwidth extension ratio defined as f_{FE}/f_{TIA}. The pairing of $\omega_z = 0.075\omega_1$ and $\beta_1 = 0.25$ is chosen as it achieves a good noise performance as well as a good quality of the output eye. The corresponding frequency response is plotted in Figure 8a, where the EMA introduces 5 dB of peaking and extends the bandwidth by a factor of 2.2×. The gain peaking in the overall frequency response is less than 0.1 dB. Figure 8b shows the pulse response at the output of the FE. To quantify the vertical and horizontal eye openings, the output pulse is sampled at a bit rate clock relative to its peak. The pulse is sampled at both the rising and falling edges of the clock. The sum of the magnitude of the samples at the even clock edges (filled markers for $n \neq 0$) quantifies the ISI. The sum of the samples at the odd clock edges (hollow markers) is considered as a jitter indicator (JI). Note that the falling edges of the clock are the zero-crossing points of the data. Therefore, the defined JI includes only the deterministic jitter caused by the residual ISI or ringing in the time domain [18]. The sum of ISI and JI samples is less than 6.5% of the main cursor sample which implies that the eye has a wide internal opening area, as evident also from the eye diagram in Figure 9a, obtained through simulation. Figure 9b shows the output eye diagram when the limited bandwidth TIA is followed by a wideband MA. The comparison between the two eyes in Figure 9 demonstrates the capability of the presented technique in restoring the bandwidth without impairing the midband gain or increasing power dissipation.

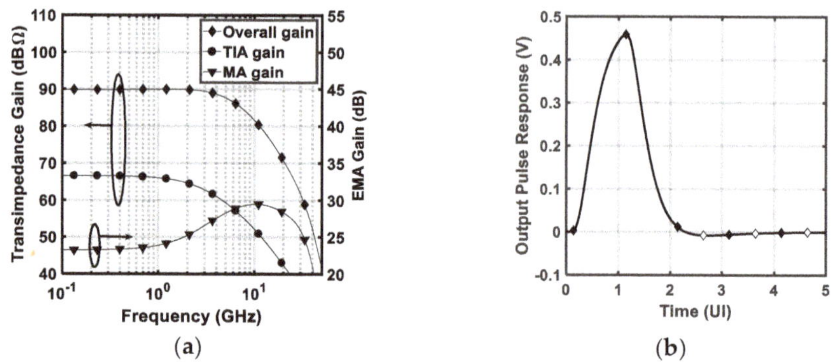

Figure 8. (**a**) Amplitude response (**b**) output response to an input current pulse with peak-to-peak value of 15 μA_{pp} and width of 100 ps. The EMA parameters are $\omega_Z/\omega_1 = 0.075$ and $\beta_1 = 0.25$.

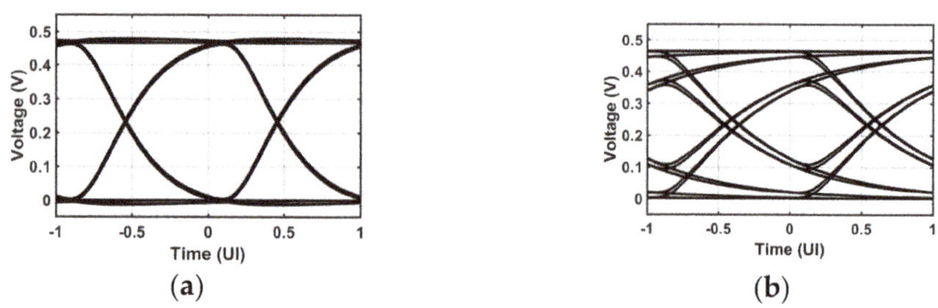

Figure 9. Matlab generated 10 Gb/s output eye diagrams when the limited-bandwidth TIA is followed by (**a**) an EMA, and (**b**) a wideband MA. The peak-to-peak value of the input current is fixed at 15 μA_{pp}.

4.3. Noise Analysis

Figure 10a shows the model used for noise analysis. The main noise sources in the Inv-TIA are the channel and feedback thermal noise, shown in Figures 2 and 10 as $I^2_{n,ch}$ and $I^2_{n,RF}$, respectively. The power spectral densities of these two sources can be expressed as: $I^2_{n,ch} = 4kT\gamma g_m$ and $I^2_{n,RF} = 4kT/R_F$ where k is the Boltzmann constant, T is the temperature in Kelvin, and γ is the excess noise factor. Under a constant gain-bandwidth product constraint, the noise-optimum FET size is $C_I = 0.7C_D$ [14]. Therefore, the transconductance of the TIA's input device can be calculated as $g_m = 2\pi f_T C_I$, where f_T is the technology transit frequency at the selected bias point. In Figure 10a, the amplifier following the TIA is modeled by $H_{post}(s)$ and its input-referred noise PSD is denoted by $V^2_{n,in} = 4kT/g_{m,post}$. $H_{post}(s)$ is given by (9) and (10) for both the proposed and conventional designs, using $\beta_{pro}(s)$ and $\beta_{con}(s)$, respectively. In simulations that follow, $g_{m,post}$, γ, and f_T are fixed at 10 mΩ^{-1}, 2, and 150 GHz, respectively.

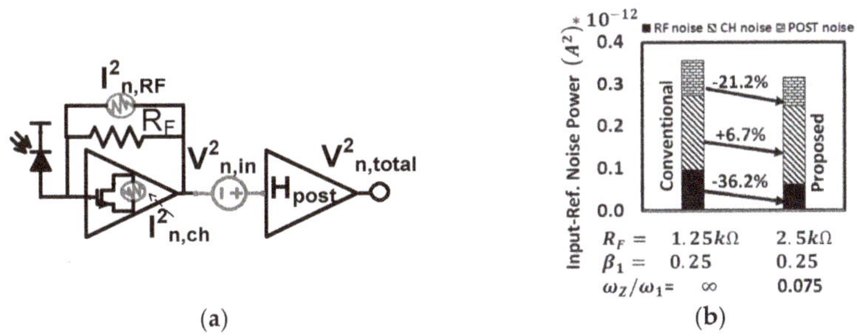

Figure 10. (**a**) Circuit model used for noise analysis (**b**) Matlab simulated noise reduction in the proposed FE compared to its conventional counterparts. The arrows indicate the amount of change for each noise component.

Linear equalization extends both the signal and the noise bandwidths [15]. Therefore, the integration of the noise power spectral density (PSD) must be performed at the receiver output to take into consideration how the equalizer processes the noise. To do so, the contribution to the output noise PSD from each noise source is first calculated. Because all noise sources are uncorrelated, the total output noise PSD is constructed by adding up all individual power spectra. The total output noise PSD is then integrated up to infinity to calculate the integrated output-referred noise power $\left(v^2_{n,total}\right)$ having units of V^2. The total integrated input-referred noise power $\left(i^2_{n,total}\right)$ in units of A^2 is then determined by dividing the $v^2_{n,total}$ by the squared effective gain $\left(Z_{TIA,eff}\right)^2$ calculated from the VEO at the output of the FE. This gain calculation accounts for the residual ISI in the signal presented to the decision circuit. The input-referred noise current is then calculated as the square root of $i^2_{n,total}$. Further discussion about the noise analysis for equalizer-based optical receivers is available in our previously published work [15].

4.4. Performance Comparison

To assess the improvement of the proposed FE versus its conventional counterpart, both FEs are simulated in Matlab. The traditional FE has the same block diagram as in Figure 7 without the pole insertion in the feedback loops. Therefore, its analysis is the same as presented earlier, but replacing each $\beta_{pro}(s)$ in (10) with $\beta_{con}(s)$. The value of the TIA's feedback resistor is tuned to set the ratio of f_{TIA}/f_{bit} to 0.57 and 0.26 for the conventional and the proposed FEs, respectively. In the latter, the values of β_1 and ω_Z are chosen to achieve an overall bandwidth of $f_{FE} = 0.56 f_{bit}$. The power consumption and the DC gain of the proposed EMA are kept equal to that of the conventional MA by fixing the values of A_1 and β_1 in both circuits. The performance of the two FEs is summarized in Table 1. Although the two FEs have approximately the same overall bandwidth, the proposed FE achieves 6 dB higher gain compared to its conventional version. This improvement in the transimpedance gain resulted from the increased value of R_F for the limited-bandwidth TIA. It is worth mentioning that this gain improvement comes without any additional power dissipation because changing R_F and ω_Z does not affect the DC power dissipation as will be shown in the practical implementation in the next section.

Table 1. Design parameters and performance summary of the proposed front-end in comparison to its conventional counterpart.

		MATLAB [(1)]		Spectre [(2)]			
		10 Gb/s		10 Gb/s		20 Gb/s [(4)]	
		Conventional	Proposed	Conventional	Proposed	Conventional	Proposed
TIA	R_F (kΩ)	1.25	2.5	0.7	1.6	0.4	0.8
	f_{TIA}/f_{bit}	0.57	0.26	0.64	0.27	0.68	0.3
MA/EMA	$\omega_Z/2\pi$ (GHz)	∞	2.25	∞	**5.25**	∞	**11.47**
	β_1	0.25	0.25	**0.14**	**0.14**	**0.15**	**0.15**
	Peaking (dB) @ f_N	0	5.05	0	4.8	0	3.5
FE	$\boldsymbol{Z_{VEO}}$ **(dBΩ)**	**83.6**	**89.98**	**79.7**	**87.1**	**71.2**	**77.2**
	f_{FE}/f_{bit}	0.57	0.56	0.6	0.61	0.59	0.54
	Peaking (dB)	0	0.084	0	0	0	0
	$\boldsymbol{i_{n,rms}}$ **(μA$_{rms}$)**	**0.598**	**0.531**	**1.2**	**0.95**	**2.41**	**1.74**
	Sensitivity Improvement (dB)						
	Noise-based	–	**0.52**	–	**1**	–	**1.4**
	PP-based [(3)]	–	**0.61**	–	**0.5**	–	**0.84**
	Total	–	**1.125**	–	**1.5**	–	**2.24**

[(1)] Simulations based on Figure 7. [(2)] Simulations based on Figure 11a. [(3)] For V_{CDR}^{PP} = 50 mV$_{pp}$. [(4)] The 20 Gb/s simulations are discussed in Section 5.4.

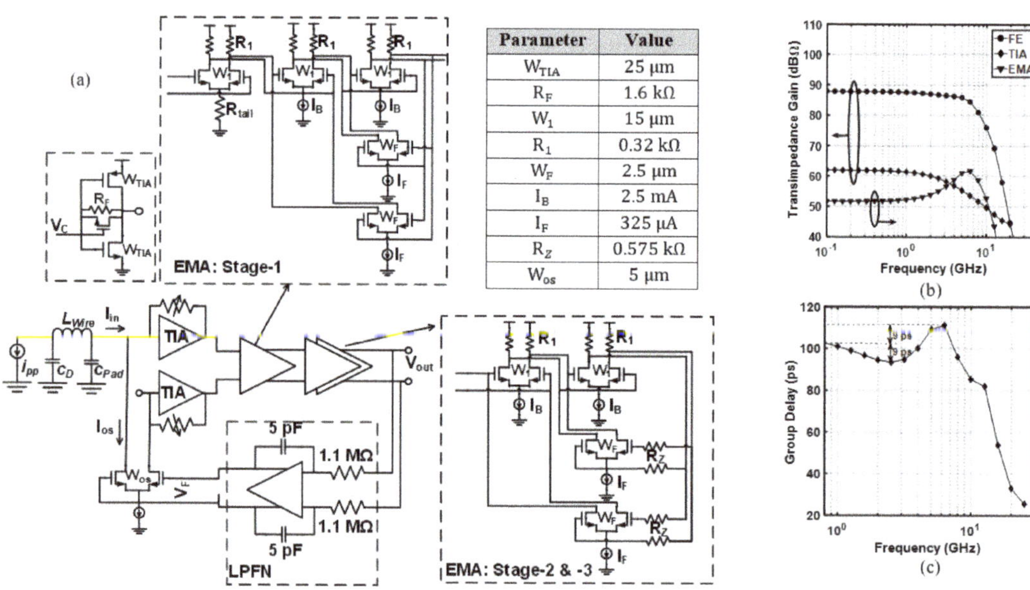

Parameter	Value
W_{TIA}	25 μm
R_F	1.6 kΩ
W_1	15 μm
R_1	0.32 kΩ
W_F	2.5 μm
I_B	2.5 mA
I_F	325 μA
R_Z	0.575 kΩ
W_{os}	5 μm

Figure 11. (**a**) Block diagram and circuitry of the implemented front-end. Parameter values for 10 Gb/s operation are tabulated. (**b**) Simulated amplitude response. (**c**) Simulated group-delay.

The input-referred noise power of both FEs is compared in Figure 10b. In the proposed FE, the feedback resistor and the post amplifier noise powers are improved compared to their counterparts in the conventional design. That is, increasing the value of R_F in the proposed FE reduces its thermal noise contribution and increases the input-referral gain which suppresses the noise from the follow-on amplifier. The channel noise is slightly increased in the proposed FE due to HFP that amplifies the high-frequency noise. Overall, the presented design technique reduces the input-referred noise current by 11.2%. The lower noise and higher gain in the presented FE led to 0.52 dB and 0.61 dB improvements in the noise-based sensitivity and the PP compared to the traditional design.

5. Circuitry and Layout of the Implemented Front-End

Figure 11a shows the block diagram and the circuitry of the implemented front-end. A replica TIA is used to provide pseudodifferential power-supply noise rejection. The TIA is followed by a three-stage EMA. A series resistor (R_Z) is inserted in the feedback loops of the second and third stages. This resistor, in combination with the parasitic capacitance of the transistor in the feedback loops, creates the zero required for bandwidth extension. Compared to Figure 7, the EMA's third stage is added to relax the gain requirements and assist in recovering the bandwidth. A low-pass feedback network (LPFN) is connected between the output of the EMA and the input of the TIA. The LPFN amplifies the difference between the DC levels at V_{Out} and returns a feedback voltage of V_F that is then converted to a current I_{os} by the transconductance of M_{os} and subtracted from the input current for offset compensation. The LPFN is a single-pole RC filter using a Miller-boosted 5 pF capacitor and a 1.1 MΩ resistor. A low cut-off frequency of 1 MHz is achieved as a trade-off between the on-chip area and the tolerable baseline wander for long runs of consecutive identical digits. The low common-mode voltage at the TIA's output prevents the use of a tail current source for the first differential pair in the EMA's first stage and therefore a poly silicon resistor is used instead.

The FE is simulated in TSMC-65 nm using a Cadence Spectre simulator. The input parasitics are modeled by a pad capacitance (C_{Pad}) of 45 fF, a photodiode capacitance (C_D) of 80 fF and a bondwire inductance (L_{wire}) of 0.5 nH. The loading from the subsequent output buffer is modeled by a load capacitance of $(C_L = 150\text{ fF})$ connected at the output of the EMA. An additional 50 fF capacitance is added to all nodes to model the wiring and layout parasitic. The receiver's output stage (not shown in Figure 11a) is a conventional differential amplifier with a load resistance of 100 Ω chosen as a trade-off between output signal amplitude and compatibility with the off-chip 50 Ω environment.

Figure 12 shows the chip layout in TSMC 65 nm CMOS technology. The chip includes two standalone FEs. One FE is the direct implementation of the circuit in Figure 11a while the other is its conventional version (i.e., R_Z is replaced by a short circuit). The total size of the chip is 1 mm × 0.7 mm. Each front-end is pad limited and occupies 665 μm × 460 μm (0.31 mm^2), including the I/O RF pads, while the active area, including the offset compensation loop, is about 0.0114 mm^2. The high-speed RF input and output probing pads are differential G-S-G-S-G since each FE has differential inputs and outputs. The TIA, the MA/EMA, and the output buffer are powered by different supplies.

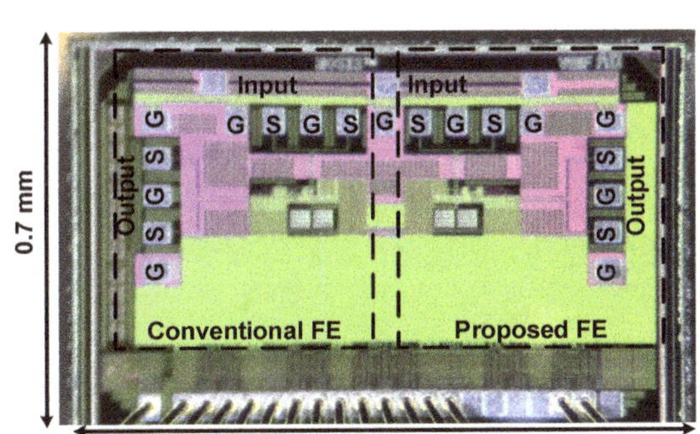

Figure 12. Chip layout.

5.1. Validation of Bandwidth Extension

Similar to the previous section, both the proposed and the conventional FEs are simulated and compared. The proposed FE's TIA bandwidth is 27% of the targeted 10 Gb/s data rate. The tail current source in the feedback pair I_F sets the feedback gain β_1 and is chosen to satisfy the power penalty condition. The series resistor R_Z is then chosen to achieve the required bandwidth extension. The device dimensions and component values are tabulated in Figure 11a for nominal 10 Gb/s operation. All transistors in the signal and feedback paths use minimum length. Current sources, however, employ transistors with longer than minimum length. The corresponding amplitude responses are shown in Figure 11b. The EMA introduces a peaking of 4.8 dB at the Nyquist frequency and restores the bandwidth by a factor of 2.28×, achieving an overall bandwidth of 6.1 GHz.

The simulated group-delay is also shown in Figure 11c where the GDV is within ±10% of the unit interval over the frequency range of interest. Figure 13a,b shows the 10 Gb/s eye diagrams at the output of the FE when the limited-bandwidth TIA is followed by a wideband MA or by the EMA, respectively. The eye diagrams obtained through simulation demonstrate the capability of the proposed peaking technique in restoring the bandwidth without impairing the low-frequency gain. The bandwidth extension improves the VEO by a factor of 1.7×. Figure 13c shows the eye diagram of the traditional FE. In this simulation, R_Z is shorted and R_F is reduced to widen the TIA's bandwidth while the current sources (I_F and I_B) are unchanged. Comparing Figure 13b,c shows that the presented design technique improves the effective gain by a factor of 2.34×. Interestingly, for the proposed design, the gain is improved by almost the same amount as the TIA's bandwidth is reduced. This emphasizes the linear relation between the gain and the bandwidth in the single-stage Inv-TIA. Table 1 summarizes the simulated performance of the two FEs where the presented FE shows 1.5 dB better sensitivity compared to its conventionally designed counterpart.

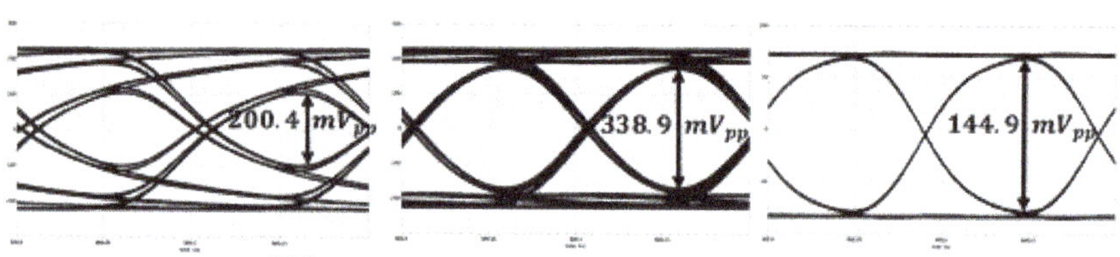

(a) (b) (c)

Figure 13. Simulation results for the 10 Gb/s output eye diagrams when the limited-bandwidth TIA is followed by (**a**) a wideband MA and (**b**) the proposed EMA. In (**c**), the TIA's bandwidth is widened, and a wideband MA is employed. The input current is fixed at 15 μA_{pp} for all simulations.

5.2. Sensitivity to Process and Temperature Variations

Figure 14 shows the simulated performance of the presented receiver under process and temperature variations. Figure 14a shows that the EMA exhibits more peaking at a lower temperature. For a given temperature, the peaking can vary by up to 6.5 dB over different process corners. The FE gain and bandwidth in Figure 14b can vary up to 13.5 dB and 3.4 GHz over different corners, respectively. The gain and bandwidth variations relative to their values at room temperature reach up to 24.3% and 22.5%, respectively, as the temperature varies from 20 °C to 80 °C. This performance variation is mainly caused by the constant current sources used in this design and can be counteracted by employing temperature-compensated or constant-gm biasing techniques [19]. Adaptation techniques can be also employed to continuously monitor the output eye diagram and set the circuit parameters accordingly to maintain the best quality for the equalized eye [20]. In the implemented prototype, the TIA's feedback resistor and current sources in the forward and feedback paths are made variable. This allows for post-fabrication control on peaking frequency, peaking magnitude, and the TIA's high-frequency roll-off. Therefore, the amplitude responses of both the EMA and the TIA track each other to achieve the targeted bandwidth with minimal GDV.

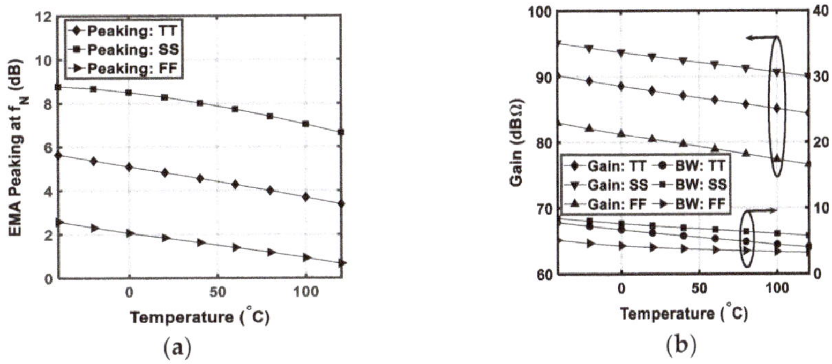

Figure 14. Simulated performance under process and temperature variations: (**a**) EMA's peaking at Nyquist frequency; (**b**) gain and bandwidth of the overall FE.

To prove that the proposed technique works despite the PT variations, Figure 15 shows the simulated 10 Gb/s eye diagram at the SS process corner and −20 °C. The uncompensated eye (left) shows a significant distortion. By carefully adjusting the circuit parameters, a clean eye is obtained (right) with an internal opening similar to that obtained under nominal operations. To generate the eye on the right, the circuit parameters are changed as follows: I_B is reduced from 2.5 mA to 1.65 mA, I_F is increased from 0.325 mA to 0.4 mA, and R_Z is reduced from 0.575 kΩ to 0.445 kΩ. The tunability range of all circuit parameters are limited to less than 35% of their nominal values which is feasible for realization. Further, the capacitance introduced by the configurable current sources appears at tail nodes and therefore does not alter the signal path.

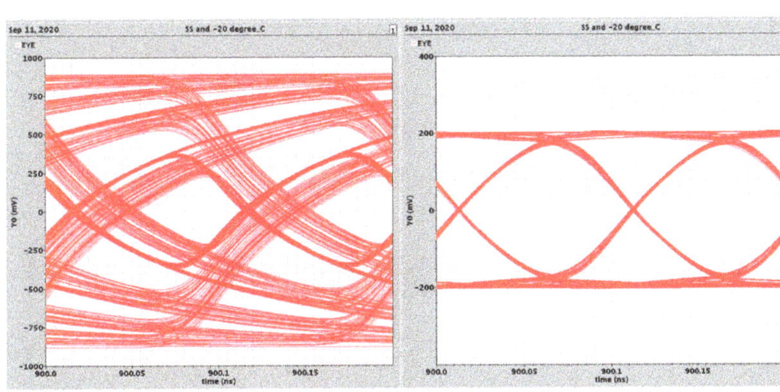

Figure 15. Simulated 10 Gb/s eye diagrams under SS process corner and −20 °C (**left**) uncompensated, (**right**) compensated.

5.3. Stability

In the presence of a complex feedback and high amplitude peaking in the EMA, the stability of the presented FE becomes an important consideration. The pole-zero simulation in Figure 6a shows that a pair of complex poles (P_A) moves toward the y-axis as ω_z is reduced. ω_z is the frequency of the introduced zero that ideally cancels the bandwidth-limiting pole created by the low-bandwidth TIA. As a result, the TIA's 3-dB bandwidth cannot be made arbitrarily small to avoid the EMA's pole pair travelling to the right-hand plane (RHP). Further, for a given ω_z, the poles P_A may enter the RHP at excessively large feedback gain β_1. However, the values of β_1 that lead to RHP poles are far from those in the proposed design. For example, in the FE in Figure 7, when ω_z is set to $2\pi f_{bit}/4$, the poles P_A do not travel to the RHP until after $\beta_1 > 6$ and $\beta_1 > 5.5$ for f_{bit} of 10 Gb/s and 20 Gb/s, respectively, while β_1 is typically limited to less than 0.3.

5.4. Discussion and Comparison to Prior Work

The performance of the proposed FE is compared to other 10 Gb/s high-gain receivers in the literature as shown in Table 2. Although thorough circuit simulations are sufficient to prove the concept behind our design, the absence of optical measurements complicates the comparison with prior art. The work in [8] consists of an Inv-TIA followed by three stages of an Inv-based Cherry-Hooper voltage amplifier. In this architecture, active interleaving feedback and local positive feedback are applied to extend the bandwidth. The circuit is implemented in a single-ended structure and measured with electrical and optical inputs for various data rates. Only electrical measurements at 10 Gb/s are listed in Table 2. The work in [8] is measured for two modes of operation denoted on Table 2 by best sensitivity mode and lowest power mode (see Figure 18 in [8]). The average of these two modes shows approximately 2× better sensitivity and 2.3× better energy efficiency compared to the work presented here. The reason for this better performance is mainly because of the single-ended structure in [8] that reduces the power dissipation and thermal noise sources compared to the differential structure used in this work. Further, the single-ended implementation enabled measurements at low supply voltages, which are not available in this work due to the DC biasing requirements on differential amplifiers. The proposed design has a much higher output peak-to-peak amplitude at the sensitivity level than [8], which is not optimized for high-gain operation and incurs a significant PP when the receiver is followed by a practical decision circuit.

Table 2. Performance comparison with published 10 Gb/s receivers.

Performance Parameter	[9]	[12]	[8]		[21]	This Work (4)
			Lowest Power	Best Sens.		
RX topology	Diff.	Diff.	Sing.	Sing.	Diff.	Diff.
Passive inductor	No	No	No	No	Yes	No
CMOS tech. (nm)	130	65	65	65	40	65
f_T (GHz)	85	150	150	150	250	150
Data rate (Gb/s)	10	10	10	10	10	10
C_{PD} (fF)	NA	50	60 (2)	60 (2)	100 (1)	120
PRBS length	31	31	7	7	7	11
Sensitivity (μA_{pp})	–	13	–	–	23.9 (3)	24.4
Output voltage (mV_{pp})	175	400	15.85 (4)	53.55 (3)	136	339
Energy efficiency (pJ/b)	**18.9**	**2.3**	**0.6**	**1.6**	**7.5**	**2.4**

(1) On-chip capacitor is added to consider the effect of the PD junction capacitance. (2) Calculated from the average input-referred noise current. (3) Calculated from measured eye diagrams that are not shown in [8]. (4) Circuit simulation with parasitic capacitances taken into consideration.

The presented receiver shows better energy efficiency than [21] which is implemented in a more advanced technology node and a comparable energy efficiency to [12] which is implemented in the same technology. The combination of multistage shunt-feedback TIA and the noiseless DFE in [12] has resulted in an excellent sensitivity at the cost of more complexity and power dissipation on the equalizer that consumes 74% of the total power. Therefore, a design that incorporates the high-gain FE in [12] with our proposed equalization technique with no additional power dissipation could lead to significant improvement on the energy-efficiency of the receiver while maintaining a good sensitivity. The work presented here shows comparable voltage sensitivity to the limiting amplifier introduced in [9], built by applying an active interleaving feedback to third-order gain cells. Finally, our work shows the largest output voltage amplitude for an input set to the sensitivity limit which makes it suitable to drive the subsequent clock and data recovery (CDR) circuit with negligible power penalty.

5.5. Operation at Higher Data Rate

The circuit in Figure 11a is also examined for 20 Gb/s operation with the same simulation setups described in Section 5.1 First, the TIA's bandwidth is set to 6 GHz (30% of the targeted data rate) by employing a feedback resistor of 800 Ω. Then, the limited-bandwidth TIA is followed by a wideband MA and the EMA, one at a time. Both amplifiers have the same value of I_B and I_f and therefore they consume the same DC power. The MA has a flat amplitude response with a bandwidth of 18.7 GHz. However, the overall bandwidth of the combined TIA/MA is dominated by the TIA's bandwidth. The EMA, on the other hand, introduces 3.5 dB of amplitude peaking at 10 GHz that extends the overall bandwidth of the combined TIA/EMA to 10.9 GHz. Figure 16a,b shows the simulation results for the output eye diagram for both scenarios. The internal eye opening improves by 1.6× when the EMA is employed compared to the case in which the wideband MA is used, demonstrating the capability of the presented technique in restoring the targeted bandwidth. The eye diagram in Figure 16c is obtained from the FE that includes TIA/MA after extending the TIA's bandwidth to 13.5 GHz by reducing its feedback resistor to 400 Ω, achieving an overall bandwidth of 11.8 GHz. Comparing Figure 16b,c) emphasizes that the presented design technique improves the effective gain compared to its conventional wide-bandwidth counterpart. The performance of the proposed FE at 20 Gb/s in comparison to its conventional counterpart is summarized in Table 1.

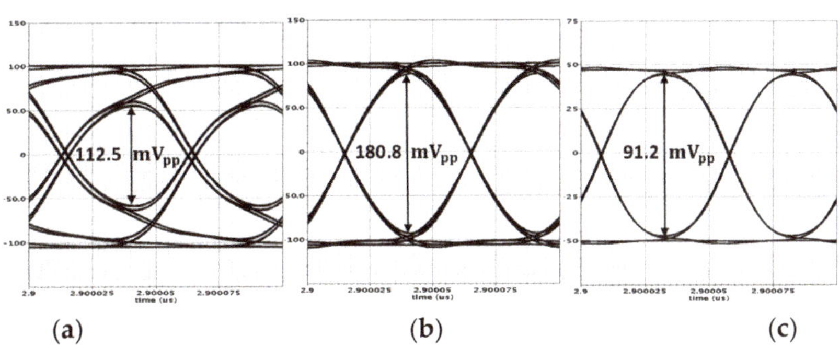

Figure 16. Simulation results for the 20 Gb/s output eye diagrams when the limited-bandwidth TIA is followed by (**a**) a wideband MA and (**b**) the proposed EMA (**b**). In (**c**), the TIA's bandwidth is widened, and a wideband MA is employed. The input current is fixed at 25 μA_{pp} for all simulations.

5.6. Operation with Large Input Signal

The presented analysis assumes that the gain cells are in linear operation. In reality, the circuit performance is strongly affected by the signal amplitude. As the signal propagates through cascaded stages, the latter gain cells start to saturate as a result of the increased voltage swing. Eventually, these cells act as unity-gain buffers and consequently the loop-gain falls below unity due to the presence of the active feedback. This in turn reduces the bandwidth. The impact of large input levels on the bandwidth of the active feedback-based structure is observed in [9] and an inverse scaling technique [22] is proposed as a potential solution for the problem. However, inverse scaling complicates the system analysis especially in the presence of interleaving feedback.

Alternatively, a straightforward automatic gain control similar to that presented in [6] can be employed. The technique has three steps: (1) aggressively reducing the TIA's gain at the cost of introducing a severe peaking in its amplitude response; (2) re-configure one of the MA stages to act as a low-pass filter to suppress the TIA's peaking and set the receiver bandwidth; (3) increasing the transconductance of the active feedback cell in the remaining MA stages to reduce their gain. In other words, at very high inputs, the TIA and the EMA interchange their roles. That is, the TIA introduces a high-frequency peaking that is then suppressed by the subsequent low-bandwidth amplifier. Figure 17 shows the simulation results for output eye diagrams when the input is set to 1 mA_{pp} at 10 Gb/s and 20 Gb/s. To generate these eyes, the TIA's feedback resistor is reduced to 60 Ω and the LPFs are removed from the EMA circuit. Despite the 7 dB of peaking in the TIA's amplitude response, the overall FE shows a flat amplitude response and a bandwidth of 12 GHz. The eye is fully open at 10 Gb/s. At 20 Gb/s, the internal eye opening is better than 60% of the maximum value. At both data rates, the eye opening is larger than it was at the sensitivity level. The widened eyes demonstrate the capability of the circuit to handle large input signals.

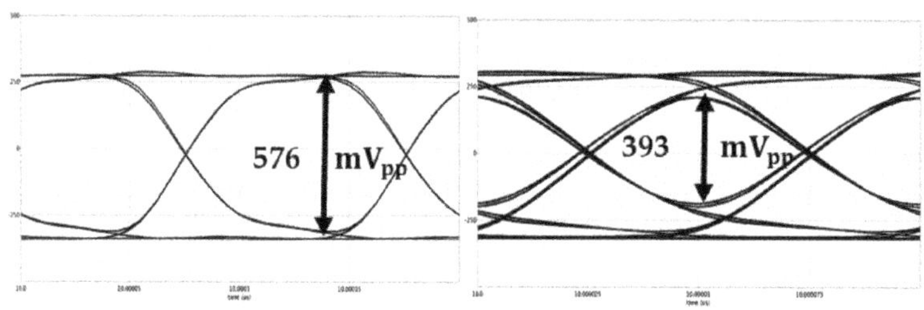

Figure 17. Simulation results for the output eye diagram when the input current is set to 1 mA_{pp} at (**a**) 10 Gb/s and (**b**) 20 Gb/s.

6. Conclusions

A design technique that relaxes the trade-off between gain and bandwidth in CMOS multi-stage amplifiers has been presented. To improve gain and reduce noise, the transimpedance amplifier is designed with a larger feedback resistor and its bandwidth limitation is compensated by a follow-on equalizing main amplifier (EMA). The EMA leverages the improved performance of state-of-the-art active-feedback main amplifier designs, but with the added benefit of high-frequency peaking. By embedding the equalizer stage in the gain stage, the overall circuit attains the improved performance of traditional equalizer-based designs, while achieving better energy efficiency due to the elimination of the standalone equalizer stage. The proposed front-end outputs an eye diagram with vertical openings of 338.9 mV_{pp} and 180 mV_{pp} at 10 Gb/s and 20 Gb/s, respectively. The vertical eye openings are doubled compared those of the conventional wide band front-end that operates at the same data rate and dissipates the same power, demonstrating the capability of the proposed technique to drive a subsequent decision circuit with a negligible power penalty. Simulation results also verify that the presented FE functions properly with large input signals and exhibits a robust performance against process and temperature variations.

Author Contributions: Conceptualization, D.A.; methodology, D.A.; software, D.A.; validation, D.A. and C.W.; formal analysis, D.A.; investigation, D.A.; resources, O.L.-L. and G.E.R.C.; writing—original draft preparation, D.A.; writing—review and editing, O.L.-L. and G.E.R.C.; visualization, D.A.; supervision, O.L.-L. and G.E.R.C.; project administration, O.L.-L. and G.E.R.C.; funding acquisition, O.L.-L. and G.E.R.C. All authors have read and agreed to the published version of the manuscript.

Funding: This work was supported in part by the Natural Sciences and Engineering Research Council of Canada through the Strategic Project Grant Program (Grant number is STPGP 494385-16) and in part by the Faculty of Engineering and Computer Science at Concordia University.

Institutional Review Board Statement: Not applicable.

Informed Consent Statement: Not applicable.

Data Availability Statement: Not applicable.

Acknowledgments: The authors would like to thank CMC Microsystems for providing access to CAD tools, fabrication, and test equipment. They would like to thank T. Obuchowicz for CAD support. They would also like to thank A. I. Abbas and R. Priti for assistance in testing the receiver prototype.

Conflicts of Interest: The authors declare no conflict of interest.

References

1. Li, H.; Hsu, C.M.; Sharma, J.; Jaussi, J.; Balamurugan, G. A 100-Gb/s PAM-4 Optical Receiver With 2-Tap FFE and 2-Tap Direct-Feedback DFE in 28-nm CMOS. *IEEE J. Solid-State Circuits* **2022**, *57*, 44–53. [CrossRef]
2. Li, H.; Sakib, M.; Dosunmu, O.; Liu, A.; Balamurugan, G.; Rong, H.; Jaussi, J.; Casper, B. A 112 Gb/s PAM4 CMOS Optical Receiver with Sub-pJ/bit Energy Efficiency. In Proceedings of the IEEE Optical Interconnects Conference (OI), Santa Fe, NM, USA, 24–26 April 2019.
3. Daneshgar, S.; Li, H.; Kim, T.; Balamurugan, G. A 128 Gb/s PAM4 Linear TIA with 12.6 pA/$\sqrt{\text{Hz}}$ Noise Density in 22 nm FinFET CMOS. In Proceedings of the IEEE Radio Frequency Integrated Circuits Symposium (RFIC), Atlanta, GA, USA, 7–9 June 2021.
4. Säckinger, E. *Broadband Circuits for Optical Fiber Communication*; John Wiley & Sons: Hoboken, NJ, USA, 2005.
5. Kim, J.; Buckwalter, J.F. Bandwidth Enhancement With Low Group-Delay Variation for a 40-Gb/s Transimpedance Amplifier. *IEEE Trans. Circuits Syst. I Reg. Pap.* **2010**, *57*, 1964–1972.
6. Li, D.; Minoia, G.; Repossi, M.; Baldi, D.; Temporiti, E.; Mazzanti, A.; Svelto, F. A Low-Noise Design Technique for High-Speed CMOS Optical Receivers. *IEEE J. Solid-State Circuits* **2014**, *49*, 1437–1447. [CrossRef]
7. Chen, T.; Chan, C.; Sheen, R.R. Transimpedance Limit Exploration and Inductor-Less Bandwidth Extension for Designing Wideband Amplifiers. *IEEE Trans. Very Large Scale Integr. (VLSI) Syst.* **2016**, *24*, 348–352. [CrossRef]
8. Fard, M.M.P.; Liboiron-Ladouceur, O.; Cowan, G.E.R. 1.23-pJ/bit 25-Gb/s Inductor-Less Optical Receiver With Low-Voltage Silicon Photodetector. *IEEE J. Solid-State Circuits* **2018**, *53*, 1793–1805. [CrossRef]
9. Huang, H.; Chien, J.; Lu, L. A 10-Gb/s Inductorless CMOS Limiting Amplifier With Third-Order Interleaving Active Feedback. *IEEE J. Solid-State Circuits* **2007**, *42*, 1111–1120. [CrossRef]
10. Ray, S.; Hella, M.M. A 53 dBΩ 7-GHz Inductorless Transimpedance Amplifier and a 1-THz + GBP Limiting Amplifier in 0.13-μm CMOS. *IEEE Trans. Circuits Syst. I Reg. Pap.* **2018**, *65*, 2365–2377. [CrossRef]
11. Pan, Q.; Wang, Y.; Lu, Y.; Yue, C.P. An 18-Gb/s Fully Integrated Optical Receiver With Adaptive Cascaded Equalizer. *IEEE J. Sel. Top. Quantum Electron.* **2016**, *22*, 361–369. [CrossRef]
12. Ahmed, M.; Talegaonkar, M.; Elkholy, A.; Shu, G.; Elmallah, A.; Rylyakov, A.; Rylyakov, P. A 12-Gb/s -16.8-dBm OMA Sensitivity 23-mW Optical Receiver in 65-nm CMOS. *IEEE J. Solid-State Circuits* **2018**, *53*, 445–457. [CrossRef]
13. Sackinger, E. The Transimpedance Limit. *IEEE Trans. Circuits Syst. I Reg. Pap.* **2010**, *57*, 1848–1856. [CrossRef]
14. Sackinger, E. On the Noise Optimum of FET Broadband Transimpedance Amplifiers. *IEEE Trans. Circuits Syst. I Reg. Pap.* **2012**, *59*, 2881–2889. [CrossRef]
15. Abdelrahman, D.; Cowan, G.E.R. Noise Analysis and Design Considerations for Equalizer-Based Optical Receivers. *IEEE Trans. Circuits Syst. I Reg. Pap.* **2019**, *66*, 3201–3212. [CrossRef]
16. Radice, F.; Bruccoleri, M.; Mammei, E.; Bassi, M.; Mazzanti, A. A low-noise programmable-gain amplifier for 25 Gb/s multi-mode fiber receivers in 28 nm CMOS FDSOI. In Proceedings of the 41st European Solid-State Circuits Conference (ESSCIRC), Graz, Austria, 14–18 September 2015.
17. Ray, S.; Chowdhury, A.; Hella, M.M. Enhancing the Stability of Broadband Amplifiers Using Third Order Nested Feedback. In Proceedings of the IEEE International Symposium on Circuits and Systems (ISCAS), Florence, Italy, 27–30 May 2018.
18. Shahramian, S.; Yasotharan, H.; Carusone, A.C. Decision Feedback Equalizer Architectures With Multiple Continuous-Time Infinite Impulse Response Filters. *IEEE Trans. Circuits Sys. II Express Briefs* **2012**, *59*, 326–330. [CrossRef]
19. Liu, C.; Yan, Y.; Goh, W.; Xiong, Y.; Zhang, L.; Madihian, M. A 5-Gb/s Automatic Gain Control Amplifier With Temperature Compensation. *IEEE J. Solid-State Circuits* **2012**, *47*, 1323–1333. [CrossRef]
20. Kim, W.; Seong, C.; Choi, W. A 5.4-Gbit/s Adaptive Continuous-Time Linear Equalizer Using Asynchronous Undersampling Histograms. *IEEE Trans. Circuits Syst. II Express Briefs* **2012**, *59*, 553–557. [CrossRef]
21. Chien, Y.; Fu, K.; Liu, S. A 3–25 Gb/s Four-Channel Receiver With Noise-Canceling TIA and Power-Scalable LA. *IEEE Trans. Circuits Syst. II Express Briefs* **2014**, *61*, 845–849. [CrossRef]
22. Sackinger, E.; Fischer, W.C. A 3-GHz 32-dB CMOS limiting amplifier for SONET OC-48 receivers. *IEEE J. Solid-State Circuits* **2000**, *35*, 1884–1888. [CrossRef]

Journal of
*Low Power Electronics
and Applications*

MDPI

Article

Towards Integration of a Dedicated Memory Controller and Its Instruction Set to Improve Performance of Systems Containing Computational SRAM

Kévin Mambu *, Henri-Pierre Charles *, Maha Kooli * and Julie Dumas *

CEA, LIST, Université Grenoble Alpes, F-38000 Grenoble, France
* Correspondence: kevin.mambu@cea.fr (K.M.); henri-pierre.charles@cea.fr (H.-P.C.); maha.kooli@cea.fr (M.K.); julie.dumas@cea.fr (J.D.)

Abstract: In-memory computing (IMC) aims to solve the performance gap between CPU and memories introduced by the memory wall. However, it does not address the energy wall problem caused by data transfer over memory hierarchies. This paper proposes the data-locality management unit (DMU) to efficiently transfer data from a DRAM memory to a computational SRAM (C-SRAM) memory allowing IMC operations. The DMU is tightly coupled within the C-SRAM and allows one to align the data structure in order to perform effective in-memory computation. We propose a dedicated instruction set within the DMU to issue data transfers. The performance evaluation of a system integrating C-SRAM within the DMU compared to a reference scalar system architecture shows an increase from ×5.73 to ×11.01 in speed-up and from ×29.49 to ×46.67 in energy reduction, versus a system integrating C-SRAM without any transfer mechanism compared to a reference scalar system architecture.

Keywords: in-memory computing; energy modeling; non-von neumann; instruction set; compilation; stencils; convolutions; sram; energy wall; memory wall

Citation: Mambu, K.; Charles, H.-P.; Kooli, M.; Dumas, J. Towards Integration of a Dedicated Memory Controller and Its Instruction Set to Improve Performance of Systems Containing Computational SRAM. *J. Low Power Electron. Appl.* **2022**, *12*, 18. https://doi.org/10.3390/jlpea12010018

Academic Editors: Alex Serb and Adnan Mehonic

Received: 15 December 2021
Accepted: 14 February 2022
Published: 16 March 2022

1. Introduction

Von Neumann architectures are limited by the performance bottleneck characterized by the "memory wall", i.e., the performance limitation of memory units compared to CPU, and the "energy wall", i.e., the gap between the energies consumed for computation and data transfers between different system components.

Figure 1a exposes the energy discrepancy between each component of a standard von Neumann architecture. We note that the energy increases by ×100 between the CPU and the cache memory, and by ×10,000 between the CPU and the DRAM memory [1]. In-memory computing (IMC) is a solution to implement non-von Neumann architectures and mitigate the memory wall by moving computation directly into memory units [2]. It allows the reduction of data transfers and thus energy consumption. However, the efficiency of IMC depends on the proper arrangement of data structures. Indeed, to be correctly computed in the memory, data should be arranged to respect a precise order (e.g., aligned in memory rows) imposed by IMC hardware design constraints.

While various state-of-the-art works propose IMC solutions, very few take into account their integration to complete computer systems while describing efficient methods to transfer data from IMC to high-latency memories or peripherals. This lack of consideration can be explained by the majority of IMC architectures being currently specialized for a few use cases, i.e., AI and big data, which limits their efficiency for general-purpose computing. We propose a data-locality management unit (DMU), a transfer block presented in Figure 1b, coupled to an SRAM-based IMC unit to generate efficient data transfer and reorganization through a dedicated instruction set. As IMC architecture, we consider the computational SRAM (C-SRAM), an SRAM-based bit-parallel IMC architecture detailed in [2–4], and able to perform logical and arithmetical operations in-parallel thanks to an arithmetic and logic unit (ALU) in its periphery. We integrate it within a CPU and a DRAM as main memory.

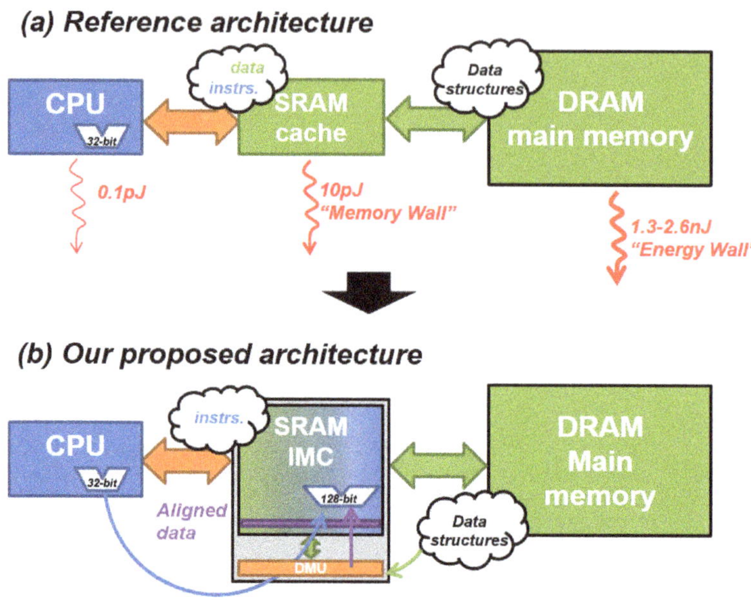

Figure 1. (**a**) Performance bottlenecks of von Neumann architecture; energy costs are based on [1]. (**b**) Proposed architecture with IMC to mitigate the "memory wall" and a "DMU" block to mitigate the "energy wall" between IMC and low-latency memory.

The system evaluation of C-SRAM and DMU on three applications—frame differencing, Sobel filter and matrix multiplication—versus a system integrating C-SRAM without a dedicated transfer mechanism shows an increase from ×5.73 to ×11.01 in speed-up and from ×29.49 to ×46.67 in energy reduction, according to a baseline scalar architecture.

The rest of this paper is organized as follows. Section 2 presents related works. Section 3 details our proposed solution. Section 4 evaluates the DMU on three applications and discusses the gains obtained through its integration. Finally, Section 5 concludes the paper and exposes future works.

2. Related Work

2.1. In-Memory Computing (IMC)

IMC architectures of the state-of-the-art can be differentiated by their technology and programming model [5]. Volatile memory-based IMC architectures include DRAM and SRAM technologies. DRAM-based IMC architectures propose to enhance DRAM memories with bulk-bitwise computation operators. These solutions offer cost and area efficiency and large parallelism, although their arithmetic support is limited to logical or specialized operators [6,7]. SRAM-based IMC architectures are less scalable than DRAM-based solutions in terms of design, but they implement more elaborated computation operators, either through strict IMC using bit-lines and sense amplifiers, or through near-memory computing by using an arithmetic logic unit (ALU) in the periphery of the bit-cell array [3,8,9]. Other approaches using emerging technologies such as MRAM or ReRAM [10,11] have been explored. They present interesting opportunities in terms of access latency and nonvolatile capability, but have drawbacks in terms of cycle-to-cycle variability and analog-to-digital conversion of input data.

Despite all these technological differences, a common challenge regarding system integration is the specific memory management required to properly map data to be able to perform effective computations. A study of the literature presents three main mapping strategies of existing IMC architectures. First, bit-parallel mapping requires each operand of a given operation to be aligned on the same bit-cells to be computed, which induces the logical memory alignment of data [3,6]. Then, in bit-serial data mapping, operands are aligned on the same rows, and all bits of a given operand are aligned on the same bit-cell [8]. This requires data to be physically transposed in the array to be effectively computed. Third, crossbar mapping is a 2D mapping scheme used to perform computation in the form of convolutions between analog data driven through different word-lines and data along each bit-line [11]. Bit-serial and crossbar mapping schemes show advantages in terms of computation capabilities, but are difficult to integrate due to their data management constraints (data transposing for bit-serial and digital-to-analog conversion and rearrangement for crossbar) which conflict with the inherent bit-parallel nature of modern computer systems.

In this paper, we focus on an SRAM-based bit-parallel architecture, which shows a better compromise between computation capability, system integration and data compatibility [2–4].

2.2. Data Management Solutions for IMC

In this subsection, we focus on data management solutions devised by previous works for IMC architectures. First, IMC architectures based on emerging technologies evaluate their solutions while not considering the cost of external data transfers with high-latency memories or peripherals, as their architectures are not yet mature for system integration. We only retain two state-of-the-art solutions of data management when focusing on conventional technologies, e.g., DRAM and SRAM-based.

"PIM-enabled instructions" [12] (PEI) establishes the specifications for a generic IMC architecture while also defining its integration to the memory hierarchy. PEI computation units (PCU) are integrated within a hybrid memory cube, each managing a DRAM module to perform NMC on the main memory while a local PCU is coupled to the host CPU to perform NMC at the cache-level, e.g., on SRAM. While the paper proposes hardware and software mechanisms to ensure data coherency and locality management, the complexity of this solution makes this architecture difficult to implement and program. Moreover, the cost of data movements for PCUs computing on DRAM means that data management needs to be statically scheduled to reduce dynamic rearrangement as often as possible. Finally, PEI does not define specialized instructions for memory accesses, limiting the charge of data management to the host CPU and extending it to DRAM memory.

The duality cache [8] is designed to be implemented in the last-level cache (LLC). Its data management mechanism is based on the cache controller of the LLC to transfer data to different data banks. Other mechanisms such as the transpose memory unit (TMU) are implemented to transpose data from bit-parallel-in-DRAM to bit-serial-in-IMC. While the hardware and software environment of the duality cache make for an effective data management mechanism for its target applications (GPU kernels, especially AI), it is not explicitly programmable and it is constraining for developers. This limitation implies the static allocation and management of data before computation, and according to the hardware mechanisms of the duality cache.

The transfer mechanism we describe in this paper proposes a dedicated instruction set, integrated as an extension of the IMC ISA to offer explicit management on dynamically allocated data to developers. Compared to previous solutions, our interest is compatibility with general-purpose computing.

3. DMU Specification

3.1. Overview

In this section, we present a DMU, a memory controller architecture to provide memory access instructions to IMC to efficiently transfer and reorganize data before computation. We implement in the DMU the control of source and destination offsets to enable fine-grain data reorganization in IMC as well as DRAM memory to address alignment constraints necessary for certain applications, and the implementation of two different operating modes makes online data padding available. Finally, the DMU implements a dedicated instruction set to program data transfers in a single clock cycle, compared to classical DMA solutions. This instruction set is implemented as a subset of the C-SRAM instruction set architecture. The DMU controller is proposed to be tightly coupled in the periphery of the IMC unit, as shown in Figure 2. This means that there is a direct interface between the DMU and IMC without going through the system bus, which is one of the main difference compared to existing DMA controllers.

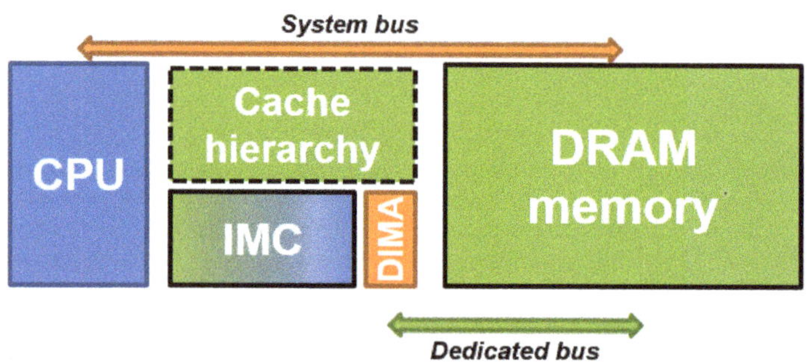

Figure 2. The integration of DMU to IMC offers an instruction set for efficient data transfers as well as a dedicated transfer bus with the main memory.

3.2. Instruction Set Architecture

Table 1 shows the proposed instructions defined for the DMU. All instructions are nonblocking for the host architecture, and a `BLOCKING_WAIT` instruction is implemented to ensure synchronization between it and the C-SRAM.

Table 1. Summary of DMU instructions.

Operation	Parameters
`SET_SRC_DRAM_REGION` (Nonblocking)	DRAM base address, region width, element size
`SET_DST_DRAM_REGION` (Nonblocking)	DRAM base address, region width, element size
`READ_TRANSFER` (MEM→IMC, nonblocking)	Source X position, source Y position, dest. IMC address, length, source offset, dest offset, operating mode
`COPY` (IMC→IMC, nonblocking)	Source IMC address, dest IMC address, source offset, dest offset, operating mode
`WRITE_TRANSFER` (MEM→IMC, nonblocking)	Dest X position, dest Y position, source IMC address, length, source offset, dest. offset, operating mode
`BLOCKING_WAIT`	None

`SET_SRC_DRAM_REGION` and `SET_DST_DRAM_REGION` take as parameters 2D regions in DRAM memory space, characterized by: their base address A and their width and height, respectively, W and H. The number of bits required for A is architecture-dependent while the numbers of bits required to encode H and W are implementation-dependent. By defining in advance the source and destination data structures in DRAM, `READ_TRANSFER` and `WRITE_TRANSFER` instructions only require, respectively, the source and destination positions (X, Y) in the target region, which can be encoded in $log_2(W)$ and $log_2(H)$ bits. This indirect addressing effectively reduces the number of bits required for both instructions, which are the most frequently called instructions of the instruction set. Figure 3 illustrates this mechanism from the developer's point of view.

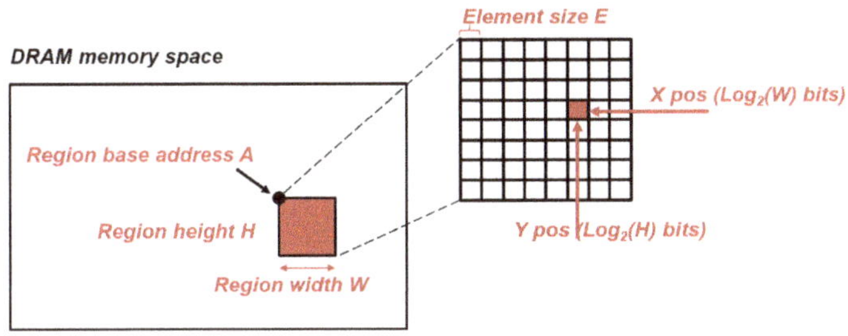

Figure 3. Representation of a 2D region in the DRAM memory space, defined by `SET_DRAM_REGION`, and a single element in said region. This decomposition allows reducing the number of bits required in transfer instructions for DRAM address.

`READ_TRANSFER`, `WRITE_TRANSFER` and `COPY` can operate to transfer data and perform online reorganization. For example, the parameterizing of source and destination offsets allow the data to be padded upon arrival in the C-SRAM. To cover most use-cases induced with the configuration of the destination offset, we implement in the DMU two operating modes through the `transfer_start` register, illustrated in Figure 4. A *zero-padding mode* fills the blanks in between destination data with zeros to perform unsigned byte extension, while an *overwriting mode* preserves the data present in the destination C-SRAM row and updates only relevant bytes. The former is destructive but enables online byte extension to perform higher-precision arithmetic for workloads such as image processing or machine learning, while the latter is more suitable for nondestructive data movements. Since most iterative codes such as convolutions induce strong data redundancy, `COPY` can be used to duplicate data and mitigate accesses with the DRAM for better energy efficiency.

Algorithm 1 describes the side effects generated by the `READ_TRANSFER` instruction, according to its parameters and the offset mechanism described in Figure 4.

Algorithm 1: `READ_TRANSFER` description.

```
Data: x, y, dst_addr, length, src_off, dst_off, op_mode
src_addr = @DRAM[A + E × (src_y × W + src_x)];
forall i ∈ [0, length[ do
    for j ∈ [0, E[ do
    |   dst_addr[E × i × dst_off + j] = src_addr[E × i × src_off + j];
    end
    if op_mode == DMU_ZERO_PAD then we overwrite the gaps between written data with zeroes
        for j ∈ [1, dst_off[ do
            for k ∈ [0, E[ do
            |   dst_addr[E × (i × dst_off + j) + k] = 0;
            end
        end
    else we preserve data in gaps between written data
    |   continue;
    end
end
```

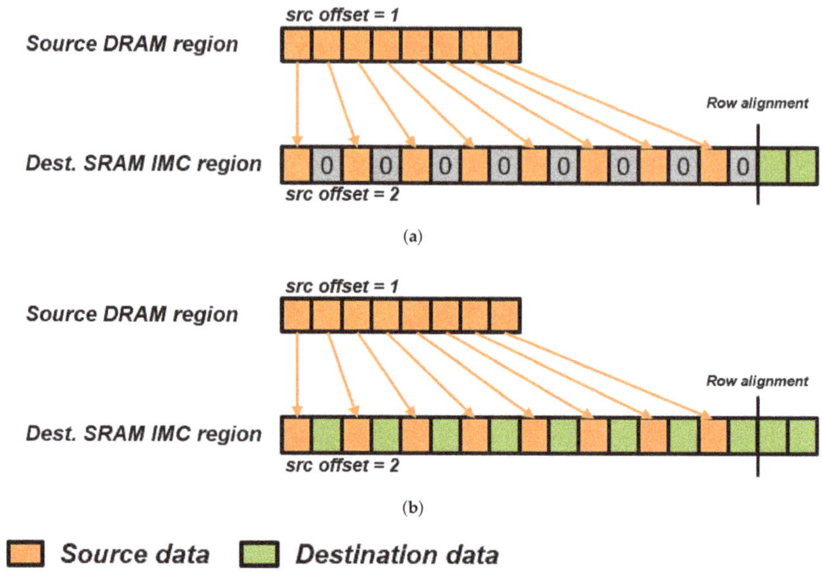

Figure 4. DMU operating modes and their impact on destination memory, here an SRAM IMC memory. (**a**) Zero-padding mode. (**b**) Overwriting mode.

4. Evaluation and Results

4.1. Experimental Methodology

Figure 5a,b present, respectively, the reference and the experimental architecture for the evaluation. The reference architecture is a 1 GHz in-order scalar CPU with a single-level cache hierarchy of 16 kB data and instruction caches, and a 512 MB LP-DDR main memory. Our architecture substitutes the 16 kB data cache with an 8 kB data cache, an 8 kB C-SRAM unit and our DMU controller. Both architectures are equivalent in terms of memory capacity but differ by their computation capabilities and the usage of each of their memory units. For each architecture, the cache memories implement a write-through policy to ensure that the written output data are present in the main memory.

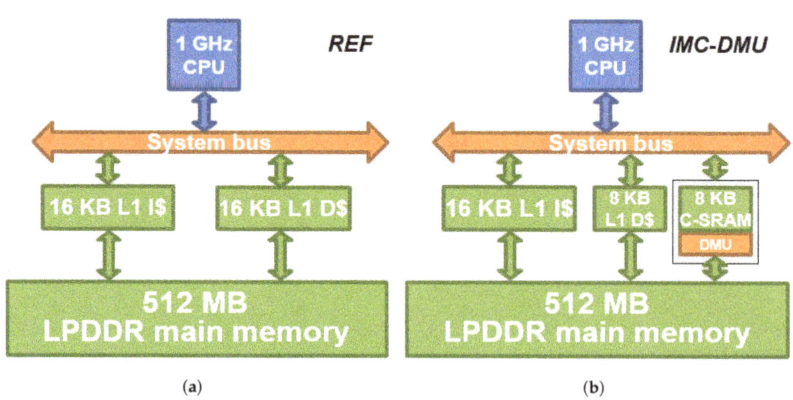

Figure 5. Experimental memory architectures for the evaluation. All cache units have write-through policy. Our proposed architecture substitutes the 16 kB L1 D with an 8 kB L1 D, an 8 kB C-SRAM and a DMU. (**a**) Reference architecture; (**b**) proposed architecture.

Our simulation methodology is based on QEMU in order to perform system-level modeling and evaluation while performing ISA exploration for IMC as well as our DMU instruction [13]. We describe and generate events to model the energy and latency costs of our architecture, from CPU and IMC to the main memory and the cache hierarchy. Table 2 shows the memory parameters of, respectively, the reference architecture and our proposed IMC architecture. These parameters were extracted using CACTI [14] and the characterizations from [2]. Though the specifications of the C-SRAM architecture do not define fixed dimensions, e.g., a fixed row size and number of rows, we consider a 128-bit C-SRAM architecture for the rest of this paper.

Table 2. Memory parameters of the reference and proposed architecture, used for the experimental evaluation.

Component	Operational Latency	Operational Energy Cost
Reference architecture (REF), using CACTI [14]:		
16 kB L1 I/D	1 ns	7.01 pJ
512 MB LP-DDR	17 ns	1.067 nJ
Our evaluated architecture (IMC-DMU), using [2,14]:		
16 kB L1 I	1 ns	7.01 pJ
8 kB L1 D	1 ns	4.93 pJ
8 kB L1 C-SRAM	3 ns	7.94 pJ
512 MB LP-DDR	17 ns	1.067 nJ

4.2. Applications

We consider three applications to evaluate our proposed architecture (IMC-DMU) versus the reference scalar architecture (REF):

- *Frame differencing* is used in computer vision to perform motion detection [15], and performs saturated subtraction between two (or more) consecutive frames in a video stream to highlight pixel differences. It has linear complexity in both computing and memory.
- A *Sobel filter* applies two 3×3 convolution kernels on an input image to generate its edge-highlighted output. It is a standard operator in Image processing as well as computer vision to perform edge detection [16]. It has linear arithmetic complexity and shows constant data redundancy (2×9 reads per input pixel, on average).
- *Matrix-matrix multiplication* is used in various domains such as signal processing or physics modeling, and is a standard of linear algebra as the `gemm` operator [17]. It has cubic ($\mathcal{O}(n^3)$) complexity in computing and memory and shows quadratic ($\mathcal{O}(n^2)$) data redundancy.

4.3. DMU Programming and Data Optimization

Iterative codes using complex patterns such as convolution windows often require streams of consecutive and redundant data to be transferred and aligned before computation according to the mapping scheme of IMC. To reduce the overall transfer latency to populate IMC with said data, we use `READ_TRANSFER` to copy nonpresent data from the DRAM to IMC and `COPY` to directly duplicate redundant data instead of accessing the DRAM memory.

Figure 6 shows the memory access patterns performed for convolution windows at stride 1, each element but the very first and last on each row of the image during the visit is read from the DRAM only once. The rest at some point can be directly read from the C-SRAM to reduce the overall transfer latency. Figure 7 implements the data management described in Figure 6. This technique can be applied to other iterative codes such as stencils, though the appropriate address generations require effort from developers. We are currently developing a mechanism to automatically generate data transfers and duplication from programmable memory access patterns to effortlessly achieve quasi-optimal energy efficiency. We will soon publish our specifications and our results.

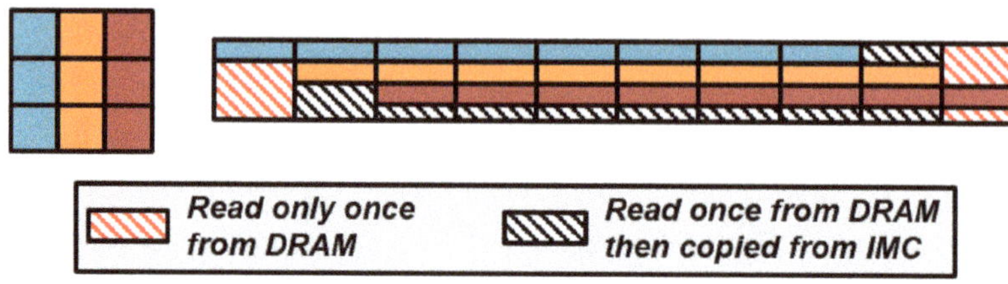

Figure 6. A stream of convolution windows, transferred row-major, shows opportunities of data duplication to reduce transfer latency.

```
/*
 * The following code transfers image data by eight convolution windows to C-SRAM,
 * the convolution data are aligned in order to perform vector computation
 */
// CSRAM_NBYTES & CSRAM_ROW_NBYTES are implementation-specific
#define NB_UINT16_PER_ROW (CSRAM_ROW_NBYTES / sizeof(uint16_t))
#define KERN_STRIDE 1
// _csram_16b is the memory mapping of C-SRAM at 16-bit granularity
volatile uint16_t \_csram_16b[CSRAM_NBYTES / CSRAM_ROW_NBYTES][NB_UINT16_PER_ROW];
for(int i = 1; i < img_height-1; i += 1)
{
  for(int j = 1; j < img_width-1; j += NB_UINT16_PER_ROW)
  {
    int nrow = 0;
    for(int iker = -1; iker <= 1; iker += 1)
    {
      READ_TRANSFER(i+iker, j-1, &_csram_16b[nrow], NB_UINT16_PER_ROW, KERN_STRIDE, sizeof(uint16_t), ZERO_PAD);
      COPY(&_csram_16b[nrow ][1], &_csram_16b[nrow+1], NB_UINT16_PER_ROW-1, KERN_STRIDE, sizeof(uint16_t), ZERO_PAD);
      READ_TRANSFER(i+iker, j , &_csram_16b[nrow+1][NB_UINT16_PER_ROW-1], 1, KERN_STRIDE, sizeof(uint16_t), ZERO_PAD);
      COPY(&_csram_16b[nrow+1][1], &_csram_16b[nrow+2], NB_UINT16_PER_ROW-1, KERN_STRIDE, sizeof(uint16_t), ZERO_PAD);
      READ_TRANSFER(i+iker, j+1, &_csram_16b[nrow+2][NB_UINT16_PER_ROW-1], 1, KERN_STRIDE, sizeof(uint16_t), ZERO_PAD);

      nrow += 3;
    }
  }
}
```

Figure 7. Example code transferring convolution windows to C-SRAM using our DMU instruction set.

4.4. Results and Discussion

We evaluated the performance of our proposed architecture, according to a reference architecture, in terms of speed-ups and energy reductions. We considered three scenarios: (1) using the C-SRAM without DMU (C-SRAM-only), (2) using the C-SRAM with the proposed DMU controller to fetch input data strictly from the main memory and (C-SRAM+DMU) and (3) using the C-SRAM with the proposed DMU controller to perform data transfers and data reuse whenever possible. In case 1, the data are transferred from the L1 data cache to the C-SRAM by the CPU, while in cases 2 and 3, the CPU issues data transfers directly between the main memory and the C-SRAM using the DMU. Case 3 is particularly relevant to the Sobel filter, which presents data redundancy due to the application of the convolution filters on the input images.

Figure 8 shows the energy reduction and speed-up for the three applications, compared to the reference scalar architecture. The X-axis represents the size of the inputs, and Y-axis represents the improvement factors evaluated for each application (higher is better). Table 3 shows the average of the maximum speed-up and energy reductions evaluated for each implementation across all applications. While the C-SRAM-only implementation shows improvement compared to the scalar system, the integration of the DMU to the C-SRAM improves the speed-up and energy reduction, respectively, from ×5.73 and ×11.01 to ×29.49 and ×46.67.

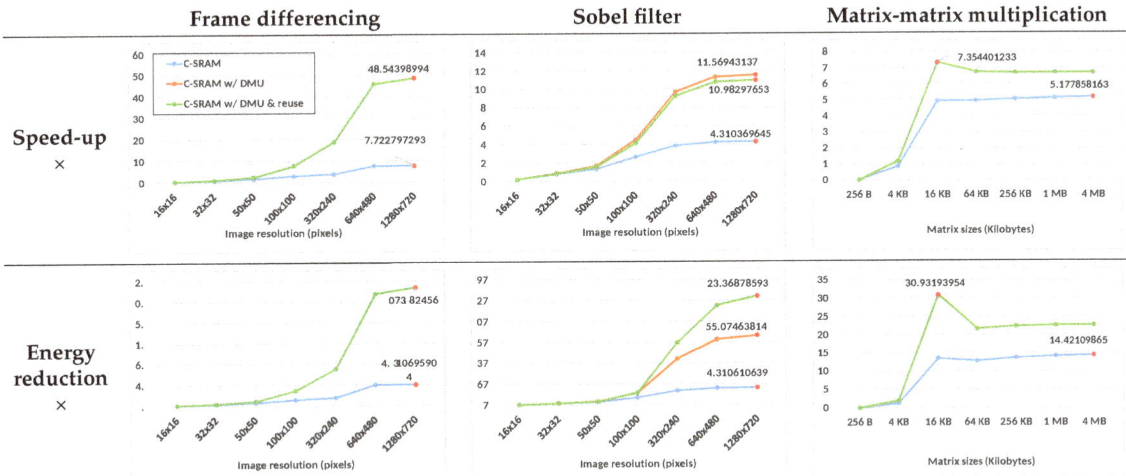

Figure 8. Energy reduction and speed-up for all applications compared to the reference scalar architecture. The X and Y axes of the plots are, respectively, the data sizes and the improvement factors, i.e., higher is better.

Table 3. Average maximum speed-up and energy reduction per evaluated implementation.

	Average Maximum Speed-Up	Average Maximum Energy Reduction
C-SRAM-only	×5.73	×11.01
C-SRAM + DMU	×29.49	×46.67

4.4.1. Frame Differencing

We observe that both the speed-up and the energy reduction are similar without and with the DMU for small resolutions, but are significantly improved by using the DMU starting 100 × 100 image resolution, with peak factors achieved of, respectively, ×48.54 and ×57.09, compared to ×7.72 and ×10.35 when using only the C-SRAM. Because the DMU offers a direct transfer bus between the main memory and C-SRAM, the entirety of the input data can be fetched directly from the C-SRAM without generating as much conflict misses on the L1 data cache. Our hypothesis is supported by the performance gap between the C-SRAM-only and the C-SRAM + DMU implementation getting wider as the images get larger starting 100 × 100 resolution. The input and output sizes become too large for the L1 data cache.

4.4.2. Sobel Filter

Similar to frame differencing, the speed-up and energy consumption of the Sobel filter is consistently improved by the DMU. In addition, the Sobel filter offers the opportunity to perform data reuse between consecutive convolution windows by reading part of their components directly from the C-SRAM instead of the main memory. By using the DMU without reusing data, the peak speed-up and energy reduction compared to the reference scalar architecture is ×11.57 and ×33.41, while the data reuse implementation shows factors of ×10.98 and ×52.22, which shows a trade-off between the execution performance of C-SRAM + DMU and its energy efficiency. This trade-off is explained by the overhead of the address generation algorithm required for IMC data duplication. The overhead might change depending on the implementation of the algorithm and the CPU performance.

4.4.3. Matrix-Matrix Multiplication

The overall performance of matrix-matrix multiplication is almost similar with and without DMU on small matrices. However, for both implementations the speed-up and energy reduction compared to the reference architecture improve significantly starting from 16 kB matrices, to, respectively, ×7.35 and ×30.93 for the C-SRAM + DMU implementation. The improvement factors are higher for C-SRAM + DMU at this input size because only one of the two operand matrices is transferred through the L1 data cache in this implementation, contrary to the C-SRAM-only implementation, which transfers all input and output matrices through the L1 data cache. Furthermore, the performance drops for both implementations at a 64 kB input size, because the capacity of the L1 data cache has been attained on the experimental architecture to store the first operand matrix. The fact that this drop is more significant for the C-SRAM + DMU implementation compared to the C-SRAM-only variant shows the impact of the cache hierarchy on the performance of IMC architectures without any dedicated transfer mechanism.

5. Conclusions

We presented the DMU, a programmable memory controller architecture to efficiently transfer and reorganize data between the SRAM IMC memory and the main memory. We integrated the DMU in a C-SRAM architecture and evaluated the energy reduction and speed-up for three applications, compared to a reference scalar architecture. The integration of the DMU to C-SRAM improved the speed-up and energy reduction, respectively, from ×5.73 and ×11.01 to ×29.49 and ×46.67.

Our future works include the physical implementation of the DMU on a test chip for the validation of our experiments and the compiler support of its ISA to implement an efficient programming model at the language level. We also plan to describe the specifications of a more elaborate instruction set, able to transfer complex data structures such as stencil kernels and convolution windows using pattern descriptors, in order to automate transfer optimizations at the hardware level.

Author Contributions: Conceptualization, K.M., H.-P.C., M.K. and J.D.; methodology, all authors; software, all authors; validation, all authors; formal analysis, all authors; investigation, all authots; resources, all authors; data curation, all authots; writing—original, K.M.; writing—review and editing, H.-P.C., M.K. and J.D.; visualization, all authors; supervision, H.-P.C. and M.K.; project administration, H.-P.C. and M.K.; funding acquisition, H.-P.C. and M.K. All authors have read and agreed to the published version of the manuscript.

Funding: This work was supported by the EU H2020 project 955606 "DEEPSEA" – Software for Exascale Architectures.

Data Availability Statement: Not applicable.

Conflicts of Interest: The authors declare no conflict of interest.

References

1. Horowitz, M. 1.1 Computing's energy problem (and what we can do about it). In Proceedings of the 2014 IEEE International Solid-State Circuits Conference Digest of Technical Papers (ISSCC), San Francisco, CA, USA, 9–13 February 2014; pp. 10–14.
2. Noel, J.P.; Pezzin, M.; Gauchi, R.; Christmann, J.F.; Kooli, M.; Charles, H.P.; Ciampolini, L.; Diallo, M.; Lepin, F.; Blampey, B.; et al. A 35.6 TOPS/W/mm^2 3-Stage Pipelined Computational SRAM with Adjustable Form Factor for Highly Data-Centric Applications. *IEEE Solid-State Circuits Lett.* **2020**, *3*, 286–289. [CrossRef]
3. Kooli, M.; Charles, H.P.; Touzet, C.; Giraud, B.; Noel, J.P. Smart Instruction Codes for In-Memory Computing Architectures Compatible with Standard SRAM Interfaces. In Proceedings of the 2018 Design, Automation & Test in Europe Conference & Exhibition (DATE), Dresden, Germany, 19–23 March 2018; p. 6.
4. Gauchi, R.; Egloff, V.; Kooli, M.; Noel, J.P.; Giraud, B.; Vivet, P.; Mitra, S.; Charles, H.P. Reconfigurable tiles of computing-in-memory SRAM architecture for scalable vectorization. In Proceedings of the ACM/IEEE International Symposium on Low Power Electronics and Design, Boston, MA, USA, 10–12 August 2020; pp. 121–126.
5. Bavikadi, S.; Sutradhar, P.R.; Khasawneh, K.N.; Ganguly, A.; Pudukotai Dinakarrao, S.M. A Review of In-Memory Computing Architectures for Machine Learning Applications. In Proceedings of the 2020 on Great Lakes Symposium on VLSI, Virtual, China, 7–9 September 2020; pp. 89–94.

6. Seshadri, V.; Lee, D.; Mullins, T.; Hassan, H.; Boroumand, A.; Kim, J.; Kozuch, M.A.; Mutlu, O.; Gibbons, P.B.; Mowry, T.C. Ambit: In-Memory Accelerator for Bulk Bitwise Operations Using Commodity DRAM Technology. In Proceedings of the 2017 50th Annual IEEE/ACM International Symposium on Microarchitecture (MICRO), Boston, MA, USA, 14–17 October 2017; pp. 273–287.
7. Deng, Q.; Zhang, Y.; Zhang, M.; Yang, J. LAcc: Exploiting Lookup Table-based Fast and Accurate Vector Multiplication in DRAM-based CNN Accelerator. In Proceedings of the 56th Annual Design Automation Conference, Las Vegas, NV, USA, 2–6 June 2019; pp. 1–6.
8. Fujiki, D.; Mahlke, S.; Das, R. Duality Cache for Data Parallel Acceleration. In Proceedings of the 46th International Symposium on Computer Architecture, Phoenix, AZ, USA, 22–26 June 2019; pp. 1–14.
9. Lee, K.; Jeong, J.; Cheon, S.; Choi, W.; Park, J. Bit Parallel 6T SRAM In-memory Computing with Reconfigurable Bit-Precision. In Proceedings of the 2020 57th ACM/IEEE Design Automation Conference (DAC), San Francisco, CA, USA, 20–24 July 2020; pp. 1–6.
10. Bhattacharjee, D.; Devadoss, R.; Chattopadhyay, A. ReVAMP: ReRAM based VLIW architecture for in-memory computing. In Proceedings of the Design, Automation & Test in Europe Conference & Exhibition (DATE), Lausanne, Switzerland, 27–31 March 2017; pp. 782–787.
11. Ezzadeen, M.; Bosch, D.; Giraud, B.; Barraud, S.; Noel, J.P.; Lattard, D.; Lacord, J.; Portal, J.M.; Andrieu, F. Ultrahigh-Density 3-D Vertical RRAM with Stacked Junctionless Nanowires for In-Memory-Computing Applications. *IEEE Trans. Electron Devices* **2020**, *67*, 4626–4630. [CrossRef]
12. Ahn, J.; Yoo, S.; Mutlu, O.; Choi, K. Pim-enabled instructions: A low-overhead, locality-aware processing-in-memory architecture. In Proceedings of the 2015 ACM/IEEE 42nd Annual International Symposium on Computer Architecture (ISCA), Portland, OR, USA, 13–17 June 2015; pp. 336–348.
13. Mambu, K.; Charles, H.-P.; Dumas, J.; Kooli, M. Instruction Set Design Methodology for In-Memory Computing through QEMU-Based System Emulator. 2021. Available online: https://hal.archives-ouvertes.fr/hal-03449840/document (accessed on 14 December 2021).
14. Muralimanohar, N.; Balasubramonian, R.; Jouppi, N.P. *Cacti 6.0: A Tool to Model Large Caches*; HP Laboratories: Palo Alto, CA, USA, 2009.
15. Zhang, H.; Wu, K. A Vehicle Detection Algorithm Based on Three-Frame Differencing and Background Subtraction. In Proceedings of the 2012 Fifth International Symposium on Computational Intelligence and Design, Hangzhou, China, 28–29 October 2012; Volume 1, pp. 148–151.
16. Khronos Vision Working Group. The OpenVX Specification Version 1.2. Available online: https://www.google.co.th/url?sa=t&rct=j&q=&esrc=s&source=web&cd=&cad=rja&uact=8&ved=2ahUKEwjB65aM0sn2AhUkxzgGHa-sDlMQFnoECAgQAQ&url=https%3A%2F%2Fwww.khronos.org%2Fregistry%2FOpenVX%2Fspecs%2F1.2%2FOpenVX_Specification_1_2.pdf&usg=AOvVaw18k11o92s0PEjGw7rZ5Sm8 (accessed on 11 October 2017).
17. Basic Linear Algebra Subprograms Technical (BLAST) Forum Standard, Basic Linear Algebra Subprograms Technical Forum. Available online: https://www.google.co.th/url?sa=t&rct=j&q=&esrc=s&source=web&cd=&cad=rja&uact=8&ved=2ahUKEwj6orG00sn2AhU_zjgGHcvDADoQFnoECAgQAQ&url=http%3A%2F%2Fwww.netlib.org%2Futk%2Fpeople%2FJackDongarra%2FPAPERS%2F135_2002_basic-linear-algebra-subprograms-techinal-blas-forum-standard.pdf&usg=AOvVaw3GQ_wyRmkgT9TG9mwuCS0I (accessed on 21 August 2001).

Journal of
Low Power Electronics and Applications

MDPI

Article

Silicon-Compatible Memristive Devices Tailored by Laser and Thermal Treatments

Maria N. Koryazhkina [1,*], Dmitry O. Filatov [1], Stanislav V. Tikhov [1], Alexey I. Belov [1], Dmitry S. Korolev [1], Alexander V. Kruglov [1], Ruslan N. Kryukov [1], Sergey Yu. Zubkov [1], Vladislav A. Vorontsov [1], Dmitry A. Pavlov [1], David I. Tetelbaum [1], Alexey N. Mikhaylov [1], Sergey A. Shchanikov [2], Sungjun Kim [3] and Bernardo Spagnolo [1,4]

1 Research and Education Center "Physics of Solid State Nanostructures", National Research Lobachevsky State University of Nizhny Novgorod, 603022 Nizhny Novgorod, Russia; dmitry_filatov@inbox.ru (D.O.F.); tikhov@phys.unn.ru (S.V.T.); belov@nifti.unn.ru (A.I.B.); dmkorolev@phys.unn.ru (D.S.K.); krualex@yandex.ru (A.V.K.); kriukov.ruslan@yandex.ru (R.N.K.); zubkov@phys.unn.ru (S.Y.Z.); vladislav.vorontsov1@gmail.com (V.A.V.); pavlov@unn.ru (D.A.P.); tetelbaum@phys.unn.ru (D.I.T.); mian@nifti.unn.ru (A.N.M.); bernardo.spagnolo@unipa.it (B.S.)

2 Department of Information Technologies, Vladimir State University, 600000 Vladimir, Russia; seach@inbox.ru

3 Division of Electronics and Electrical Engineering, Dongguk University, Seoul 04620, Korea; sungjun@dongguk.edu

4 Dipartimento di Fisica e Chimica "Emilio Segrè", Group of Interdisciplinary Theoretical Physics, Università degli Studi di Palermo and CNISM, Unità di Palermo, I-90128 Palermo, Italy

* Correspondence: mahavenok@mail.ru

Abstract: Nowadays, memristors are of considerable interest to researchers and engineers due to the promise they hold for the creation of power-efficient memristor-based information or computing systems. In particular, this refers to memristive devices based on the resistive switching phenomenon, which in most cases are fabricated in the form of metal–insulator–metal structures. At the same time, the demand for compatibility with the standard fabrication process of complementary metal–oxide semiconductors makes it relevant from a practical point of view to fabricate memristive devices directly on a silicon or SOI (silicon on insulator) substrate. Here we have investigated the electrical characteristics and resistive switching of SiO_x- and SiN_x-based memristors fabricated on SOI substrates and subjected to additional laser treatment and thermal treatment. The investigated memristors do not require electroforming and demonstrate a synaptic type of resistive switching. It is found that the parameters of resistive switching of SiO_x- and SiN_x-based memristors on SOI substrates are remarkably improved. In particular, the laser treatment gives rise to a significant increase in the hysteresis loop in *I–V* curves of SiN_x-based memristors. Moreover, for SiO_x-based memristors, the thermal treatment used after the laser treatment produces a notable decrease in the resistive switching voltage.

Keywords: memristor; silicon oxide; silicon nitride; SOI technology; resistive switching; electrical characteristics; laser treatment; thermal treatment

Citation: Koryazhkina, M.N.; Filatov, D.O.; Tikhov, S.V.; Belov, A.I.; Korolev, D.S.; Kruglov, A.V.; Kryukov, R.N.; Zubkov, S.Y.; Vorontsov, V.A.; Pavlov, D.A.; et al. Silicon-Compatible Memristive Devices Tailored by Laser and Thermal Treatments. *J. Low Power Electron. Appl.* **2022**, *12*, 14. https://doi.org/10.3390/jlpea12010014

Academic Editors: Alex Serb and Adnan Mehonic

Received: 21 October 2021
Accepted: 28 February 2022
Published: 2 March 2022

1. Introduction

A memristor is a two-terminal nanoelectronic element that changes and remembers its resistance depending on the applied voltage and the charge flowing through it. Its main difference from semiconductor memory elements, which implement a binary code and two stable states, is the multilevel, synaptic nature of the conduction switching [1]. It is believed that this will make it possible to create next-generation computers (with a non-von-Neumann architecture) and neuromorphic artificial intelligence systems on the basis of memristors [2–12]. The main disadvantages of memristors fabricated in the form of metal–insulator–metal (MIM) or metal–insulator–semiconductor (MIS) structures are the reproducibility of resistive switching (RS) parameters, which is insufficient for practical use (stochasticity), high values of RS voltages, and the complexity of integration into a standard complementary metal–oxide–semiconductor (CMOS) fabrication process. Currently, approaches to solving these problems are being developed: the use of new materials and various interfaces [13–16], the use of signals with a special shape for RS [17], the use of optical radiation [18] or noise [19–21] as parameters controlling the switching dynamics, programming the amplitudes and durations of switching pulses [22,23], etc. Indeed, the wider application of memristors is limited by their insufficient stability, the high variability of RS parameters, and a lack of understanding of the drift–diffusion processes responsible [24]. One of the fundamental origins of the instability of the memristor parameters is the essentially stochastic nature of RS [20]. Furthermore, the noise sources can induce new ordered dynamical structures and cause new phase transition phenomena [25,26]. Therefore, of all these approaches, the one based on the constructive role of both internal (thermal) and external noise sources is most promising due to the intrinsic stochastic nature of resistive switching in memristor devices [19–21].

Transition metal oxides (e.g., HfO_x [27,28], TaO_x [29,30], ZrO_x [31,32], TiO_x [33,34], and more complex compounds such as perovskites [35]), as well as SiO_x, and GeO_x, are considered promising insulator materials for memristors. Recently, intensive research has also been carried out on memristive structures based on SiN_x [36,37]. This is of practical interest due to their compatibility with the standard technology for creating modern integrated circuits. The use of SiO_x and SiN_x insulator films involves a number of practical advantages. For example, the authors of [38] carried out a comprehensive comparison of the RS parameters of memristive structures based on HfO_x and SiO_x and showed a lower variability in resistance in different states of SiO_x-based memristors. In turn, the authors of [39] demonstrated the absence of changes in the value of currents in resistive states of SiN_x-based memristors irradiated with As^+ during 10^5 cycles of RS and the minimum variability of switching voltages. It should be noted that SiO_x- and SiN_x-based memristors have a filamentary resistive switching mechanism [40,41].

The use of a semiconductor as one of the electrodes of a memristive structure is also important from the point of view of integrating memristors into a standard CMOS fabrication process [42,43]. One of the electrodes is silicon, which simplifies the technological process and allows the memory to be integrated monolithically on a single platform with a transistor [44,45]. In these articles, bulk silicon was used as an electrode. However, in the preparation of most semiconductor devices and microcircuits, preference is given to "silicon-on-insulator" (SOI) substrates due to its advantages over bulk silicon: lower power consumption and the higher performance and density of elements [46]. Therefore, from a practical point of view, it is advisable to implement memristive structures on SOI substrates. Despite the significant number of published studies of memristive structures with a bulk silicon electrode, structures on SOI substrates must be studied independently due to the peculiarities of morphology and structure of the latter. Thus, the development and investigation of memristive structures, in which the SOI substrate acts as an electrode, is of considerable theoretical and practical interest. However, such data are nearly absent in the literature, except for several separate reports on the use of SOI in memristive devices (see, for example, [47–50]).

Despite the practical advantages of using a semiconductor as an electrode in a memristive structure, one should not forget the presence of surface states (SS) at the insulator/semiconductor interface, which is undesirable from the point of view of creating memristive structures. These states make a significant contribution to the total serial resistance of the structure [51]. A decrease in their density leads to a redistribution of the external voltage, so that the electric field strength in the insulator increases, thereby stimulating resistive switching. Thermal treatment (TT) is a widely used method of dealing with such defects. Laser treatment (LT) can be used for the same purposes. In the latter case, the effect is achieved due to heating of the substrate because of the absorption of laser radiation in it. In addition, LT is used to modify the charge state of an insulator in a flash memory device, which is used to completely erase information in memory elements [52].

Therefore, LT and TT can be effectively used to change the electrical characteristics of memristive structures.

We propose a comprehensive approach to improving the parameters of RS: namely, increasing the resistance ratio in extreme resistive states and decreasing the RS voltages of memristive structures based on promising and accessible insulator layers—SiO_x and SiN_x, fabricated under industrial conditions on SOI substrates. This approach is based not only on the use of materials that are standard for the CMOS fabrication process, but also on the use of LT and TT, which are widely used in the microelectronic industry to control the electrical parameters of devices. In addition, the investigation of the frequency dependences of electrical characteristics of memristive structures carried out in this work makes it possible to obtain the necessary detailed information about the processes occurring in the insulator film and about the state of insulator/semiconductor interfaces in different resistive states [53]. Data in this paper are presented in the same order in which they were obtained, so that the reader can unambiguously determine the contribution of LT and TT to the change in resistive switching parameters.

To the best of our knowledge, such a comprehensive study of SiO_x- and SiN_x-based memristive structures fabricated on SOI substrates, including the influence of LT and TT on their electrical characteristics (RS parameters), has not been carried out previously.

2. Materials and Methods

SiO_x and SiN_x films (with a nominal thickness of 13 nm each) were deposited on commercial SOI substrates with a device layer thickness of 360 nm by plasma-enhanced chemical vapor deposition under the following conditions:

- SiO_x—using 5% SiH_4/N_2 (160 sccm), N_2O (1500 sccm) and N_2 (240 sccm) at a pressure of 550 mTorr and high frequency (HF) power of 60 W, with a deposition rate of 200 Å/min;
- SiN_x—using 5% SiH_4/N_2 (800 sccm), NH_3 (10 sccm) and N_2 (1200 sccm) at a pressure of 580 mTorr and HF-power 60 W, with a deposition rate of 100 Å/min.

Top Au electrodes (20 nm) with a Zr sublayer (8 nm) with an area of $S\sim10^{-2}$ (in this study) and 10^{-3} cm^2 were deposited on the surface of insulators by magnetron sputtering at a temperature of 473 K. A schematic representation of the fabricated structures is shown in Figure 1. The devices were prepared in the form of a metal–insulator–semiconductor sandwich with a common bottom electrode (SOI) and local top (Au with a Zr sublayer) electrodes. Figure 2 is an optical image of a fragment of the device showing two top electrodes of a small area and one of a larger area. The optical image was obtained using a Leica DM 4000 M optical microscope (Wetzlar, Germany).

The electrical characteristics were measured using a semiconductor device parameter analyzer, Agilent B1500A (Santa Rosa, CA, USA). The sign of voltage across the structures corresponded to the potential of the top electrode relative to the potential of the bottom electrode. *I–V* curves and the small-signal *C–f*, *G–f*, and *R–f* characteristics of memristors were measured in parallel and series capacitor equivalent resistor–capacitor circuits (see Figure 1 for explanation) [54] in the frequency range 10^3–2×10^6 Hz. The values of parallel capacitance (C_p), parallel conductance loss (G_p/ω), dielectric loss tangent (tgδ), parallel (R_p), and series (R_s) resistances were determined. The parameters of parallel capacitor equivalent circuit are determined by the electronic phenomena in an insulator, while the parameters featuring a serial capacitor equivalent circuit are determined by the resistance of electrodes and that of the transition layer between the electrode and insulator film [54].

The information on relaxation processes in the insulator was obtained by analyzing the Cole–Cole diagrams—the dependences of G_p/ω on C_p, which were obtained from corresponding frequency dependences [55]. As shown below, the obtained diagrams were either a circular arc or a semicircle. Thus, in the first case, the spectrum of SS at the insulator/semiconductor interface was continuous, while the second case indicates the presence of a mono-level of SS. An analysis of the Cole–Cole diagrams makes it possible to estimate the effective density of SS at the Fermi level (N_{ss}). In the case of a continuous spectrum of SS, for such an estimate, one can use the following equation [56]:

$$N_{ss} = \frac{[G_p/\omega]_{max}}{0.4q^2 S}, \tag{1}$$

where $[G_p/\omega]_{max}$ is the maximum value of parallel conductance loss, q is the electron charge, and S is the structure area (i.e., the area of the top electrode). In the case of a mono-level of SS, one can use the following equation [56]:

$$N_{ss} = \frac{8kT[G_p/\omega]_{max}}{q^2 S}, \tag{2}$$

where k is the Boltzmann constant and T is the temperature.

In addition, measurements of capacitance–voltage and conductance–voltage characteristics were carried out in a parallel capacitor equivalent resistor–capacitor circuit at a small test signal frequencies of 10 and 100 kHz.

It should be noted that the investigated memristive structures initially had a conductive state. The investigations of the electrical characteristics of memristive structures were carried out in initial state (IS), in low-resistance state (LRS), and in high-resistance state (HRS).

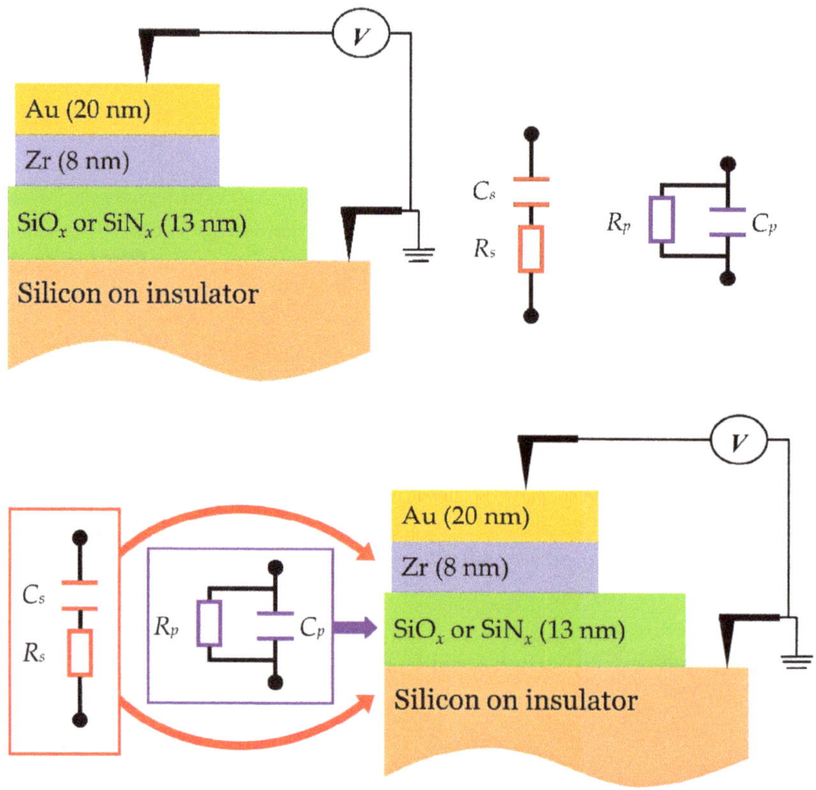

Figure 1. Schematic representation of SiO_x- and SiN_x-based memristors and the simplest capacitor equivalent resistor–capacitor circuits.

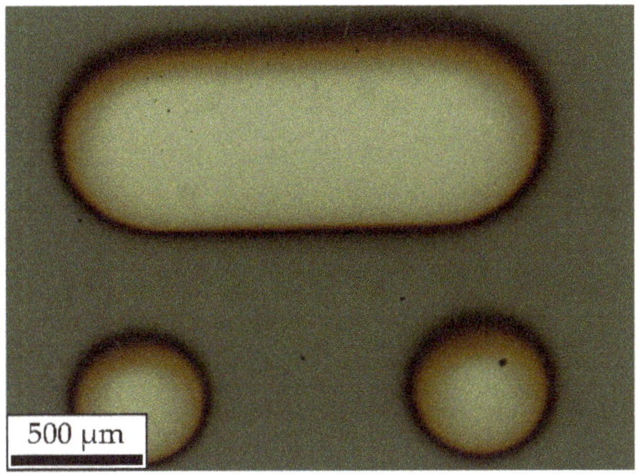

Figure 2. Optical image of a fragment of the device.

As mentioned above, LT can be used to change the electrical characteristics of memristive structures, which determine RS parameters. It is assumed that LT will make it possible to reduce the built-in charge in insulator films and to lower the density of SS at the insulator/semiconductor interface in memristive structures. Therefore, some memristive structures were subjected to LT. For this, a semiconductor laser with a power of 1.5 W and a wavelength of 460 nm, which corresponds to a photon energy of 2.7 eV, was used in the continuous mode. Irradiation was carried out through the top electrode for 10 min. It should be noted that the top Au electrodes with a thickness of 20 nm were semitransparent for the laser wavelength used [57]. Under the influence of laser radiation, the structure heats up to ≈473 K.

Thermal treatment is a widely used method for changing the density of SS at the insulator/semiconductor interface. Therefore, in order to improve the state of this interface, some of the SiO_x-based memristive structures were subjected to TT. For this purpose, memristive structures were placed in a hermetically closed metal thermostat, which was slowly heated at a rate of 13.5 K/min using an electric heater or cooled with liquid nitrogen. Investigations of electrical characteristics were carried out in a temperature range of 77–600 K in an atmosphere dried with silica gel. The temperature was maintained with an accuracy of 1 K.

Structural investigations of the SiO_x and SiN_x films and memristive structures based on them were carried out by X-ray photoelectron spectroscopy (XPS) and transmission electron microscopy (TEM). The profiling of samples using the XPS method implies the use of ion etching. The question arises of the correct determination of the etching rate and the existence of an error in determining the depth. If the rate can be determined using calibration samples, then the error is determined for each sample separately. A large contribution to the error when determining the depth is made by irregularities on the surface of the sample, due to which shading occurs during the etching process [58]. For correct interpretation of the data, information on the roughness obtained by atomic force microscopy (AFM) was used.

3. Results and Discussion

3.1. SiO_x-Based Memristive Structures on SOI Substrates

According to the AFM data (Figure 3a), the root mean square roughness of the SiO_x film is 1.8 nm. Figure 3b presents XPS data for SiO_x films before and after annealing at 550 K. The stoichiometry of the SiO_x film barely changes between before and after annealing, and is $x \approx 1.8$. One can also notice a transition layer at the SiO_x/SOI interface, the thickness of which is ~15 nm.

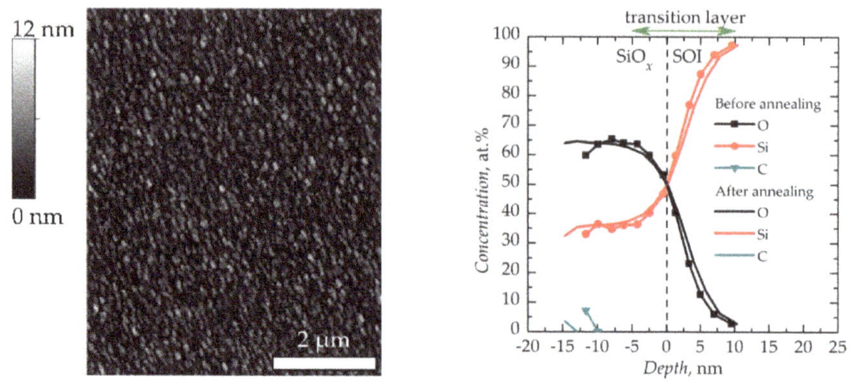

(**a**) (**b**)

Figure 3. (**a**) AFM image of SiO_x film surface; (**b**) distribution of chemical elements over depth of SiO_x film before and after annealing at 550 K. The origin of the coordinates along the abscissa coincides with the SiO_x/SOI interface.

Figure 4 shows TEM images of a cross section of a SiO_x-based memristive structure after LT and TT. According to Figure 4, the SiO_x film has an amorphous structure. At the same time, Si (area 4), ZrO (areas 1 and 3) and ZrO_2 (area 2) nanocrystallites were found in the Zr sublayer and at the interface with the insulator. The structure of the observed nanocrystallites was determined by comparing the interplanar spacing in TEM images with the literature data. This means partial oxidation of Zr electrode and silicon oxide reduction in contact with this electrode during treatments.

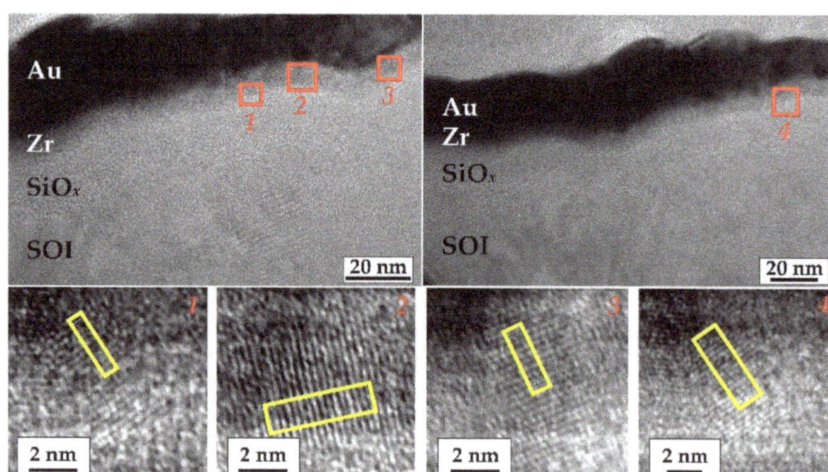

Figure 4. High-resolution TEM images of two cross-sectional regions of a SiO_x-based memristive structure after LT and TT. The inset shows scaled images of the nanocrystallites (parts that were used to determine the interplanar spacing are highlighted by yellow rectangles).

The SiO_x-based memristive structures before LT and TT did not require electroforming [59], since initially they had a conductive state (Figure 5a, curve 1). When a voltage of −6 V was applied, the memristive structure switched from LRS to HRS (Figure 5a, curve 2). Subsequent application of voltage of +6 V did not lead to switching of the structure (Figure 5a, curve 3). In the absence of switching (Figure 5a, curves 1 and 3), the values of the current through the device in the forward and reverse directions of the voltage sweep hardly differed.

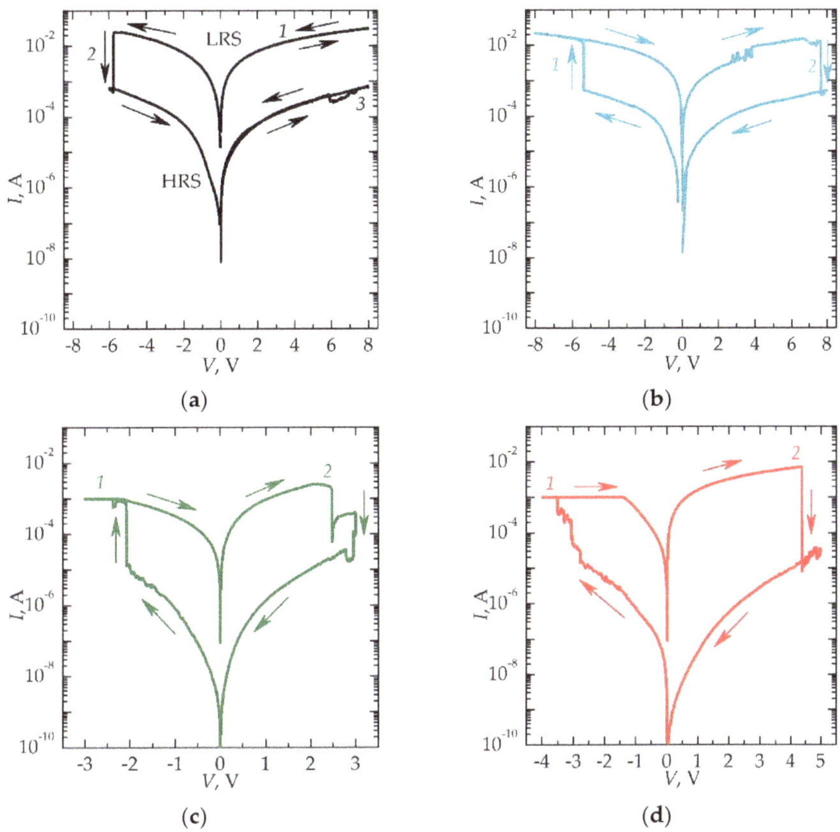

Figure 5. I–V curves of SiO_x-based memristive structure (**a**) before LT and TT, (**b**) after LT, (**c**) after TT and (**d**) after multiple RS. In the absence of switching (Figure 5a, curves 1 and 3), the values of the current through the device in the forward and reverse directions of the voltage sweep almost did not differ. The direction of the voltage sweep is shown by arrows.

The frequency dependences of the parameters of equivalent circuit of memristive structures in IS (i.e., for curve 1 in Figure 5a) and HRS (i.e., for curve 2 in Figure 5a) are shown in Figure 6. The structure in IS is characterized by large ohmic losses at a low frequencies (Figure 6, curves 2, 3) and a low parallel resistance R_p shunting the structure (Figure 6, curve 5). After switching into HRS, the losses decreased by three orders of magnitude (Figure 6, curves 7, 8), and the value of R_p, respectively, increased by three orders of magnitude (Figure 6, curve 10).

Note that the values of the relative permittivity of SiO_x films calculated from the value of C_p by the equation for a parallel plate capacitor at a frequency of 1 kHz do not change with RS, while the value of tgδ changes by three orders of magnitude. This behavior of low-signal HF parameters indicates the filamentary mechanism of RS [60]. In this case, the active part of the film impedance changes locally, i.e., on a small (compared to the total electrode area) memristor area, while the resistance and dielectric losses remain almost unchanged for the rest of the film under the electrode.

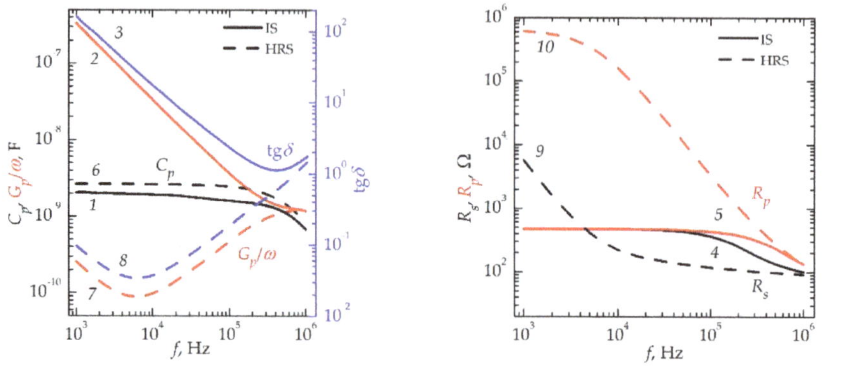

(**a**) (**b**)

Figure 6. Frequency dependences of (**a**) C_p (1, 6), G_p/ω (2, 7), tgδ (3, 8) and (**b**) R_s (4, 9), R_p (5, 10) obtained for memristive structure in IS (1–5) and HRS (6–10).

Dependences of C_p and G_p/ω on V (Figure 7) show that the semiconductor corresponds to the n-type, since the capacitance at a frequency of 100 kHz (see inset in Figure 7b) is in the form of a step with an increase towards the voltage $V > 0$ [56]. The concentration of equilibrium electrons in a silicon electrode can be estimated using the following equation [61]:

$$N_D = \frac{2\left(2\varphi_0 - \frac{kT}{q}\right)}{\varepsilon_s \varepsilon_0 q} \cdot \left(\frac{\frac{C_{ox}}{C_{min}} - 1}{C_{ox}} \right)^{-2}, \tag{3}$$

where N_D is the donor concentration in the semiconductor, φ_0 is the height of the potential barrier at the insulator/semiconductor interface, ε_s is the relative permittivity of the semiconductor, ε_0 is the vacuum permittivity, C_{ox} is the oxide capacity, equal to the maximum value of capacity in Figure 7b in the dark, and C_{min} is the minimum value of capacity in Figure 7b in the dark. The obtained value varied in the range of ~3×10^{19}–3×10^{20} cm^{-3}. This variation is associated with strong fluctuations in capacitance due to the nonuniform distribution of impurities over the thickness of the silicon electrode.

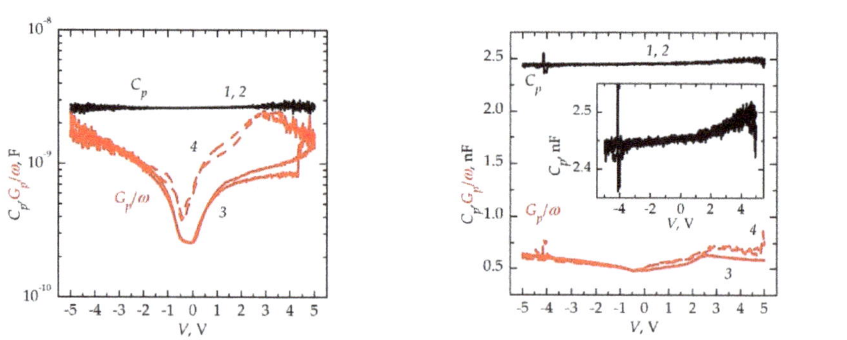

(**a**) (**b**)

Figure 7. Dependences of C_p (1, 2) and G_p/ω (3, 4) on V measured at a frequency of a small test signal (**a**) 10 and (**b**) 100 kHz and in the dark (1, 3) or under laser radiation (2, 4). Voltage sweep from −5 V to +5 V and vice versa.

The maxima in the dependences of G_p/ω on V (Figure 7a, curve 4 and Figure 7b, curves 3 and 4) in the theory of MIS structures are usually associated with SS at the insulator/semiconductor interface. If one assumes a quasicontinuous SS distribution, the N_{ss} value can be estimated using Equation (1). The value of N_{ss} is, under laser radiation, $-3.6{\cdot}10^{12}$ cm^{-2}eV^{-1} (at a frequency of 10 kHz) and 1.1×10^{12} cm^{-2}eV^{-1} (at a frequency of 100 kHz), and in the dark -1×10^{12} cm^{-2}eV^{-1} (at a frequency of 100 kHz). Thus, the density of SS on the conductive Si electrode is large and increases with decreasing frequency and under laser radiation. The capture of carriers to these states should decrease the response time of memristors in the same way as a large series resistance.

Figure 5b shows the I–V curves of a memristive structure after LT. It can be seen that LT leads to a change in the polarity of RS: applying a negative voltage leads to the switching of the structure in LRS, and applying a positive voltage leads to the switching of the structure in HRS. Similar behavior was observed in AZO/CeO_2/ITO/glass memory devices [62]. The effect can be explained in terms of the change in the active electrode of the structure, which plays the main role in the formation and oxidation of the filament; however, this requires additional investigation. The obtained I–V curves demonstrate a ratio of currents in LRS and HRS of more than 2 orders of magnitude.

The results of the effect of LT on electrical characteristics of memristive structure are shown in Figure 8. Frequency dependences of the parameters of equivalent circuit of the structure in LRS (i.e., after curve 1 in Figure 5b) and HRS (i.e., after curve 2 in Figure 5b) are shown. These data also indicate a change in the polarity of RS after LT and an almost unchanged value of the resistance of the bottom semiconductor electrode (~100 Ω). In addition, higher values of tgδ in HRS at a low frequency, as compared to structures before LT (Figure 6a), indicate incomplete oxidation of filaments.

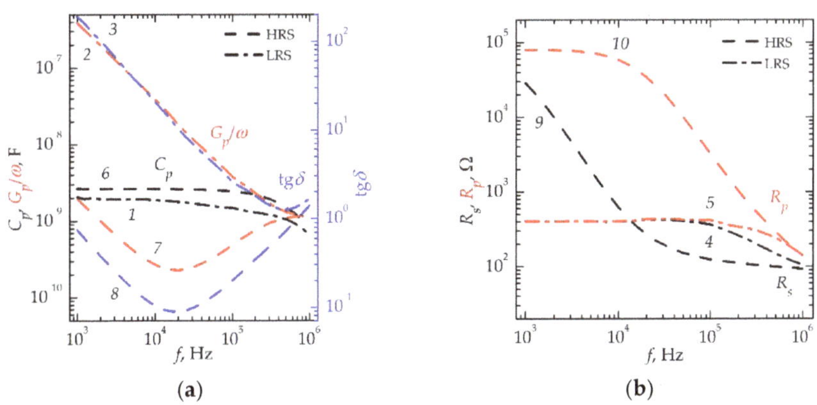

Figure 8. Frequency dependences of (**a**) C_p (1, 6), G_p/ω (2, 7), tgδ (3, 8) and (**b**) R_s (4, 9), R_p (5, 10) obtained for memristive structure in LRS (1–5) and HRS (6–10). The data were obtained after LT.

TT in a dried atmosphere at 540 K in a hermetically closed metal thermostat also changes the electrical characteristics of SiO_x-based memristive structures. This is evidenced by the frequency dependences of the parameters of equivalent circuit shown in Figure 9. Nonstandard behavior of dielectric losses and the value of parallel resistance with an increase in temperature from 77 to 540 K are noteworthy. Namely, usually, with an increase in the temperature, the concentration of free carriers in the insulator increases, so the values of tgδ [63] increase and those of R_p decrease. However, in this case, the values show the opposite tendency. The observed behavior is unusual for insulators and is probably associated with an irreversible change in the properties of the insulator because of TT. The polarity of RS after TT corresponds to the polarity after LT (Figure 5c). It should be noted that the RS voltage decreases after the TT of the structures.

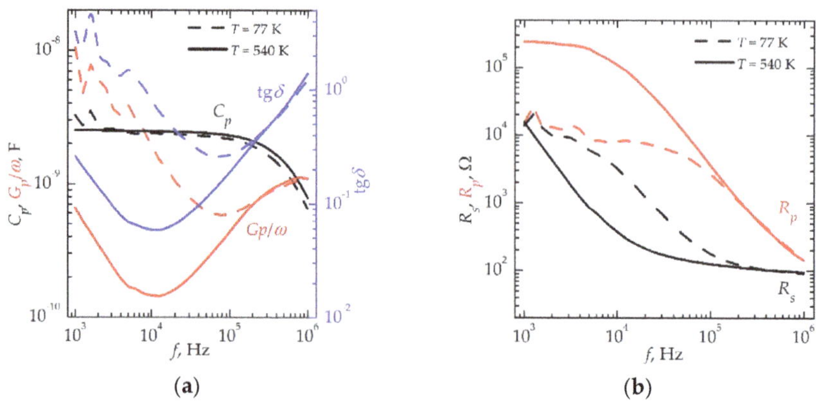

Figure 9. Frequency dependences of (**a**) C_p, G_p/ω, tgδ and (**b**) R_s, R_p of memristive structure in HRS obtained at a temperature of 77 (dashed line) and 540 K (solid line).

Figure 10 shows the results of studying the stability of the parameters of equivalent circuit and RS parameters of memristive structures after TT under multiple switching in the mode of *I–V* curve, shown in Figure 5c. The experiment was carried out as follows. After switching the structure in LRS (i.e., for curve 1 in Figure 5c), the frequency dependences of the parameters of equivalent circuit were measured. Furthermore, after switching the structure in HRS (i.e., for curve 2 in Figure 5c), the frequency dependences of the parameters of equivalent circuit were measured again. Thus, multiple (within ~2 h) switching of memristive structure from LRS to HRS and vice versa occurred, with sequential measurement of the parameters of equivalent circuit. The times for which the parameters of equivalent circuit were measured were significantly shorter than the time intervals between switches. Therefore, changes in the parameters during the testing of structures could be neglected. Thus, the observed changes in parameters occur due either to the stochasticity of RS processes, or, less likely, to changes in structures in the intervals between switching.

Figure 10a,b shows that the parameters of equivalent circuit after switching into HRS are relatively reproducible in comparison with the parameters obtained after switching into LRS; this is indicated by the weak time dependence of C_{p0}, R_{p0}, and tgδ_0 (Figure 10a,b, curves 2, 4, 6). When switching into LRS, the time dependences of C_{p0}, R_{p0}, and tgδ_0 are characterized by non-monotonic behavior, which is reflected in significant (by more than two orders of magnitude for R_{p0} and tgδ_0) chaotic changes (Figure 10a,b, curves 1, 3, and 5). The last result can be interpreted as follows. The selected mode of switching into HRS allows each time to destroy the active filament, and each switching into LRS leads to the formation of different (in terms of shape and location) filaments. It should be noted that, with multiple switching, regardless of the sign of the switching voltage and the state of the memristive structure, a monotonic decrease in the series resistance $R_{s\infty}$ from ~650 Ω to ~160 Ω was observed (Figure 10b, curve 7). It should be recalled that the value of series resistance is determined by the resistance of the semiconductor electrode. The observed behavior indicates the occurrence of electrochemical reactions on the semiconductor electrode and the accumulation of a positive charge on its surface during the recharging of the memristive structure.

Figure 10c,d shows the results of a statistical study for 10 *I–V* curves of memristive structures after TT. It can be seen that the currents through the structure in LRS and HRS differ by at least one order of magnitude (Figure 10c), and the voltages for RESET (switching from LRS to HRS, V_{RESET}) and SET (switching from HRS to LRS, V_{SET}) processes have a value in the selected range (Figure 10d).

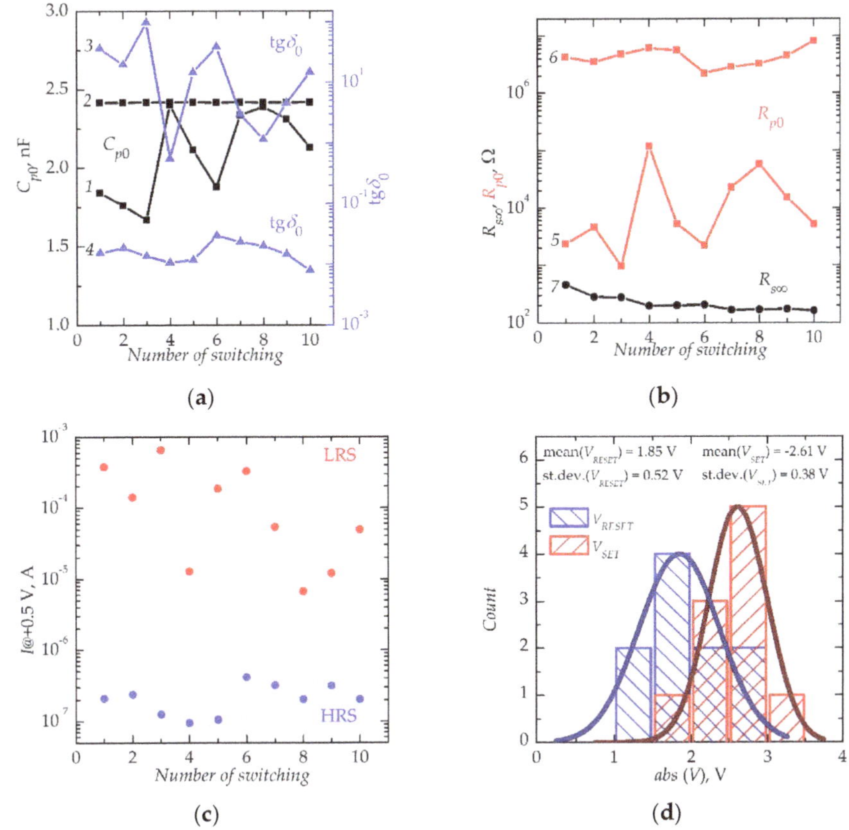

Figure 10. (a,b) The parameters of equivalent circuit of memristive structure after TT and switching into LRS (1, 3, 5) and HRS (2, 4, 6) obtained at a frequency of a small test signal of 1 kHz (C_{p0}, tgδ_0, R_{p0}) and 2 MHz ($R_{s\infty}$); (c) dependences of the currents (at a reading voltage of +0.5 V) of memristive structure in LRS (red) and HRS (blue) after TT on the number of RS cycles; (d) distribution of voltages of SET (red) and RESET (blue) processes of memristive structure after TT.

Figure 5d shows the I–V curves of memristive structures after multiple RS. An increase in the voltage values of RESET and SET processes is seen, which indicates a significant change in electrical characteristics of the structure under multiple RS.

Figure 11 shows the frequency dependences of the parameters of equivalent circuit obtained for the memristive structure in HRS (i.e., for curve 2 in Figure 5d) after multiple switching. It should be emphasized that, in contrast to the dependencies shown above, these data indicate the complete oxidation of filaments when the structure is switched in HRS. This fact is seen, in particular, from a comparison with the frequency dependence of the tgδ structure in HRS in a low-frequency region. An increase in the tgδ and G_p/ω of the structures (Figure 11a) can be explained by the fact that there is no shunting of memristor by the value of R_p and the implementation of a series connection of the capacitance C_p and memristor electrodes. In this case, the losses at low frequencies are small, the parallel capacitance is equal to the series capacitance, and the series resistance is determined by the resistance of the semiconductor electrode (~170 Ω).

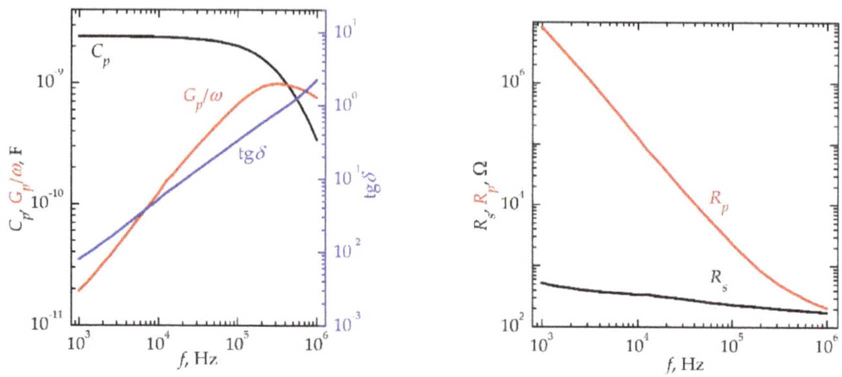

(a) (b)

Figure 11. Frequency dependences of (**a**) C_p, G_p/ω, tgδ and (**b**) R_s, R_p of memristive structure in HRS after multiple RS.

For a quantitative comparison, Table 1 shows the values of C_p, G_p/ω, tgδ, R_p, and R_s, obtained at the frequencies of 1 and 100 kHz (indices 0 and ∞, respectively). Data were obtained for the SiO_x-based memristor in different resistive states before LT and TT, after LT, during TT, and after multiple RS.

Table 1. SiO_x-based memristor equivalent circuit parameters.

Treatment	Resistive State	C_{p0}, nF	$C_{p\infty}$, nF	G_{p0}/ω, nF	$G_{p\infty}/\omega$, nF	tgδ_0	tgδ_∞	R_{p0}, Ω	$R_{p\infty}$, Ω	R_{s0}, Ω	$R_{s\infty}$, Ω
Before LT and TT	IS	2.05	1.60	334	3.68	163	2.30	476	433	476	364
	HRS	2.65	2.46	0.26	0.47	0.10	0.19	621,296	3400	5732	119
After LT	LRS	2.06	1.49	390	3.78	189	2.54	409	422	409	365
	HRS	2.65	2.45	1.98	0.49	0.75	0.20	80,309	3253	28,806	125
During TT	HRS at 77 K	3.24	2.17	11	0.59	3.24	0.27	15,124	2701	13,812	179
	HRS at 540 K	2.52	2.31	0.66	0.44	0.26	0.19	241,206	3637	15,457	126
After multiple RS	HRS	2.42	2.02	0.02	0.66	0.01	0.33	8,259,960	2396	525	234

Figure 12 shows the Cole–Cole diagrams obtained for SiO_x-based memristive structures in HRS. The data were obtained from the frequency dependences of the G_p/ω and C_p of memristive structure before LT and TT (see Figure 6a), after LT (see Figure 8a), and after TT and multiple switching (Figure 11a). It can be seen that all diagrams have a circular arc shape, i.e., the spectrum of SS at insulator/semiconductor interface is continuous in all cases. The values of N_{ss} were estimated using Equation (1) and are 1.9×10^{12}, 1.8×10^{12}, and 1.5×10^{12} cm^{-2}eV^{-1}, respectively.

Thus, only the LT of the Au/Zr/SiO_x/SOI memristive structures is not sufficient for a significant change in the value of the density of SS. Additional use of TT leads to a decrease in this value by a factor of ~1.3. Nevertheless, the combined effect of LT and TT on the Au/Zr/SiO_x/SOI memristive structures results in a decrease in RS voltages of almost 2-fold. The effect is probably associated with the annealing of SS, which, in turn, leads to a decrease in the resistance of the structure. According to the model [64], the appearance of SS is associated with the disordering of silicon subsurface near the interface with the insulator. From this point of view, annealing promotes a decrease in the density of SS due to the relaxation of this disorder. However, one should also consider that annealing can lead to a change in the concentration of electrically active impurities in both the semiconductor and the insulator. As a result, the Fermi level at the insulator/semiconductor interface can shift towards a lower density of states.

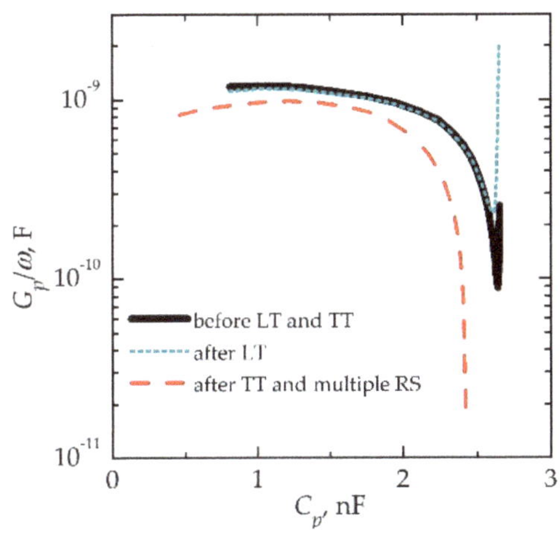

Figure 12. The Cole–Cole diagrams obtained for SiO_x-based memristive structures in HRS. The data were obtained before LT and TT, after LT, and after TT and multiple RS.

It should be noted that, along with the abovementioned influence of treatments on the density of SS, they can be responsible for the occurrence of RS. It can be assumed that the nanocrystallites observed by TEM are responsible for the initial conductive state of SiO_x-based memristors. According to the estimates from TEM images, nanocrystallites reach diameters of ~7 nm. Note that TEM studies were carried out after LT and TT. Thus, probably, such treatments led to significant and irreversible oxidation of nanocrystallites and, before treatments, the sizes of nanocrystallites could be comparable to the thickness of the insulator film. The latter could lead to shunting the devices. This explanation is indirectly confirmed by the unusual behavior of dielectric losses and the value of parallel resistance, with an increase in temperature from 77 to 540 K.

3.2. SiN_x-Based Memristive Structures on SOI Substrates

According to AFM data (Figure 13a), the root mean square roughness of the SiN_x film is 1.9 nm. In Figure 13b, XPS data for SiN_x film before and after annealing at 550 K are reported. It is shown that the stoichiometry of SiN_x film before and after annealing hardly changes and $x \approx 1.25$. One can also notice the presence of a transition layer at the SiN_x/SOI interface, the thickness of which is ~18 nm.

In Figure 14, the TEM images of a cross section of SiN_x-based memristive structures after LT are shown. According to Figure 14, the SiN_x film has an amorphous structure. At the same time, the presence of ZrN (areas 1, 3, 5–7) and Si (area 4) nanocrystallites is confirmed inside amorphous SiN_x. ZrO_2 (area 2), ZrO (area 8), and ZrN (area 9) nanocrystallites are found in the Zr sublayer and at the interface with the insulator. The presence of Si_3N_4 nanocrystallites should also be noted (area 10). The structure of the observed nanocrystallites was determined by comparing the interplanar spacing in TEM images with the literature data. Like for the SiO_x-based memristive structures, the SiN_x-based structures considered in this section initially had a conductive state. It should be noted that SiN_x-based memristive structures did not demonstrate RS before LT (Figure 15a, curve 1).

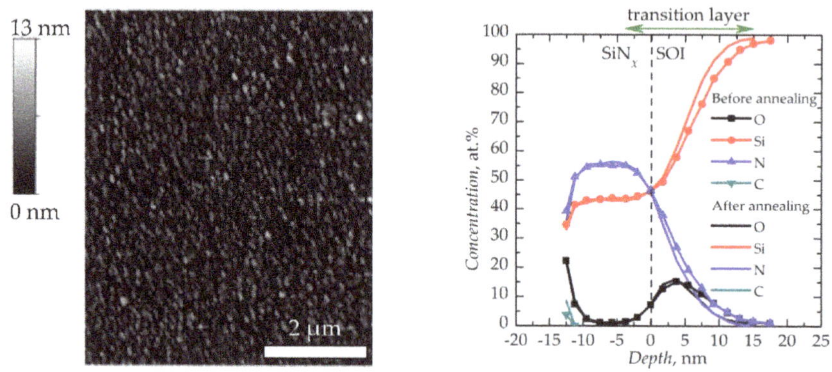

(**a**) (**b**)

Figure 13. (**a**) AFM image of SiN_x film surface; (**b**) distribution of chemical elements over depth of SiN_x film before and after annealing at 550 K. The origin of coordinates along the abscissa coincided with the SiN_x/SOI interface.

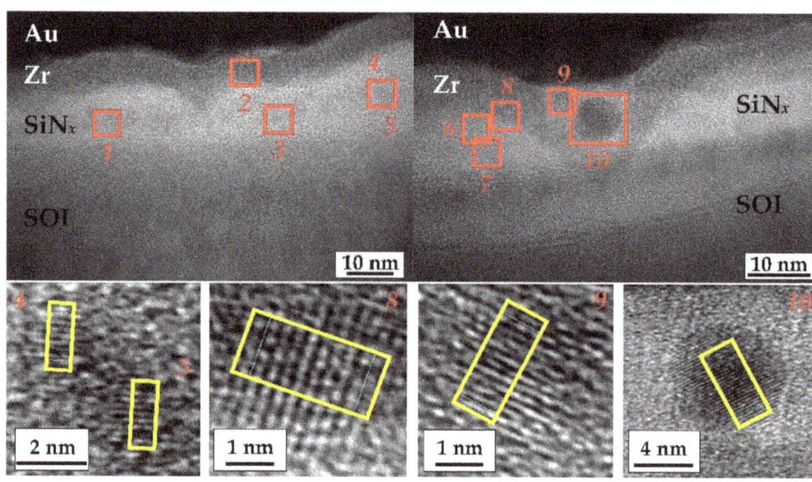

Figure 14. High-resolution TEM images of two cross-sectional regions of SiN_x-based memristive structure after LT. The inset shows scaled images of the nanocrystallites (parts that were used to determine the interplanar spacing are highlighted by yellow rectangles).

In Figure 16, the frequency dependences of the parameters of equivalent circuit of memristive structure before LT are shown. The series resistance in the structure at a high frequency, which, as determined by the resistance of memristor electrodes, is ~110 Ω.

In the theory of MIS structures, using high-frequency *C–V*, it is possible to determine the type of dopant: as a DC sweep voltage is applied to the metal, a positive slope of $1/C^2$ vs. *V* indicates acceptors and a negative slope indicates donors [65,66]. The $1/C^2$ value increases with increasing absolute voltage value (Figure 17), which indicates the *n*-type conductivity of the semiconductor film. The nonlinearity of this dependence can be a consequence of inhomogeneous doping of the semiconductor film.

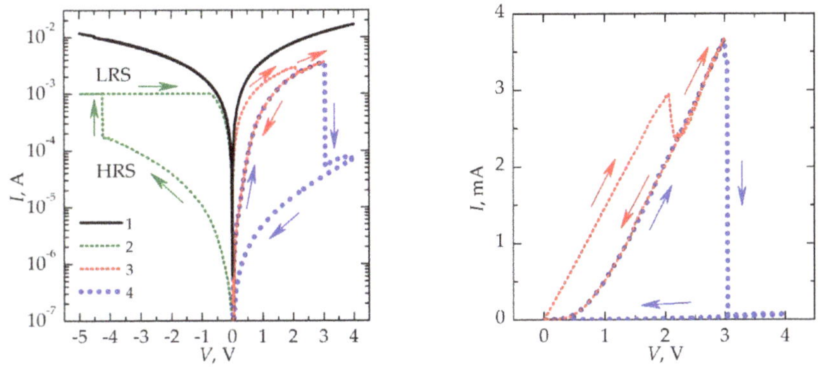

(**a**) (**b**)

Figure 15. (**a**) *I–V* curves of SiN_x-based memristive structure before (solid line) and after (dotted line) LT in semi-log plot. Curves 3 and 4 in linear plot (**b**). The direction of the voltage sweep is shown by arrows.

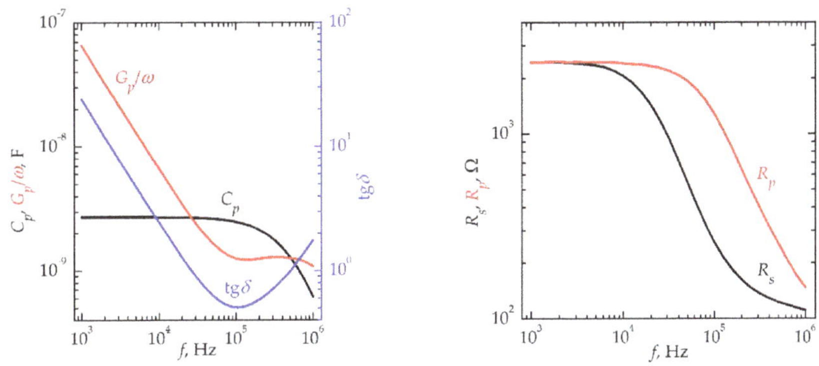

(**a**) (**b**)

Figure 16. Frequency dependences of (**a**) C_p, G_p/ω, tgδ and (**b**) R_s, R_p obtained for memristive structure before LT.

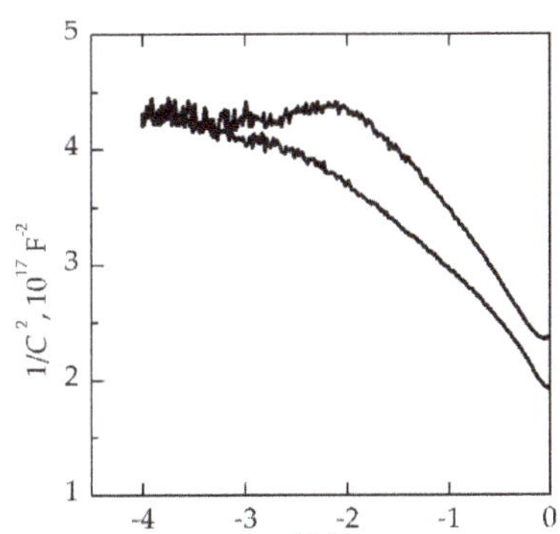

Figure 17. Dependence of nonequilibrium capacitance on voltage in coordinates $1/C^2$–V obtained for memristive structure before LT. Data were measured at a small test signal frequency of 100 kHz.

The donor concentration N_D can be estimated using the slope of the straight line, which extrapolates the data in Figure 17, and the following equation [56]:

$$N_D = \frac{2}{\varepsilon_s \varepsilon_0 q S^2} \frac{\Delta V}{\Delta \frac{1}{C^2}}. \quad (4)$$

The N_D value is ~5×10^{18} cm^{-3}. It should be noted that the obtained value is probably underestimated due to the presence of horizontal areas in the dependence.

The frequency dependence of the parameters of the equivalent circuit shows almost no changes when measured in the dark and under short-term laser radiation. This indicates the presence of an electron-enriched layer at the insulator/semiconductor interface. Therefore, like for the SiO_x-based memristive structures considered above, the structures based on SiN_x were subjected to LT in air in order to change the charge state of the traps in SiN_x. Figure 15a (curves 2–4) shows I–V curves demonstrating a significant increase in the hysteresis loop (change in the current by ~3 orders of magnitude) after LT of the structures. It should be noted that the structures also demonstrated a synaptic nature of switching (Figure 15b) [67,68]. After LT, memristive structures showed an increased value of the relative permittivity (before LT, it was 4; after, it was −4.85). This value was calculated using the equation for a parallel plate capacitor at a frequency of 1 kHz. This behavior indicates the contribution of the space charge region in the semiconductor electrode to the capacitance of the capacitor before LT.

The effect of LT on the electrical characteristics of the memristive structure is illustrated in Figure 18. Frequency dependences of the parameters of equivalent circuit of the structure in LRS (i.e., for curve 2 in Figure 15a) and in HRS (i.e., for curve 4 in Figure 15a) after LT are shown.

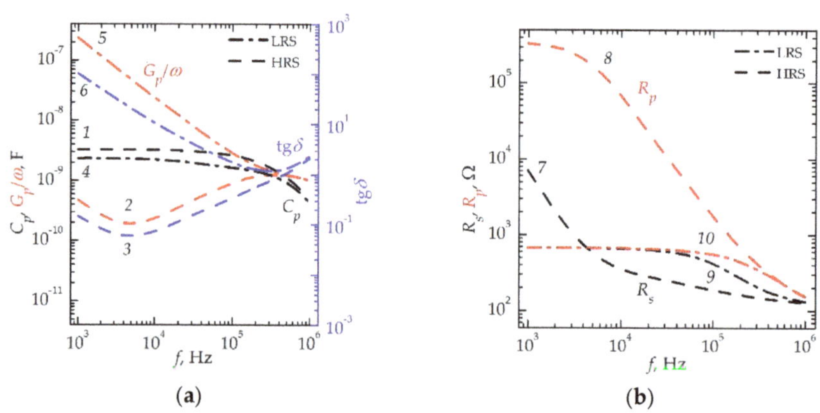

Figure 18. Frequency dependences of (**a**) C_p (1, 4), G_p/ω (2, 5), tgδ (3, 6) and (**b**) R_s (7, 9), R_p (8, 10) obtained for memristive structure in LRS (4, 5, 6, 9, 10) and HRS (1, 2, 3, 7, 8). The data were obtained after LT.

It is worth noting the presence of large losses in the structures after switching into LRS with a voltage of −5 V, which is probably due to the presence of filaments; at the same time, the losses in the structure were significantly reduced (at a low frequency up to two orders of magnitude) after switching into HRS with a voltage of +4 V. However, there was no complete destruction of filaments. This was indicated by the presence of losses at a frequency of <10^4 Hz, which are characterized for losses due to leakage currents at low frequencies [55]. Also, in the memristive structure in HRS at a low frequency, the parallel resistance increased (up to 2 orders of magnitude), shunting it. In this case, the resistance of the silicon electrode remained almost unchanged.

For a quantitative comparison, Table 2 shows the values of C_p, G_p/ω, tgδ, R_p, and R_s obtained at the frequencies of 1 and 100 kHz (indices 0 and ∞, respectively). Data were obtained for the SiN_x-based memristor in different resistive states before and after LT.

Table 2. SiN_x-based memristor equivalent circuit parameters.

Treatment	Resistive State	C_{p0}, nF	$C_{p\infty}$, nF	G_{p0}/ω, nF	$G_{p\infty}/\omega$, nF	tgδ_0	tgδ_∞	R_{p0}, Ω	$R_{p\infty}$, Ω	R_{s0}, Ω	$R_{s\infty}$, Ω
Before LT	IS	2.71	2.48	65	1.25	24	0.50	2436	1276	2432	258
After LT	LRS	2.30	1.64	236	2.86	102	1.74	675	558	675	419
	HRS	3.26	2.61	0.48	0.90	0.15	0.35	334,200	1764	6992	188

Figure 19 shows the Cole–Cole diagrams obtained for SiN_x-based memristive structures. The data were obtained from the frequency dependences of the G_p/ω and C_p of memristive structure before LT (see Figure 16a) and after LT (see Figure 18a). Note that the memristive structure did not demonstrate resistive switching before laser treatment; therefore, the diagram for this case was obtained in the initial highly conductive state of memristive structure. At the same time, after LT, the two resistive states of memristive structure became distinguishable; therefore, the diagram for the second case was obtained under the conditions of HRS of the memristive structure.

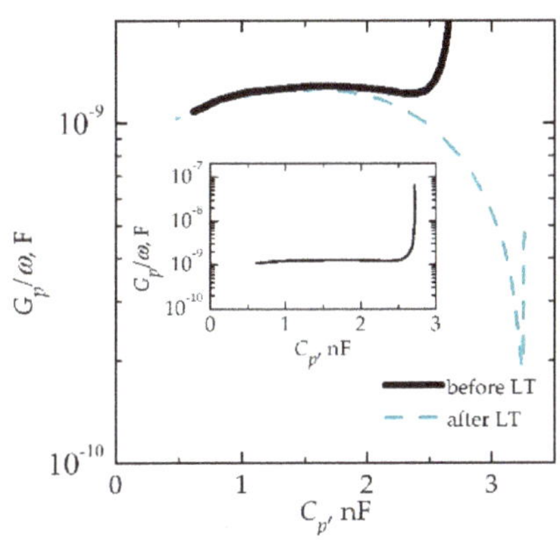

Figure 19. The Cole–Cole diagrams obtained for SiN_x-based memristive structures. The data were obtained before and after LT. Inset: same Cole–Cole diagram as before LT, but at full scale.

In the first case (before LT), the diagram had a circular arc shape, which indicates a uniform spectrum of SS at the insulator/semiconductor interface. The value of N_{ss} was estimated using Equation (1) and is equal to 1.6×10^{12} cm^{-2}eV^{-1}. SS with such a high-density value can reduce response times and contribute to the variability in RS voltage values. The sharp increase in G_p/ω at high values of C_p is due to the presence of conductive channels (see inset in Figure 19). In the second case (after LT), the shape of the diagram is close to a semicircle, which indicates the presence of a mono-level of SS. The value of N_{ss}, estimated using Equation (2), was 1.5×10^{11} cm^{-2}, which is an order of magnitude lower than before LT.

Figure 20a,b shows the results of a statistical study for 10 I–V curves of memristive structures after LT. It can be seen that the currents through the structure in LRS and HRS differ by at least 8-fold (Figure 20a), and the voltages for RESET and SET processes have a value in the selected range (Figure 20b).

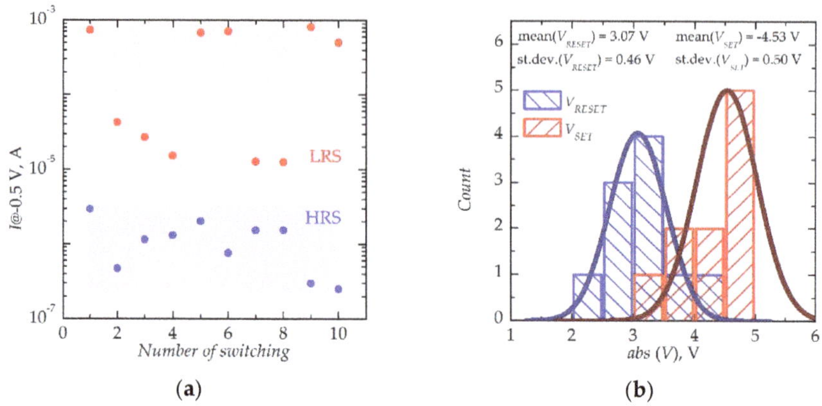

Figure 20. (**a**) Dependences of the currents (at a reading voltage of −0.5 V) of memristive structure in LRS (red) and HRS (blue) after LT on the number of RS cycles; (**b**) distribution of voltages of V_{SET} (red) and V_{RESET} (blue) processes of memristive structure after LT.

Thus, one can conclude that LT leads to a change in the spectrum of SS at the SiN_x/SOI interface. This is probably due to the more significant, in comparison with SiO_x-based structures, effect of LT on the charge state of traps in SiN_x, which determine the conductivity with the optical activation energy (for $SiN_{x<4/3}$, this value is equal to 2.6 eV [69]). It was reported in [70] that these traps can play a decisive role in the rupture and restoration of filaments during switching in SiN_x-based memristors. Therefore, LT is an effective method for changing RS parameters in metal/SiN_x/semiconductor memristive structures.

It should be noted that, along with the abovementioned influence of LT on the spectrum of SS, it can be responsible for the occurrence of RS. It can be assumed that the nanocrystallites observed by TEM are responsible for the initial conductive state of SiN_x-based memristors. According to the estimates from TEM images, nanocrystallites reach diameters of ~5–10 nm. Note that TEM studies were carried out after LT. Thus, probably, such treatment led to significant and irreversible oxidation of nanocrystallites and, before LT, the sizes of nanocrystallites could be comparable to the thickness of the insulator film. The latter could lead to shunting the devices.

4. Conclusions

This work demonstrates the robustness of the memristive phenomenon in thin-film structures based on promising and accessible insulator layers—SiO_x and SiN_x—fabricated on SOI substrates and subjected to additional laser and thermal treatments. It was shown that laser treatment leads to a significant increase in the hysteresis loop in *I–V* curves of the Au/Zr/SiN_x/SOI memristive structures. The effect was explained by the positive charging of traps in the insulator and a decrease in the density of surface states at the insulator/semiconductor interface (by an order of magnitude). Moreover, laser treatment of the Au/Zr/SiO_x/SOI memristive structures was not sufficient to produce a significant change in the value of the density of the surface states. Additional use of thermal treatment led to a decrease in this value by a factor of ~1.3. Furthermore, the combined effect of laser treatment followed by thermal treatment on the Au/Zr/SiO_x/SOI memristive structures led to a near doubling of the resistive switching voltages. The effect was, probably, associated with the annealing of surface states, which, in turn, led to a decrease in the resistance of the structure.

The CMOS compatibility of memristive devices in our study was provided by two factors. First, it is a SOI substrate, which is used in the technology of integrated circuits, including radiation-resistant ones. Secondly, it is a switching layer material, which is also fabricated using industrial technology. In this sense, the top electrode is of no fundamental importance, since in the framework of BEOL (back-end-of-line) integration it does not affect the basic FEOL (front-end-of-line) process. We chose a composite Au/Zr electrode, since it had previously proven itself well in MIM devices based on SiO_x [71] and is semitransparent, which is important for laser treatment. However, as part of the further optimization of these devices, other combinations of oxidizable and inert metals can be selected and tested.

It should be emphasized that the device layer of silicon in the SOI structure can differ greatly from bulk silicon in terms of structure and surface quality. The latter significantly affects the surface state, which can play an important role in the resistive switching mechanism. Therefore, the use of a SOI substrate in combination with specific switching insulators and additional treatment methods is of fundamental importance.

Author Contributions: Conceptualization, S.V.T. and A.N.M.; methodology, S.V.T. and D.S.K.; software, A.I.B.; validation, M.N.K., D.O.F., D.I.T., A.N.M., S.K. and B.S.; formal analysis, M.N.K., D.O.F. and S.V.T.; investigation, S.V.T., A.I.B., A.V.K., R.N.K., S.Y.Z., V.A.V. and D.A.P.; resources, A.N.M. and S.K.; data curation, M.N.K., D.O.F., S.V.T., A.I.B., D.I.T. and A.N.M.; writing—original draft preparation, M.N.K., D.O.F., S.V.T. and A.I.B.; writing—review and editing, D.I.T., A.N.M., S.A.S., S.K. and B.S.; visualization, M.N.K., S.V.T. and A.I.B.; supervision, D.O.F., D.I.T., A.N.M., S.A.S., S.K. and B.S.; project administration, A.N.M. and S.A.S.; funding acquisition, A.N.M. and S.A.S. All authors have read and agreed to the published version of the manuscript.

Funding: This work was supported by the Ministry of Science and Higher Education of Russian Federation (Project No. 13.2251.21.0098) and a National Research Foundation of Korea (NRF) grant funded by the Ministry of Science and ICT (2021K1A3A1A49098073). The studies were performed using the hardware resources of the shared use center: Research and Education Center "Physics of Solid State Nanostructures", Lobachevsky State University of Nizhny Novgorod.

Data Availability Statement: The data that support the findings of this study are available from the corresponding author upon reasonable request.

Conflicts of Interest: The authors declare no conflict of interest.

References

1. Zhang, W.; Gao, B.; Tang, J.; Li, X.; Wu, W.; Qian, H.; Wu, H. Analog-type resistive switching devices for neuromorphic computing. *Phys. Status Solidi RRL* **2019**, *13*, 1900204. [CrossRef]
2. Wang, Z.; Joshi, S.; Savel'ev, S.E.; Jiang, H.; Midya, R.; Lin, P.; Hu, M.; Ge, N.; Strachan, J.P.; Li, Z.; et al. Memristors with diffusive dynamics as synaptic emulators for neuromorphic computing. *Nat. Mater.* **2017**, *16*, 101–108. [CrossRef] [PubMed]
3. Slipko, V.A.; Pershin, Y.V. Metastable memristive lines for signal transmission and information processing applications. *Phys. Rev. E* **2017**, *95*, 042213. [CrossRef]
4. Wang, Z.; Joshi, S.; Savel'ev, S.; Song, W.; Midya, R.; Li, Y.; Rao, M.; Yan, P.; Asapu, S.; Zhuo, Y.; et al. Fully memristive neural networks for pattern classification with unsupervised learning. *Nat. Electron.* **2018**, *1*, 137–145. [CrossRef]
5. Pershin, Y.V.; Slipko, V.A. Dynamical attractors of memristors and their networks. *Europhys. Lett.* **2019**, *125*, 20002. [CrossRef]
6. Pershin, Y.V. A demonstration of implication logic based on volatile (diffusive) memristors. *IEEE Trans. Circuits Syst. II Express Briefs* **2019**, *66*, 1033–1037. [CrossRef]
7. Mehonic, A.; Sebastian, A.; Rajendran, B.; Simeone, O.; Vasilaki, E.; Kenyon, A.J. Memristors—From in-memory computing, deep learning acceleration, and spiking neural networks to the future of neuromorphic and bio-inspired computing. *Adv. Intell. Syst.* **2020**, *2*, 2000085. [CrossRef]
8. Serb, A.; Corna, A.; George, R.; Khiat, A.; Rocchi, F.; Reato, M.; Maschietto, M.; Mayr, C.; Indiveri, G.; Vassanelli, S.; et al. Memristive synapses connect brain and silicon spiking neurons. *Sci. Rep.* **2020**, *10*, 2590. [CrossRef]
9. Demin, V.A.; Nekhaev, D.V.; Surazhevsky, I.A.; Nikiruy, K.E.; Emelyanov, A.V.; Nikolaev, S.N.; Rylkov, V.V.; Kovalchuk, M.V. Necessary conditions for STDP-based pattern recognition learning in a memristive spiking neural network. *Neural Netw.* **2021**, *134*, 64–75. [CrossRef]
10. Johnson, B.A.; Brahim, K.; Balanov, A.G.; Savel'ev, S.; Borisov, P. Transition from noise-induced to self-sustained current spiking generated by a NbO_x thin film threshold switch. *Appl. Phys. Lett.* **2021**, *118*, 023502. [CrossRef]
11. Ushakov, Y.; Akther, A.; Borisov, P.; Pattnaik, D.; Savel'ev, S.; Balanov, A.G. Deterministic mechanisms of spiking in diffusive memristors. *Chaos Solitons Fractals* **2021**, *149*, 110997. [CrossRef]
12. Du, N.; Zhao, X.; Chen, Z.; Choubey, B.; Di Ventra, M.; Skorupa, I.; Bürger, D.; Schmidt, H. Synaptic plasticity in memristive artificial synapses and their robustness against noisy inputs. *Front. Neurosci.* **2021**, *15*, 696. [CrossRef] [PubMed]

13. Mikhaylov, A.; Belov, A.; Korolev, D.; Antonov, I.; Kotomina, V.; Kotina, A.; Gryaznov, E.; Sharapov, A.; Koryazhkina, M.; Kryukov, R.; et al. Multilayer metal-oxide memristive device with stabilized resistive switching. *Adv. Mater. Technol.* **2020**, *5*, 1900607. [CrossRef]
14. Nikiruy, K.E.; Iliasov, A.I.; Emelyanov, A.V.; Sitnikov, A.V.; Rylkov, V.V.; Demin, V.A. Memristors based on nanoscale layers $LiNbO_3$ and $(Co_{40}Fe_{40}B_{20})_x(LiNbO_3)_{100-x}$. *Phys. Solid State* **2020**, *62*, 1732–1735. [CrossRef]
15. Matsukatova, A.N.; Emelyanov, A.V.; Minnekhanov, A.A.; Sakharutov, D.A.; Vdovichenko, A.Y.; Kamyshinskii, R.A.; Demin, V.A.; Rylkov, V.V.; Forsh, P.A.; Chvalun, S.N.; et al. Memristors based on poly(*p*-xylylene) with embedded silver nanoparticles. *Tech. Phys. Lett.* **2020**, *46*, 73–76. [CrossRef]
16. Sun, K.; Chen, J.; Yan, X. The future of memristors: Materials engineering and neural networks. *Adv. Funct. Mater.* **2021**, *31*, 2006773. [CrossRef]
17. La Torre, C.; Fleck, K.; Starschich, S.; Linn, E.; Waser, R.; Menzel, S. Dependence of the SET switching variability on the initial state in HfO_x-based ReRAM. *Phys. Status Solidi A* **2016**, *213*, 316–319. [CrossRef]
18. Ungureanu, M.; Zazpe, R.; Golmar, F.; Stoliar, P.; Llopis, R.; Casanova, F.; Hueso, L.E. A light-controlled resistive switching memory. *Adv. Mater.* **2012**, *24*, 2496–2500. [CrossRef]
19. Patterson, G.A.; Fierens, P.I.; Grosz, D.F. On the beneficial role of noise in resistive switching. *Appl. Phys. Lett.* **2013**, *103*, 074102. [CrossRef]
20. Mikhaylov, A.N.; Guseinov, D.V.; Belov, A.I.; Korolev, D.S.; Shishmakova, V.A.; Koryazhkina, M.N.; Filatov, D.O.; Gorshkov, O.N.; Maldonado, D.; Alonso, F.J.; et al. Stochastic resonance in a metal-oxide memristive device. *Chaos Solitons Fractals* **2021**, *144*, 110723. [CrossRef]
21. Ntinas, V.; Rubio, A.; Sirakoulis, G.C.; Aguilera, E.S.; Pedro, M.; Crespo-Yepes, A.; Martin-Martinez, J.; Rodriguez, R.; Nafria, M. Power-efficient noise-induced reduction of ReRAM cell's temporal variability effects. *IEEE Trans. Circuits Syst. II Express Briefs* **2021**, *68*, 1378–1382. [CrossRef]
22. Ielmini, D.; Nardi, F.; Cagli, C. Resistance-dependent amplitude of random telegraph-signal noise in resistive switching memories. *Appl. Phys. Lett.* **2010**, *96*, 053503. [CrossRef]
23. Marchewka, A.; Waser, R.; Menzel, S. Physical simulation of dynamic resistive switching in metal oxides using a Schottky contact barrier model. In Proceedings of the International Conference on Simulation of Semiconductor Processes and Devices (SISPAD), Washington, DC, USA, 9–11 September 2015; IEEE: New York, NY, USA, 2015; pp. 297–300. [CrossRef]
24. Strukov, D.B.; Alibart, F.; Stanley Williams, R. Thermophoresis/diffusion as a plausible mechanism for unipolar resistive switching in metal-oxide-metal memristors. *Appl. Phys. A* **2012**, *107*, 509–518. [CrossRef]
25. Guarcello, C.; Valenti, D.; Spagnolo, B. Phase dynamics in graphene-based Josephson junctions in the presence of thermal and correlated fluctuations. *Phys. Rev. B* **2015**, *92*, 174519. [CrossRef]
26. Carollo, A.; Spagnolo, B.; Valenti, D. Uhlmann curvature in dissipative phase transitions. *Sci. Rep.* **2018**, *8*, 9852. [CrossRef] [PubMed]
27. Jiang, H.; Han, L.; Lin, P.; Wang, Z.; Jang, M.H.; Wu, Q.; Barnell, M.; Yang, J.J.; Xin, H.L.; Xia, Q. Sub-10 nm Ta channel responsible for superior performance of a HfO_2 memristor. *Sci. Rep.* **2016**, *6*, 28525. [CrossRef] [PubMed]
28. Lu, K.; Li, Y.; He, W.F.; Chen, J.; Zhou, Y.X.; Duan, N.; Jin, M.M.; Gu, W.; Xue, K.H.; Sun, H.J.; et al. Diverse spike-timing-dependent plasticity based on multilevel HfO_x memristor for neuromorphic computing. *Appl. Phys. A* **2018**, *124*, 438. [CrossRef]
29. Lian, X.; Wang, M.; Rao, M.; Yan, P.; Yang, J.J.; Miao, F. Characteristics and transport mechanisms of triple switching regimes of TaO_x memristor. *Appl. Phys. Lett.* **2017**, *110*, 173504. [CrossRef]
30. Choi, S.; Jang, S.; Moon, J.H.; Kim, J.C.; Jeong, H.Y.; Jang, P.; Lee, K.J.; Wang, G. A self-rectifying TaO_y/nanoporous TaO_x memristor synaptic array for learning and energy-efficient neuromorphic systems. *NPG Asia Mater.* **2018**, *10*, 1097–1106. [CrossRef]
31. Abbas, Y.; Han, I.S.; Sokolov, A.S.; Jeon, Y.R.; Choi, C. Rapid thermal annealing on the atomic layer-deposited zirconia thin film to enhance resistive switching characteristics. *J. Mater. Sci. Mater. Electron.* **2020**, *31*, 903–909. [CrossRef]
32. Upadhyay, N.K.; Sun, W.; Lin, P.; Joshi, S.; Midya, R.; Zhang, X.; Wang, Z.; Jiang, H.; Yoon, J.H.; Rao, M.; et al. A memristor with low switching current and voltage for 1S1R integration and array operation. *Adv. Electron. Mater.* **2020**, *6*, 1901411. [CrossRef]
33. Gambuzza, L.V.; Samardzic, N.; Dautovic, S.; Xibilia, M.G.; Graziani, S.; Fortuna, L.; Stojanovic, G.; Frasca, M. A data driven model of TiO_2 printed memristors. In Proceedings of the 8th International Conference on Electrical and Electronics Engineering (ELECO), Bursa, Turkey, 28–30 November 2013; pp. 1–4. [CrossRef]
34. Kim, M.; Yoo, K.; Jeon, S.P.; Park, S.K.; Kim, Y.H. The effect of multi-layer stacking sequence of TiO_x active layers on the resistive-switching characteristics of memristor devices. *Micromachines* **2020**, *11*, 154. [CrossRef] [PubMed]
35. Cvejin, K.; Mojić, B.; Samardžić, N.; Srdić, V.V.; Stojanović, G.M. Dielectric studies of barium bismuth titanate as a material for application in temperature sensors. *J. Mater. Sci. Mater. Electron.* **2013**, *24*, 1243–1249. [CrossRef]
36. Vasileiadis, N.; Karakolis, P.; Mandylas, P.; Ioannou-Sougleridis, V.; Normand, P.; Perego, M.; Komninou, P.; Ntinas, V.; Fyrigos, I.A.; Karafyllidis, I.; et al. Understanding the role of defects in silicon nitride-based resistive switching memories through oxygen doping. *IEEE Trans. Nanotechnol.* **2021**, *20*, 356–364. [CrossRef]
37. Vasileiadis, N.; Ntinas, V.; Fyrigos, I.A.; Karamani, R.E.; Ioannou-Sougleridis, V.; Normand, P.; Karafyllidis, I.; Sirakoulis, G.C.; Dimitrakis, P. A new 1P1R image sensor with in-memory computing properties based on silicon nitride devices. In Proceedings of the IEEE International Symposium on Circuits and Systems (ISCAS), Daegu, Korea, 22–28 May 2021; IEEE: New York, NY, USA, 2021; pp. 1–5. [CrossRef]

38. Ambrosi, E.; Bricalli, A.; Laudato, M.; Ielmini, D. Impact of oxide and electrode materials on the switching characteristics of oxide ReRAM devices. *Faraday Discuss.* **2019**, *213*, 87–98. [CrossRef]
39. Yen, T.-J.; Chin, A.; Gritsenko, V. Improved device distribution in high-performance SiN_x Resistive Random Access Memory via Arsenic ion implantation. *Nanomaterials* **2021**, *11*, 1401. [CrossRef]
40. Duchamp, M.; Migunov, V.; Tavabi, A.H.; Mehonic, A.; Buckwell, M.; Munde, M.; Kenyon, A.J.; Dunin-Borkowski, R.E. In situ transmission electron microscopy of resistive switching in thin silicon oxide layers. *Resolut. Discov.* **2016**, *1*, 27–33. [CrossRef]
41. Jiang, X.; Ma, Z.; Yang, H.; Yu, J.; Wang, W.; Zhang, W.; Li, W.; Xu, J.; Xu, L.; Chen, K.; et al. Nanocrystalline Si pathway induced unipolar resistive switching behavior from annealed Si-rich SiN_x/SiN_y multilayers. *J. Appl. Phys.* **2014**, *116*, 123705. [CrossRef]
42. Islamov, D.R.; Gritsenko, V.A.; Chin, A. Charge transport in thin hafnium and zirconium oxide films. *Optoelectron. Instrument. Proc.* **2017**, *53*, 184–189. [CrossRef]
43. Gismatulin, A.A.; Orlov, O.M.; Gritsenko, V.A.; Kruchinin, V.N.; Mizginov, D.S.; Krasnikov, G.Y. Charge transport mechanism in the metal–nitride–oxide–silicon forming-free memristor structure. *Appl. Phys. Lett.* **2020**, *116*, 203502. [CrossRef]
44. Bishop, M.D.; Wong, H.S.P.; Mitra, S.; Shulaker, M.M. Monolithic 3-D integration. *IEEE Micro* **2019**, *39*, 16–27. [CrossRef]
45. Saylan, S.; Aldosari, H.M.; Humood, K.; Abi Jaoude, M.; Ravaux, F.; Mohammad, B. Effects of top electrode material in hafnium-oxide-based memristive systems on highly-doped Si. *Sci. Rep.* **2020**, *10*, 19541. [CrossRef] [PubMed]
46. Popov, V.P.; Antonova, A.I.; Frantsuzov, A.A.; Safronov, L.N.; Feofanov, G.N.; Naumova, O.V.; Kilanov, D.V. Properties of silicon-on-insulator structures and devices. *Semiconductors* **2001**, *35*, 1030–1037. [CrossRef]
47. Hoessbacher, C.; Fedoryshyn, Y.; Emboras, A.; Melikyan, A.; Kohl, M.; Hillerkuss, D.; Hafner, C.; Leuthold, J. The plasmonic memristor: A latching optical switch. *Optica* **2014**, *1*, 198–202. [CrossRef]
48. Puppo, F.; Doucey, M.A.; Di Ventra, M.; De Micheli, G.; Carrara, S. Memristor-based devices for sensing. In Proceedings of the IEEE International Symposium on Circuits and Systems (ISCAS), Melbourne, VIC, Australia, 1–5 June 2014; IEEE: New York, NY, USA, 2014; pp. 2257–2260. [CrossRef]
49. Li, C.; Han, L.; Jiang, H.; Jang, M.-H.; Lin, P.; Wu, Q.; Barnell, M.; Yang, J.J.; Xin, H.L.; Xia, Q. Three-dimensional crossbar arrays of self-rectifying $Si/SiO_2/Si$ memristors. *Nat. Commun.* **2017**, *8*, 15666. [CrossRef]
50. Pragnya, P.; Pinkowitz, A.; Hull, R.; Gall, D. Electrochemical memristive devices based on submonolayer metal deposition. *APL Mater.* **2019**, *7*, 101121. [CrossRef]
51. Baltakesmez, A. Improved barrier parameters and working stability of Au/*p*-GO/*n*-lnP/Au–Ge Schottky barrier diode with GO interlayer showing resistive switching effect. *Vacuum* **2019**, *168*, 108825. [CrossRef]
52. Skorobogatov, S. Local heating attacks on flash memory devices. In Proceedings of the IEEE International Workshop on Hardware-Oriented Security and Trust (HOST), San Francisco, CA, USA, 27 July 2009; IEEE: New York, NY, USA, 2009; pp. 1–6. [CrossRef]
53. Kärkkänen, I.; Shkabko, A.; Heikkilä, M.; Vehkamäki, M.; Niinistö, J.; Aslam, N.; Meuffels, P.; Ritala, M.; Leskelä, M.; Waser, R.; et al. Impedance spectroscopy study of the unipolar and bipolar resistive switching states of atomic layer deposited polycrystalline ZrO_2 thin films. *Phys. Status Solidi A* **2015**, *212*, 751–766. [CrossRef]
54. Epshtein, S.L. *Measuring of Capacitor Characteristics*; Energiya: Moscow, Russia, 1965; 236p. (In Russian)
55. Oreshkin, P.T. *Physics of Semiconductors and Dielectrics*; Vysshaya Shkola: Moscow, Russia, 1977; 448p. (In Russian)
56. Ovsyuk, V.N. *Electronic Processes in Semiconductors with Space-Charge Regions*; Nauka: Novosibirsk, Russia, 1984; 254p. (In Russian)
57. Antipov, A.; Arakelian, S.; Vartanyan, T.; Gerke, M.; Istratov, A.; Kutrovskaya, S.; Kucherik, A.; Osipov, A. Optical properties of multilayer bimetallic films obtained by laser deposition of colloidal particles. *Opt. Spectrosc.* **2016**, *121*, 765–768. [CrossRef]
58. Hofmann, S. *Auger- and X-ray Photoelectron Spectroscopy in Materials Science*; Springer: Berlin/Heidelberg, Germany, 2013; 528p. [CrossRef]
59. Wang, T.Y.; Meng, J.L.; Li, Q.X.; Chen, L.; Zhu, H.; Sun, Q.Q.; Ding, S.J.; Zhang, D.W. Forming-free flexible memristor with multilevel storage for neuromorphic computing by full PVD technique. *J. Mater. Sci. Technol.* **2021**, *60*, 21–26. [CrossRef]
60. Gorshkov, O.N.; Mikhaylov, A.N.; Kasatkin, A.P.; Tikhov, S.V.; Filatov, D.O.; Pavlov, D.A.; Belov, A.I.; Koryazhkina, M.N.; Bobrov, A.I.; Malekhonova, N.V.; et al. Resistive switching in the $Au/Zr/ZrO_2$-Y_2O_3/TiN/Ti memristive devices deposited by magnetron sputtering. *J. Phys. Conf. Ser.* **2016**, *741*, 012174. [CrossRef]
61. Gurtov, V.A. *Solid State Electronics*; Tekhnosfera: Moscow, Russia, 2008; 512p. (In Russian)
62. Ismail, M.; Kim, S. Negative differential resistance effect and dual bipolar resistive switching properties in a transparent Ce-based devices with opposite forming polarity. *Appl. Surf. Sci.* **2020**, *530*, 147284. [CrossRef]
63. Kumar, N.; Chand, S. Effects of temperature, bias and frequency on the dielectric properties and electrical conductivity of Ni/SiO_2/*p*-Si/Al MIS Schottky diodes. *J. Alloys Compd.* **2020**, *817*, 153294. [CrossRef]
64. Hasegawa, H.; Sawada, T. On the electrical properties of compound semiconductor interfaces in metal/insulator/semiconductor structures and the possible origin of interface states. *Thin Solid Films* **1983**, *103*, 119–140. [CrossRef]
65. Chaabouni, F.; Abaab, M.; Rezig, B. Characterization of *n*-ZnO/*p*-Si films grown by magnetron sputtering. *Superlattices Microstruct.* **2006**, *39*, 171–178. [CrossRef]
66. Hu, C. *Modern Semiconductor Devices for Integrated Circuits*; Pearson: London, UK, 2010; p. 351.
67. Wang, L.G.; Zhang, W.; Chen, Y.; Cao, Y.Q.; Li, A.D.; Wu, D. Synaptic plasticity and learning behaviors mimicked in single inorganic synapses of $Pt/HfO_x/ZnO_x$/TiN memristive system. *Nanoscale Res. Lett.* **2017**, *12*, 65. [CrossRef] [PubMed]

68. Krishnaprasad, A.; Choudhary, N.; Das, S.; Dev, D.; Kalita, H.; Chung, H.S.; Aina, O.; Jung, Y.; Roy, T. Electronic synapses with near-linear weight update using MoS_2/graphene memristors. *Appl. Phys. Lett.* **2019**, *115*, 103104. [CrossRef]
69. Nasyrov, K.A.; Shaĭmeev, S.S.; Gritsenko, V.A.; Han, J.H.; Kim, C.W.; Lee, J.-W. Electron and hole injection in metal-oxide-nitride-oxide-silicon structures. *J. Exp. Theor. Phys.* **2006**, *102*, 810–820. [CrossRef]
70. Gismatulin, A.A.; Gritsenko, V.A.; Yen, T.-J.; Chin, A. Charge transport mechanism in SiN_x-based memristor. *Appl. Phys. Lett.* **2019**, *115*, 253502. [CrossRef]
71. Mikhaylov, A.N.; Belov, A.I.; Guseinov, D.V.; Korolev, D.S.; Antonov, I.N.; Efimovykh, D.V.; Tikhov, S.V.; Kasatkin, A.P.; Gorshkov, O.N.; Tetelbaum, D.I.; et al. Bipolar resistive switching and charge transport in silicon oxide memristor. *Mater. Sci. Eng. B* **2015**, *194*, 48–54. [CrossRef]

Journal of
Low Power Electronics and Applications

MDPI

Article

Design of In-Memory Parallel-Prefix Adders

John Reuben

Chair of Computer Architecture, Friedrich-Alexander-Universität Erlangen-Nürnberg (FAU), 91058 Erlangen, Germany; johnreuben.prabahar@fau.de

Abstract: Computational methods in memory array are being researched in many emerging memory technologies to conquer the 'von Neumann bottleneck'. Resistive RAM (ReRAM) is a non-volatile memory, which supports Boolean logic operation, and adders can be implemented as a sequence of Boolean operations in the memory. While many in-memory adders have recently been proposed, their latency is exorbitant for increasing bit-width ($O(n)$). Decades of research in computer arithmetic have proven parallel-prefix technique to be the fastest addition technique in conventional CMOS-based binary adders. This work endeavors to move parallel-prefix addition to the memory array to significantly minimize the latency of in-memory addition. Majority logic was chosen as the fundamental logic primitive and parallel-prefix adders synthesized in majority logic were mapped to the memory array using the proposed algorithm. The proposed algorithm can be used to map any parallel-prefix adder to a memory array and mapping is performed in such a way that the latency of addition is minimized. The proposed algorithm enables addition in $O(log(n))$ latency in the memory array.

Keywords: resistive RAM (ReRAM); non-volatile memory (NVM); majority logic; memristor; 1Transistor-1Resistor (1T–1R); in-memory computing; processing-in-memory; parallel-prefix adder; logic-in-memory; memristive logic

Citation: Reuben, J. Design of In-Memory Parallel-Prefix Adders. *J. Low Power Electron. Appl.* **2021**, *11*, 45. https://doi.org/10.3390/jlpea11040045

Academic Editors: Alex Serb and Adnan Mehonic

Received: 14 October 2021
Accepted: 17 November 2021
Published: 24 November 2021

1. Introduction

Conventional computer architecture is facing an acute problem—the 'von Neumann bottleneck' or 'memory wall'. The shuffling of data between processing and memory units is energy-consuming and time-consuming and degrades the performance of contemporary computing systems [1,2]. In other words, the energy needed to move data (between memory and processing units) forms a significant portion of the computational energy. To overcome the memory wall, the processor and memory unit must be brought closer to each other. A 3D stacking of DRAM dies over logic die, often referred to as near-memory computing [3], was pursued earlier to reduce the latency and energy for data movement between processor and memory. The recent trend is to move computing to the location of the data, i.e., in-memory computing.

In in-memory computing, the data are processed at their location (i.e., in the memory array) and not moved out of the memory array to a separate processing unit. At present, diverse operations from arithmetic operations to cognitive tasks such as machine learning and pattern recognition are being explored in memory arrays [4]. This article focuses on arithmetic operations and how adders can be implemented in memory. It should be noted that in-memory computing is pursued in many memory technologies—both conventional (SRAM, DRAM) and emerging non-volatile memories (Resistive RAM, STT-MRAM, PCM, FeFET). However, in this article, we restrict our focus to Resistive RAM technology to achieve a greater focus on the design of parallel-prefix adders. Resistive RAM device is a two-terminal Metal–Insulator–Metal structure in which data can be stored as resistance. A positive voltage across the structure forms a conductive filament (low resistance state) and a negative voltage ruptures the filament (high resistance state), leading to two stable resistances. Boolean gates can be implemented in the memory array by altering the structure of the memory array, the peripheral circuitry around the array, or both. Arithmetic circuits such as adders can be implemented as a chain of such Boolean operations.

Although different in-memory adders have been proposed in the literature, the latency of in-memory adders is a severe disadvantage in in-memory computing, i.e., an addition operation needs a long sequence of Boolean operations. A poorly optimized in-memory adder may take longer to compute (add two n-bit numbers) than the combined time it takes to fetch data from memory and add them in a CMOS-based processor. In a computing system, adders constitute the basic computational unit. In-memory adders have not had their latency studied and optimized for an increasing bit-width (n-bit operand). In practice, 32-bit/64-bit in-memory adders require hundreds of cycles due to $O(n)$ latency requirements. It was originally proposed that parallel-prefix (PP) adders could bring down the latency caused by the rippling of carry in CMOS-based adders. PP adders are the fastest adders in conventional CMOS technology [5,6]. To improve the latency of in-memory adders, it is necessary to learn lessons from the decades of research on CMOS adders and adopt them for in-memory addition. Therefore, parallel-prefix adders were pursued in this work to improve the latency of in-memory adders. More specifically, we propose a generic methodology to design any PP adder in memory. As an example, we consider the Ladner–Fischer type of PP adder and demonstrate how this can be implemented in-memory in O($log(n)$ latency. The presented method requires no major modifications to the peripheral circuitry of the memory array and is also energy-efficient.

The rest of the paper is organised as follows. Section 2 reviews the state-of-the-art in-memory adders and classifies in-memory adders on the basis of state fullness, logic primitive and architecture. The review identifies the exorbitant latency of adders with increasing bit-width, as a significant issue that needs attention. Section 3 presents PP adders as a solution to the long latency incurred by the rippling of carry. Section 4.1 reviews the in-memory majority gate, which is the fundamental logic gate used in this work to implement the PP adder in the memory array. Section 4.2 elaborates how PP adders can be synthesized using majority logic. Having synthesized PP adders in majority logic, Section 4.3.3 elaborates how they can be mapped to the memory array. We present the simulation methodology in Section 5.1. In Section 5.2, we analyse how the latency of the proposed adder grows with increasing bit-width. Sections 5.3 and 5.4 analyse how the energy and area of the proposed adder grow with increasing bit-width. In Section 5.5, we compare the proposed adder with other adders reported in the literature, followed by the Conclusion in Section 6.

2. In-Memory Adders: A Brief Review

Conventionally, adders were designed using logic gates built from CMOS transistors. In contrast, an in-memory adder is designed using a 'functionally complete' Boolean logic primitive. NOR, for example, is functionally complete, since any Boolean logic can be expressed using NOR gates. Therefore, if an NOR gate can be implemented in the memory array, any arithmetic circuit can be implemented in the memory array. NAND, IMPLY + FALSE [7] and Majority + NOT [8] are other functionally complete logic primitives. In the last 5 years, several in-memory adders have been proposed. They can be classified as the following:

1. State variable used for computation—stateful or non-stateful;
2. Logic primitive used for computation—NAND or NOR or IMPLY or MAJORITY or XOR or a combination of these;
3. Adder architecture—how is the carry propagated?

Stateful in-memory adders perform an addition by logic gates, where each gate is executed by manipulating the resistance of a memristor (i.e., the internal state) rather than by a mix of resistance and voltage [9]. If voltage is also used, in addition to resistance, the logic gate and the adder are said to be non-stateful (Figure 1a). This is one of the characteristics of in-memory adders that, with certain modifications to conventional memory, a particular logic primitive can be realized and other logic primitives need to be realized in terms of this logic primitive. The NOR-based memristive logic family (MAGIC), for example, requires that all other gates (AND, OR, XOR) are expressed in terms of NOR gates and then used in the memory array. Similarly, in the NAND-based adder reported in [10], an XOR gate is implemented as NAND gates (one XOR requires four memory cycles). Figure 1b illustrates a one-bit full adder expressed solely as NOR/NAND/Majority gates (expressing a circuit using single logic primitive is preferred for in-memory implementation). Finally, an issue that is often overlooked in this emerging area is the issue of carry-propagation. The manner in which carry is propagated from LSB to MSB decides the speed of the in-memory adder. In the in-memory community, different adder architectures, from ripple carry (slowest) to parallel-prefix (fastest) adders, have been proposed.

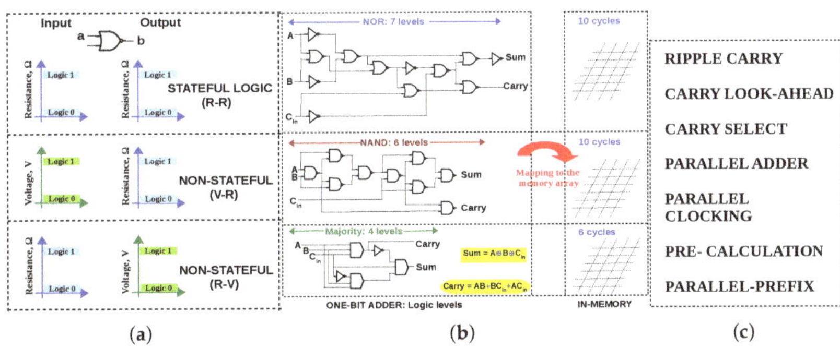

Figure 1. (**a**) An in-memory logic gate (adder) is stateful if its only state variable is resistance. Non-stateful logic gates (adder) also use voltage in conjunction with resistance. (**b**) In-memory adder implementation favors homogeneity of logic primitives; 1-bit full adder in terms of NOR gates [11], NAND gates [12] and majority gates [13]; (**c**) Different carry-propagation techniques result in different adder architectures.

Table 1 lists different in-memory adders that have been reported recently and their latency for 8-bit and n-bit. The adders are also classified based on the three characteristics we reviewed—state variable, logic primitive and adder architecture. A key observation is that logic primitive plays an important role in determining the latency. IMPLY is a weak logic primitive, and generally incurs more latency than all other logic primitives. XOR and the majority are generally stronger logic primitives than OR/AND/NOR. This is evident from the fact that XOR-based and majority-based ripple-carry adders are faster than NAND/NOR-based ripple-carry adders [13]. In other words, with the adder configuration being the same (ripple-carry), logic primitive plays an important role in determining the latency of in-memory addition. Another important finding is that the adder architecture plays a key role in deciding the latency for increasing bit-width. This is evident from the latency of OR + AND-based adders in Table 1. Both adders used the same logic primitive (OR + AND), but [14] uses ripple-carry architecture, achieving a latency of $6n + 1$ while [15] uses a parallel-prefix configuration to achieve a latency of $8log_2(n) + 13$. As an example, a 32-bit adder based on OR + AND logic will require 193 cycles and 53 cycles for ripple-carry and parallel-prefix architectures, respectively. Hence, for larger bit-widths, architecture (carry propagation technique) plays an important role in latency. In summary, both adder architecture and logic primitive influence the latency of in-memory adder. Therefore, majority logic primitive and parallel-prefix adder architecture were chosen in this work to drastically minimize the latency of in-memory adders.

Table 1. Latency of recently reported in-memory adders (8-bit and n-bit).

Stateful	Primitive	Architecture	Latency (8 bit)	Latency (n-bit)	Ref.
Yes	IMPLY	Ripple carry	58	$5n + 18$	[7]
Yes	IMPLY	Parallel-serial	56	$5n + 16$	[16]
Yes	IMPLY + OR	Ripple carry	54	$6n + 6$	[17]
Yes	IMPLY	Semi-parallel	136	$17n$	[18]
Yes	NOR	Ripple carry	83	$10n + 3$	[19]
Yes	NOR	Look-Ahead	48	$5n + 8$	[20]
No	OR + AND	Ripple carry	49	$6n + 1$	[14]
Yes	ORNOR	Parallel-clocking	31	$2n + 15$	[21]
Yes	RIMP/NIMP*	Pre-calculation	20	$2n + 4$	[22]
Yes	XOR	Ripple carry	18	$2n + 2$	[23]
No	XOR + MAJ	Ripple carry	18	$2n + 2$	[24]
Yes	XNOR/XOR	Carry-Select	9	–	[25]
No	OR + AND	Parallel-prefix	37	$8log_2(n) + 13$	[15]

In a Complementary Resistive Switch (CRS) adder, RIMP/NIMP* denotes reverse implication and inverse implication.

3. Parallel-Prefix Adders: A Solution for the Carry-Propagation Problem

When two n-bit binary numbers A ($a_{n-1}a_{n-2} \cdot a_0$) and B ($b_{n-1}b_{n-2} \cdot b_0$) are added, the sum bit S_i at the ith bit position is computed as,

$$S_i = H_i \oplus C_{i-1} \tag{1}$$

where, $H_i = A_i \oplus B_i$ and C_{i-1} is the carry computed in the previous bit position. To compute the sum bits of the next significant bit position, the incoming carry C_{i-1} is propagated to the next position. This is accomplished using carry generate bits ($G_i = A_i \cdot B_i$) and carry propagate bits ($P_i = A_i + B_i$). The carry-out (C_i) of a particular bit position is always a function of the carry from the previous bit (C_{i-1}), and they are expressed as follows:

$$C_i = G_i + P_i \cdot C_{i-1} \tag{2}$$

Thus, during the 8-bit addition of $a_7a_6a_5a_4a_3a_2a_1a_0$ and $b_7b_6b_5b_4b_3b_2b_1b_0$, sum bit $S_6 = H_6 \oplus C_5$ and C_5 is a function of a_5, b_5, C_4 according to Equations (1) and (2). In other words, S_6 cannot be computed until C_5 is computed, which recursively depends on the carry-out of the lower significant bit. This is the decades old carry-propagation problem and significantly affects the speed of n-bit addition as n grows. Ripple-carry adders are extremely slow for 32-bit/64-bit addition due to this carry propagation. To improve this situation, carry-skip adders were proposed, which allowed for carries to skip across block of bits instead of rippling through them. This was followed by Carry-lookahead adders, where carries were computed in parallel and achieved logarithmic logic depth [26]. Parallel-prefix (PP) adders improved on the carry-look ahead adder by expressing carry-propagation as a prefix computation [27]. They are the fastest family of adders [5,6] in conventional transistor-based implementations.

PP adders have a 'carry-generate block', followed by a 'sum-generate block' (Figure 2). Internally, the carry-generate block has a pre-processing stage, which computes G_i, P_i, H_i for every bit. Using them, carry bits $C_{out}C_{n-1} \cdot C_1C_0$ are computed using the prefix computation technique. This is followed by the sum-generate block, where $S_i = H_i \oplus C_{i-1}$ is computed. The reader is referred to [27,28] for a detailed explanation of the stages of a parallel-prefix adder. Kogge–Stone, Ladner–Fischer, Brent–Kung, Sklansky, Ling, etc., are examples of PP adders. According to the taxonomy of PP adders [29], these adders essentially form a compromise between logical depth, fan-out and wiring tracks. PP adders can reduce the logical depth to $O(log(n))$, for n-bit adders [30].

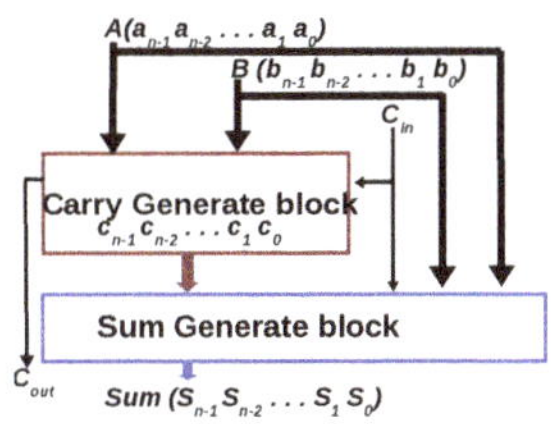

Figure 2. Generic Structure of PP adders: A 'carry-generate block' calculates carry by prefix computation and is then followed by a 'sum-generate block' ([30]).

4. In-Memory Implementation of Parallel-Prefix Adders

4.1. In-Memory Majority Gate

Before their implementation in memory, the PP adders must first be synthesized in terms of logic gates which can be implemented in memory. As stated, different logic primitives require different modifications to the memory array or its peripheral circuitry (or both). Therefore, for the in-memory implementation of adders, it is important to minimize the different types of logic primitives used. Consequently, it is beneficial to express the adder using one logic primitive, rather than four different logic primitives. Recently, an in-memory majority gate was proposed in [31,32]. The three inputs to the majority gate are the three resistances of the memory cells, and the output majority is computed as a READ operation (Figure 3a). This majority gate does not necessitate any major modifications to the peripheral circuitry of a regular memory array, and is also energy-efficient (access transistor for each memory cell minimizes sneak currents, thus lowering energy consumption when compared to other adders implemented in 1S–1R configuration). As depicted in Figure 3b, multiple majority gates can be executed in array columns, which suits PP adders with a similar structure.

4.2. Homogeneous Synthesis of Parallel-Prefix Adders

Conventionally, PP adders are synthesized in terms of AND, OR and XOR gates for CMOS implementation. Figure 4a depicts an eight-bit PP adder of the Ladner–Fischer type. Three different logic primitives are required—AND, OR and XOR. As stated, different logic primitives require different modifications to the memory array and its peripheral circuitry. If a particular Boolean logic gate cannot be implemented in the memory array, it has to be re-formulated in terms of a logic gate that can be implemented in the memory array. For example, in the NAND-based logic family reported in [10], the XOR gate cannot be implemented; therefore, it is expressed as four NAND gates. As depicted in Figure 4a, a single XOR becomes three levels of NAND logic, increasing its latency. In contrast, by expressing a PP adder purely in terms of MAJORITY+NOT gates, the PP adder can be efficiently implemented in the memory array. Furthermore, a majority-based PP adder achieves a marginal reduction in logical depth compared to conventional AND-OR-XOR implementation (Figure 4). This is due to the majority being a stronger logic primitive than NAND/NOR/IMPLY [13]. To synthesize PP adders in terms of majority gates, logic synthesis tools can be used. A logic synthesis tool is proposed in [8], which takes any AND-OR-INVERT-based logic and synthesizes it purely in terms of majority and NOT gates. Boolean logic minimization techniques such as re-shaping, push-up, node merging, etc., are used to re-synthesize and optimize conventional AND-OR-INVERT logic in terms of MAJORITY-INVERT [33–38]. Since the majority is the fundamental logic primitive for many emerging nanotechnologies, there are also works which pioneered the synthesis of PP adders solely in terms of majority gates. The reader is referred to [5,30,39] for such works. Therefore, a variety of techniques can be used to transform PP adders in terms of majority and NOT gates. Figure 4b depicts a 8-bit PP adder, synthesized solely in terms of majority and NOT gates. In addition to achieving homogeneity, the majority-based PP adder incurs one level of reduction in logical depth compared to the AND-OR-XOR-based PP adder.

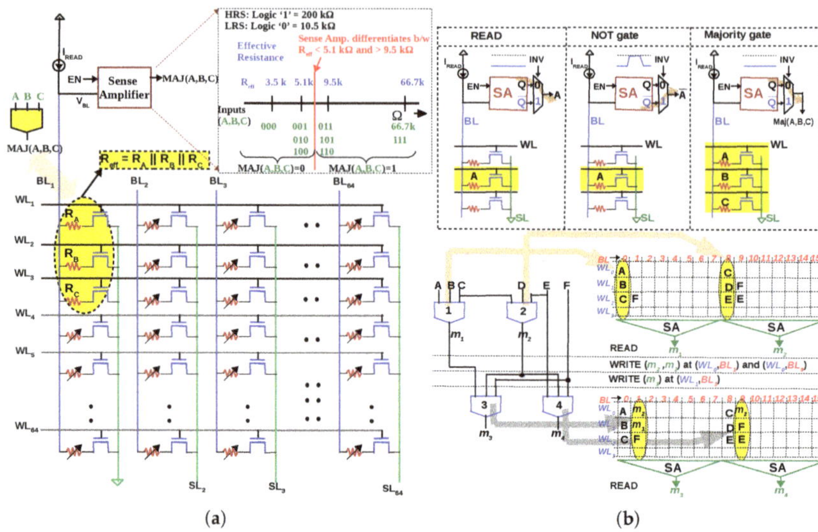

Figure 3. (**a**) In-memory Implementation of majority gate [31,32]: In a 1T-1R array, the resistances (R_A, R_B, R_C) in the three rows will be parallel if three rows are selected at the same time. (Inputs of the majority gate A, B, C are represented as resistances R_A, R_B, R_C). During READ, the effective resistance R_{eff} can accurately be sensed to implement an in-memory majority gate. (**b**) NOT operation can be implemented by inverting the output of the SA. With a majority and NOT gate implemented as a READ operation, the array can be used to execute multiple levels of logic by writing back the data, simplifying computing to READ and WRITE operations.

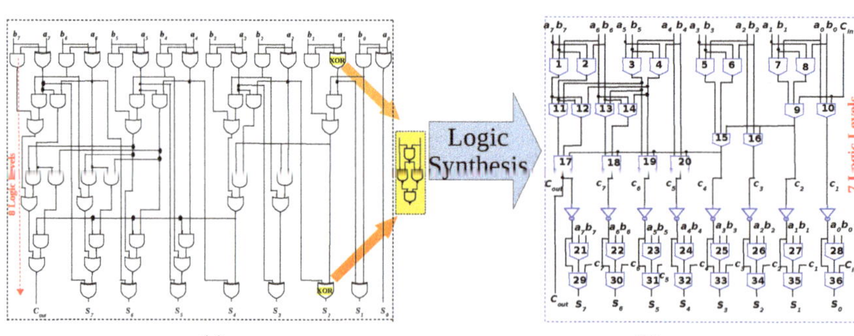

Figure 4. (**a**) Eight-bit PP adder of Ladner–Fischer type expressed in terms of AND, OR , XOR gates. (**b**) Re-synthesized and optimized in terms of MAJORITY and NOT gates [5,30].

4.3. Mapping Methodology

Having synthesized the PP adder in terms of majority and NOT gates, they can be implemented in memory using the in-memory majority gate described in Section 4.1. The NOT gate can be implemented as a simple READ operation with the output inverted. The design of in-memory PP adders presented in this paper is generic and can be used to implement any PP adder. However, in this section, the Ladner–Fischer adder of Figure 4b is chosen and the in-memory implementation (mapping) steps are elaborated.

4.3.1. In-Memory Mapping as an Optimization Problem: Objectives

The mapping of the majority-based PP adder to the memory array can be treated as an optimization problem. Any optimization problem has objectives or goals, which should be achieved in the presence of certain constraints. The objectives of in-memory mapping are as follows:

1. Latency of in-memory PP adder must be minimized (O_1);
2. Energy consumption during addition must be minimized (O_2);
3. Area of the array used during computation must be minimized (O_3).

The aforementioned objectives are no different from the objectives of any VLSI circuit. All objectives cannot be met simultaneously in this mapping, and trade-offs must be made between latency of addition (O_1) and the area of array that is used (O_3). Any arithmetic circuit implemented in memory is bound to be very slow due to the high latency of in-memory adders. The latency of in-memory adders reported in the literature grows, as $O(n)$ and 32-bit/64-bit in memory require hundreds of cycles [9]. Therefore, in this mapping, we focus on and minimize the latency. Minimizing the latency might result in the array area being compromised. However, latency is the more serious issue compared to array area in in-memory addition, for the following reasons:

1. A conventional adder (in CMOS) is devoted to addition while we re-use the existing memory array in in-memory computation. Hence, the increased array area required during addition is not a disadvantage, as long as computation can be performed in the memory array without an extra array;
2. ReRAM memory cell or memristor is a nano-device and does not significantly contribute. For example, a single 1T-1R cell in 130 nm CMOS occupies 0.2 μm^2 [40].

A significant portion of the energy consumed during in-memory addition is dissipated in the memory array. This is predominantly due to sneak-currents in the 1S–1R array. In contrast, our proposed PP adder is implemented in a transistor-accessed memory array (1T–1R); therefore, the energy dissipation in the array is negligible. The major energy consumption is the energy consumed while the cells switch states (WRITE) and the majority operation (READ). Therefore, the energy consumed during addition is minimized if latency is minimized. In other words, latency (O_1) is the most important objective to be minimized.

4.3.2. In-Memory Mapping as an Optimization Problem: Constraints

The constraints are specific to this design methodology and can be summarized as follows:

1. Majority operation must be executed at three consecutive rows (C_1);
2. Due to the bounded endurance of ReRAM devices, the number of times a cell is switched must be minimized. (C_2).

C_1 must be satisfied during mapping because, during majority operation, three rows must simultaneously be selected. In principle, the three selected rows need not be contiguous and can be in different locations in the memory array (e.g., row 5, 8, 15 of a 64 $\times$ 64 array). However, row-decoding will become complicated. For practical in-memory implementation, the mapping must be 'peripheral circuit friendly'. In [32], a triple-row decoder is proposed for triple row-activation during majority operation. To implement this decoder, multiple single-row decoders were interleaved. Furthermore, the same row-decoder must be able to perform single-row decoding and triple-row decoding. This is because, during normal memory operation, a single row must be selected and, during majority, three rows must be selected. To this end, an address translator circuit is used in the row decoder, which seamlessly switches between single-row activation and triple-row activation. The triple-row decoder [32] is designed in such a way that only three consecutive rows can be selected. Therefore, while mapping, the inputs of the majority gate (to be executed in memory in the next step) must be written in three consecutive rows.

Constraint C_2 is posed by a characteristic of non-volatile memories called endurance. A memory device's endurance refers to its ability to switch between two stable states while maintaining a sufficient resistance ratio. Experimentally reported endurances vary from 10^6 to 10^{12}. Due to this limited endurance, the number of times a memory cell is switched during addition must be minimized.

4.3.3. Algorithm

Having identified the objectives and constraints, we formulate a generic methodology to map any PP adder to the memory array. As stated, if the PP adder is available in terms of AND-OR-XOR gates, they must be re-synthesized in terms of the majority and NOT gates using logic synthesis techniques/tools. Given a majority-based PP adder, optimal in-memory implementation is an optimization problem—minimize O_1, while meeting C_1 and C_2.

The following steps implement the PP adder in the memory:

1. Start with Logic level 1;
2. Simultaneously execute all majority gates of a logic level in the columns of the array (O_1);
3. Write the outputs of the majority gates to the precise locations where they are needed in the next logic level such that all the majority gates of the following level can be executed simultaneously (O_1);
4. During Step 3, write the outputs of the majority gates to a new location and do not overwrite the existing data (C_2);
5. During Step 3, write the outputs of the majority gates to contiguous locations in the memory array (C_1);
6. Repeat Steps 2–5 for the remaining logic levels.

Figure 5 illustrates the mapping of an 8-bit PP adder to the memory array. Majority gates 1–8 of the first logic level are executed simultaneously in one memory cycle. Since we know that, at the next level, majority gates 9, 10, 11, 12, 13, 14 need to be executed, we write the outputs of the first logic level ($m_1, m_2, m_3, m_4, m_5, m_6, m_7, m_8$) to the exact location where they will be needed. When we write the output of the majority gates back to the array, they are written in consecutive rows (C_1) and are not overwritten on existing data (C_2). The in-memory steps are highlighted in yellow in Figure 5. The in-memory steps corresponding to logic levels 1 and 2 are:

1. Majority at col. (1, 9, 26, 33, 42, 49, 58, 65) rows 4–6 as a READ operation;
2. Write ($m_1 m_1 m_3 m_5 m_7$) at col. (2, 10, 34, 50, 59) , row 4;
3. Write ($m_2 m_2 m_4 m_6 m_8$) at col. (2, 10, 34, 50, 59), row 5;
4. Write ($m_3 m_4 m_3 m_4$) at col. (2, 10, 17, 25) , row 6;
5. Majority at col. (2, 10, 17, 25, 59, 73) rows 4–6 as a READ operation.
6. ....

In this manner, the seven logic levels of an 8-bit adder can be executed in memory in 18 cycles. A detailed mapping of all seven logic levels is presented in Appendix A.

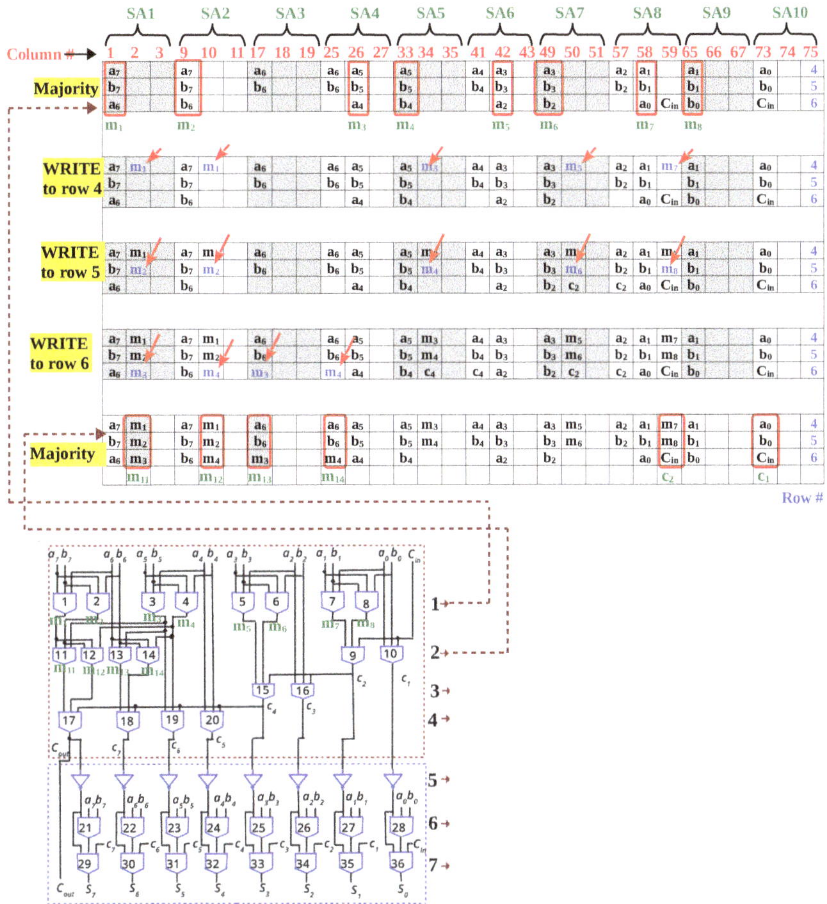

Figure 5. Illustration of mapping of the first two logic levels to a memory array. Since each majority operation is executed as a READ operation, it can be written to the exact location it is needed at the next logic level while satisfying C_1 and C_2. In the above mapping, eight columns share a sense amplifier.

5. Performance of In-Memory Parallel-Prefix Adders

5.1. Simulation Methodology

To verify the proposed in-memory adder through simulation, the 1T-1R memory array and its peripheral circuitry were designed in IHP's 130 nm CMOS process. The memory array was composed of 1T-1R cells in which the ReRAM is modelled using the Stanford-PKU model with a 130 nm NMOS transistor as access transistor. A time-based sense amplifier [9] was used to read from the array (majority operation) and an op-amp was used to simultaneously write multiple bits into the array. A triple-row decoder was designed by interleaving multiple single-row decoders. Detailed schematics of the peripheral circuitry are given in [9]. A simultaneous reading (majority operations) and writing across columns of the array was verified by simulation. As described in Section 4.3.3, the adder can be executed in memory as a sequence of READ (majority) and WRITE operations, which are orchestrated by the memory controller (the memory controller can be designed as a finite-state machine and was not designed in this work).

5.2. Latency of In-Memory PP Adders with Increasing Bit-Width

The latency of PP adders grows as $log\ (n)$. From Figure 6, one can observe that, from 8-bit to 16-bit, the number of logic levels increased by only a single level, i.e., from seven levels to eight levels. The major advantage of the PP adder lies in this $(O(log(n))$ logic levels), and we aim to extend this advantage to our in-memory implementation. For the 16-bit version, we have to add an extra level of logic to the carry-generate block to calculate the carry (sum generate block remains at three logic levels; see Figure 6). In general, the number of logic levels, l is given by

$$l = log_2 n + 4 \tag{3}$$

for the n-bit PP adder (Ladner–Fischer type) synthesized in majority logic [30]. When this 16-bit adder was mapped to the memory array following the procedure used for an 8-bit adder (Section 4.3.3), 22 cycles were incurred. As a result of the interconnections between logic levels, the number of in-memory cycles is always higher than the number of logic levels. For an 8-bit adder (Figure A1), a careful comparison of the in-memory cycles indicated that every logic level is translated into at least two cycles, i.e., $2l$ in-memory cycles. The first few logic levels of the carry–generate block required two more WRITE cycles in addition to the aforementioned WRITE cycles. This additional requirement applies for $(l-5)$ of the l levels. Consequently, the number of cycles required for l logic levels of an n-bit PP adder can be calculated as follows:

$$\begin{aligned} Cycles_{in-memory} &= (2l) + 2(l-5) \\ &= 4l - 10 \\ &= 4(log_2 n + 4) - 10 \\ &= 4 \cdot log_2 n + 6 \end{aligned} \tag{4}$$

Therefore, any PP adder can be implemented in $O(log_2 n)$ cycles, which is the fastest in-memory adder reported to date (a detailed comparison is given in Section 5.5).

5.3. Energy of In-Memory PP Adders with Increasing Bit-Width

The energy consumed during in-memory addition is composed of the actual energy consumed due to addition (switching ReRAM cells during writing; energy consumed in the SA during majority operation) and the array leakage energy. The array leakage energy is the inherent energy consumption due to sneak currents in transistor-less arrays (e.g., some works, such as [41], used a diode to suppress these sneak currents). However, the proposed adder is executed in a 1T–1R array where the sneak currents are negligible. Hence, array leakage energy can be neglected. The energy used to write into an ReRAM cell is $E_{WRITE} \approx 12$ pJ/bit for IHP's ReRAM. The energy used for majority operation is the energy consumed in the SA, and is given by, $E_{MAJ} \approx 0.63$ pJ/majority operation. As can be seen in Figure A1, during eight-bit addition, there are 36 majority operations; 8 NOT and 85 bits are written to the array. Neglecting the energy of an NOT operation (which is only 0.13 pJ/bit), the energy needed for eight-bit in-memory addition is

$$Energy_{8-bit} = 36 \times E_{MAJ} + 84 \times E_{WRITE} \tag{5}$$

Observing that E_{WRITE} is $20 \times E_{MAJ}$, the in-memory addition energy is dominated by the energy that is needed to write into the array.

$$Energy_{8-bit} \approx 84 \times E_{WRITE} \tag{6}$$

where E_{WRITE} is the energy that is needed to write to a single bit. Similarly, during 16-bit addition in memory, 180 cells are written [9]. In general, for n-bit addition, $(2n-2) \times 6$ cells are written, making the energy for n-bit addition,

$$Energy_{n-bit} \approx (2n-2) \times 6 \times E_{WRITE} \tag{7}$$

To summarize, the energy for the proposed in-memory adder grows as $\approx 12n$ times the WRITE energy/bit.

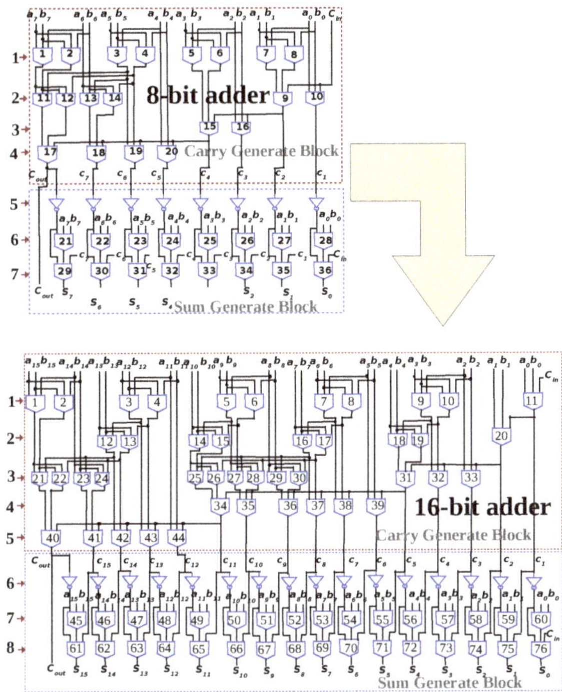

Figure 6. Eight-bit and 16-bit PP adder (Ladner–Fischer type) expressed in majority logic [5,30]. From 8-bit to 16-bit, the number of logic levels increased from 7 to 8, i.e., ($O(log(n))$) latency in terms of logic levels, before mapping to the memory array.

5.4. Area of In-Memory PP Adders with Increasing Bit-Width

In all in-memory adders, the peripheral circuitry of the array is modified to support logic operations, resulting in an increase in the CMOS peripheral circuit area. This increase is a significant factor to consider, since this increase in the silicon area is used solely to make the array 'computable'. Therefore, a holistic comparison between in-memory adders should consider both the increase in the peripheral circuitry area and the array area (occupied during addition), with the former being the more significant factor. In this work, the triple-row decoder is the only change required, while all other parts of the peripheral circuitry do not change, since computation is performed using normal memory operations (READ and WRITE). The array area used during the addition is simple to calculate—only six rows are needed, independent of the adder size (see Figure A1). In the of Figure A1 mapping, it is assumed that eight columns share a sense amplifier (this is the case when considering pitch-matching, although there are works which assume a sense amplifier for each column). For 8-bit addition, 80 columns are needed, and for n-bit addition, $8n + 16$ columns are needed. Therefore, for n-bit addition, the required array area is $6 \times (8n + 16)$.

5.5. Comparison with Other In-Memory Adders

In this section, we compare the presented in-memory PP adder design methodology with other adders and evaluate the latency with increasing bit-width. In Table 2, the latency of the proposed in-memory PP adder is compared with the latency of in-memory adders summarized in Table 1. With the exception of the two PP adders, the latency of all other adders is $O(n)$. The sklansky PP adder of [15] incurs a delay of $8log_2(n) + 13$, while the majority-based PP adder presented in this work incurs a latency of $4log_2(n) + 6$. With a PP architecture, majority logic-based implementation outperforms the OR/AND implementation of [15] in terms of latency. This proves that majority is a stronger logic primitive than OR/AND. The issue of latency becomes more evident when we observe the latency for increasing bit-width. For an 8-bit addition, the XOR-based ripple-carry adders [23,24] incur a latency of 18, which is the same latency as that incurred by the majority-based PP adder. A superficial observation may lead one to conclude that logic primitive alone plays a key role, and both XOR and MAJ are equally good, irrespective of the architecture used. However, the latency of XOR-based adders [23,24] grows to $2n + 2$, while that of majority-based PP adder grows to $4log_2(n) + 6$. In other words, for a 32-bit addition, the XOR-based adders incurs 66 cycles, while the proposed majority-based PP adder will incur only 26 cycles. This disparity further increases for 64-bit additions. In Figure 7, the latency of in-memory adders is plotted for increasing bit-width to better visualize this trend. As plotted in Figure 7, the proposed adder is one of the in-memory adders with the least latency, since it logarithmically depends on n. This latency advantage is obtained with only a minor modification to the row-decoder of a conventional memory. It must be noted that most other in-memory adders, compared in Table 2, require significant modifications to the peripheral circuitry. The energy consumption of the proposed in-memory adder is mainly due to the HRS $\leftrightarrow$ LRS switching energy of the cells during addition. The leakage energy, due to sneak-path currents (which constitutes a significant portion of the total addition energy in 1S–1R adders), is avoided by the access transistor.

Table 2. Latency comparison of in-memory adders (8-bit and n-bit).

Logic Primitive	Architecture	Latency (8 bit)	Latency (n-bit)	Ref.
IMPLY	Ripple carry	58	$5n + 18$	[7]
IMPLY	Parallel-serial	56	$5n + 16$	[16]
IMPLY + OR	Ripple carry	54	$6n + 6$	[17]
IMPLY	Semi-parallel	136	$17n$	[18]
NOR	Ripple carry	83	$10n + 3$	[19]
NOR	Look-Ahead	48	$5n + 8$	[20]
OR + AND	Ripple carry	49	$6n + 1$	[14]
ORNOR	Parallel-clocking	31	$2n + 15$	[21]
RIMP/NIMP	Pre-calculation	20	$2n + 4$	[22]
XOR	Ripple carry	18	$2n + 2$	[23]
XOR + MAJ	Ripple carry	18	$2n + 2$	[24]
XNOR/XOR	Carry-Select	9	–	[25]
OR + AND	Parallel-prefix	37	$8log_2(n) + 13$	[15]
Majority + NOT	Parallel-prefix	18	$4log_2(n) + 6$	This work

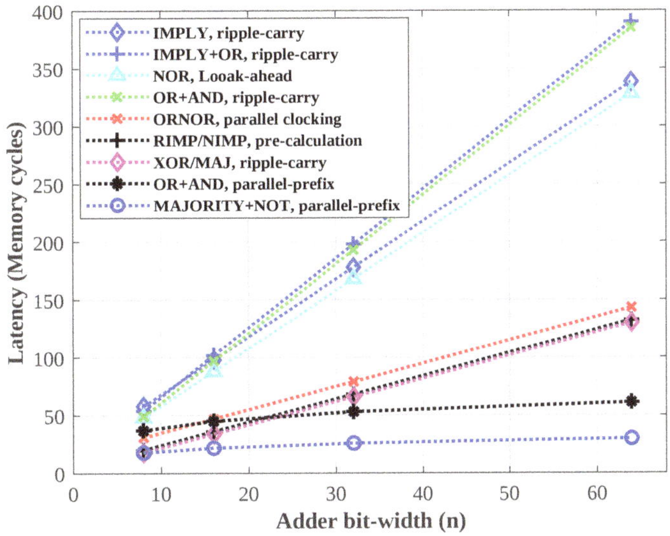

Figure 7. Latency of in-memory adders with increasing bit-width, n. An adder with $O(log(n))$ latency is required for 32-bit/64-bit addition to harness the power of in-memory computation.

6. Conclusions

The latency of in-memory adders is a severe disadvantage in in-memory computing, i.e., any adder is implemented in the memory array as a long sequence of Boolean operations. A poorly optimized in-memory adder may take longer to compute than the combined time it takes to fetch data from memory and compute in a CMOS processor. In-memory adders have not had their latency analyzed and optimized for higher bit-width, and consequently incur $O(n)$ latency for n-bit addition (32-bit/64-bit adders, typically used in microprocessors, will require hundreds of cycles). In this work, a design methodology is presented to tackle the exorbitant latency of in-memory adders. The strength of the majority logic primitive is coupled with the parallel-prefix (PP) adder architecture to achieve a latency of $4log_2(n)$+6 for parallel-prefix additions in the memory array. The main contribution of this work is a generic mapping methodology, used to map a parallel-prefix adder circuit (synthesized in majority logic) to the memory array with minimum latency. Multiple majority operations can be performed simultaneously in the columns of the array, and could achieve a $O(log(n))$ latency for any PP adder. Using the proposed design methodology, 32-bit and 64-bit adders (used in processors) can be implemented in 26 and 30 memory cycles, respectively. This can pave the way for arithmetic and similar computing tasks to be efficiently performed at the data location.

Funding: This research received no external funding.

Institutional Review Board Statement: Not applicable.

Informed Consent Statement: Not applicable.

Data Availability Statement: Not applicable.

Conflicts of Interest: The author declares no conflict of interest.

Appendix A. Mapping of 8-Bit Ladner-Fischer Adder to Memory Array

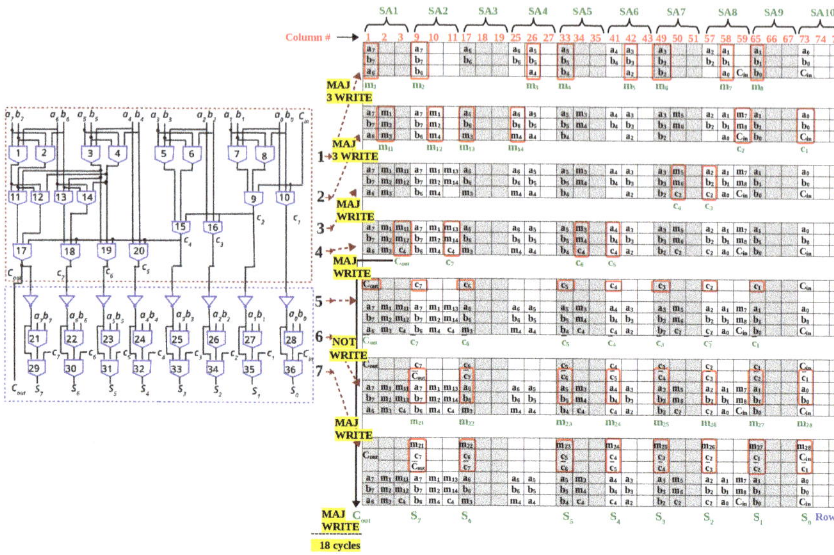

Figure A1. Mapping of the eight-bit LF adder of Figure 5 to memory array. All the majority gates in a level are simultaneously executed (red boxes). During parallel-prefix addition, m_i represents the output of the ith majority gate, and c_i is the carry (denoted in green color, since it is read as a voltage before being written into the array). 3 WRITE denotes writing cycles to 3 different rows, where more than 1 bit may be written in each row.

References

1. Horowitz, M. Computing's energy problem (and what we can do about it). In Proceedings of the 2014 IEEE International Solid-State Circuits Conference Digest of Technical Papers (ISSCC), San Francisco, CA, USA, 9–13 February 2014; pp. 10–14. [CrossRef]
2. Pedram, A.; Richardson, S.; Horowitz, M.; Galal, S.; Kvatinsky, S. Dark Memory and Accelerator-Rich System Optimization in the Dark Silicon Era. *IEEE Des. Test* **2017**, *34*, 39–50. [CrossRef]
3. Singh, G.; Chelini, L.; Corda, S.; Awan, A.J.; Stuijk, S.; Jordans, R.; Corporaal, H.; Boonstra, A. A Review of Near-Memory Computing Architectures: Opportunities and Challenges. In Proceedings of the 2018 21st Euromicro Conference on Digital System Design (DSD), Prague, Czech Republic, 29–31 August 2018; pp. 608–617.
4. Sebastian, A.; Le Gallo, M.; Khaddam-Aljameh, R.; Eleftheriou, E. Memory devices and applications for in-memory computing. *Nat. Nanotechnol.* **2020**, *15*, 529–544. [CrossRef] [PubMed]
5. Jaberipur, G.; Parhami, B.; Abedi, D. Adapting Computer Arithmetic Structures to Sustainable Supercomputing in Low-Power, Majority-Logic Nanotechnologies. *IEEE Trans. Sustain. Comput.* **2018**, *3*, 262–273. [CrossRef]
6. Ziegler, M.; Stan, M. A unified design space for regular parallel prefix adders. In Proceedings of the Proceedings Design, Automation and Test in Europe Conference and Exhibition, Paris, France, 16–20 February 2004; Volume 2, pp. 1386–1387. [CrossRef]
7. Kvatinsky, S.; Satat, G.; Wald, N.; Friedman, E.G.; Kolodny, A.; Weiser, U.C. Memristor-Based Material Implication (IMPLY) Logic: Design Principles and Methodologies. *IEEE Trans. Very Large Scale Integr. (VLSI) Syst.* **2014**, *22*, 2054–2066. [CrossRef]
8. Amarú, L.; Gaillardon, P.E.; Micheli, G.D. Majority-Inverter Graph: A New Paradigm for Logic Optimization. *IEEE Trans. Comput.-Aided Des. Integr. Circuits Syst.* **2016**, *35*, 806–819. [CrossRef]
9. Reuben, J.; Pechmann, S. Accelerated Addition in Resistive RAM Array Using Parallel-Friendly Majority Gates. *IEEE Trans. Very Large Scale Integr. (VLSI) Syst.* **2021**, *29*, 1108–1121. [CrossRef]
10. Shen, W.; Huang, P.; Fan, M.; Han, R.; Zhou, Z.; Gao, B.; Wu, H.; Qian, H.; Liu, L.; Liu, X.; et al. Stateful Logic Operations in One-Transistor-One- Resistor Resistive Random Access Memory Array. *IEEE Electron Device Lett.* **2019**, *40*, 1538–1541. [CrossRef]
11. Ben-Hur, R.; Ronen, R.; Haj-Ali, A.; Bhattacharjee, D.; Eliahu, A.; Peled, N.; Kvatinsky, S. SIMPLER MAGIC: Synthesis and Mapping of In-Memory Logic Executed in a Single Row to Improve Throughput. *IEEE Trans. Comput.-Aided Des. Integr. Circuits Syst.* **2019**, *39*, 2434–2447. [CrossRef]

12. Adam, G.C.; Hoskins, B.D.; Prezioso, M.; Strukov, D.B. Optimized stateful material implication logic for three- dimensional data manipulation. *Nano Res.* **2016**, *9*, 3914–3923. [CrossRef]
13. Reuben, J. Rediscovering Majority Logic in the Post-CMOS Era: A Perspective from In-Memory Computing. *J. Low Power Electron. Appl.* **2020**, *10*, 28. [CrossRef]
14. Ali, K.A.; Rizk, M.; Baghdadi, A.; Diguet, J.P.; Jomaah, J.; Onizawa, N.; Hanyu, T. Memristive Computational Memory Using Memristor Overwrite Logic (MOL). *IEEE Trans. Very Large Scale Integr. (VLSI) Syst.* **2020**, *28*, 2370–2382. [CrossRef]
15. Siemon, A.; Menzel, S.; Bhattacharjee, D.; Waser, R.; Chattopadhyay, A.; Linn, E. Sklansky tree adder realization in 1S1R resistive switching memory architecture. *Eur. Phys. J. Spec. Top.* **2019**, *228*, 2269–2285. [CrossRef]
16. Karimi, A.; Rezai, A. Novel design for a memristor-based full adder using a new IMPLY logic approach. *J. Comput. Electron.* **2018**, *17*, 11303–11314. [CrossRef]
17. Cheng, L.; Li, Y.; Yin, K.-S.; Hu, S.-Y.; Su, Y.-T.; Jin, M.-M.; Wang, Z.-R.; Chang, T.-C.; Miao, X.-S. Functional Demonstration of a Memristive Arithmetic Logic Unit (MemALU) for In-Memory Computing. *Adv. Funct. Mater.* **2019**, *29*, 1905660. [CrossRef]
18. Ganjeheizadeh Rohani, S.; Taherinejad, N.; Radakovits, D. A Semiparallel Full-Adder in IMPLY Logic. *IEEE Trans. Very Large Scale Integr. (VLSI) Syst.* **2020**, *28*, 297–301. [CrossRef]
19. Talati, N.; Gupta, S.; Mane, P.; Kvatinsky, S. Logic Design Within Memristive Memories Using Memristor-Aided loGIC (MAGIC). *IEEE Trans. Nanotechnol.* **2016**, *15*, 635–650. [CrossRef]
20. Kim, Y.S.; Son, M.W.; Song, H.; Park, J.; An, J.; Jeon, J.B.; Kim, G.Y.; Son, S.; Kim, K.M. Stateful In-Memory Logic System and Its Practical Implementation in a TaOx-Based Bipolar-Type Memristive Crossbar Array. *Adv. Intell. Syst.* **2020**, *2*, 1900156. [CrossRef]
21. Siemon, A.; Drabinski, R.; Schultis, M.J.; Hu, X.; Linn, E.; Heittmann, A.; Waser, R.; Querlioz, D.; Menzel, S.; Friedman, J.S. Stateful Three-Input Logic with Memristive Switches. *Sci. Rep.* **2019**, *9*, 14618. [CrossRef] [PubMed]
22. Siemon, A.; Menzel, S.; Waser, R.; Linn, E. A Complementary Resistive Switch-Based Crossbar Array Adder. *IEEE J. Emerg. Sel. Top. Circuits Syst.* **2015**, *5*, 64–74. [CrossRef]
23. TaheriNejad, N. SIXOR: Single-Cycle In-Memristor XOR. *IEEE Trans. Very Large Scale Integr. (VLSI) Syst.* **2021**, *29*, 925–935. [CrossRef]
24. Pinto, F.; Vourkas, I. Robust Circuit and System Design for General-Purpose Computational Resistive Memories. *Electronics* **2021**, *10*, 1074. [CrossRef]
25. Wang, Z.-R.; Li, Y. ; Su, Y.-T.; Zhou, Y.-X.; Cheng, L.; Chang, T.-C.; Xue, K.-H.; Sze, S.M.; Miao, X.-S. Efficient Implementation of Boolean and Full-Adder Functions With 1T1R RRAMs for Beyond Von Neumann In-Memory Computing. *IEEE Trans. Electron Devices* **2018**, *65*, 4659–4666. [CrossRef]
26. Dimitrakopoulos, G.; Papachatzopoulos, K.; Paliouras, V. Sum Propagate Adders. *IEEE Trans. Emerg. Top. Comput.* **2021**, *9*, 1479–1488. [CrossRef]
27. Knowles, S. A family of adders. In Proceedings of the 15th IEEE Symposium on Computer Arithmetic. ARITH-15 2001, Vail, CO, USA, 11–13 June 2001; pp. 277–281. [CrossRef]
28. Dimitrakopoulos, G.; Nikolos, D. High-speed parallel-prefix VLSI Ling adders. *IEEE Trans. Comput.* **2005**, *54*, 225–231. [CrossRef]
29. Harris, D. A taxonomy of parallel prefix networks. In *The Thirty-Seventh Asilomar Conference on Signals, Systems Computers*; IEEE: Pacific Grove, CA, USA, 2003; pp. 2213–2217. [CrossRef]
30. Pudi, V.; Sridharan, K.; Lombardi, F. Majority Logic Formulations for Parallel Adder Designs at Reduced Delay and Circuit Complexity. *IEEE Trans. Comput.* **2017**, *66*, 1824–1830. [CrossRef]
31. Reuben, J. Binary Addition in Resistance Switching Memory Array by Sensing Majority. *Micromachines* **2020**, *11*, 496. [CrossRef]
32. Reuben, J.; Pechmann, S. A Parallel-friendly Majority Gate to Accelerate In-memory Computation. In Proceedings of the 2020 IEEE 31st International Conference on Application-Specific Systems, Architectures and Processors (ASAP), Manchester, UK, 6–8 July 2020; pp. 93–100.
33. Wang, P.; Niamat, M.Y.; Vemuru, S.R.; Alam, M.; Killian, T. Synthesis of Majority/Minority Logic Networks. *IEEE Trans. Nanotechnol.* **2015**, *14*, 473–483. [CrossRef]
34. Chung, C.C.; Chen, Y.C.; Wang, C.Y.; Wu, C.C. Majority logic circuits optimisation by node merging. In Proceedings of the 2017 22nd Asia and South Pacific Design Automation Conference (ASP-DAC), Chiba, Japan, 16–19 January 2017; pp. 714–719. [CrossRef]
35. Riener, H.; Testa, E.; Amaru, L.; Soeken, M.; Micheli, G.D. Size Optimization of MIGs with an Application to QCA and STMG Technologies. In Proceedings of the 2018 IEEE/ACM International Symposium on Nanoscale Architectures (NANOARCH), Athens, Greece, 17–19 July 2018; pp. 1–6.
36. Devadoss, R.; Paul, K.; Balakrishnan, M. Majority Logic: Prime Implicants and n-Input Majority Term Equivalence. In Proceedings of the 2019 32nd International Conference on VLSI Design and 2019 18th International Conference on Embedded Systems (VLSID), Delhi, India, 5–9 January 2019; pp. 464–469. [CrossRef]
37. Neutzling, A.; Marranghello, F.S.; Matos, J.M.; Reis, A.; Ribas, R.P. maj-n Logic Synthesis for Emerging Technology. *IEEE Trans. Comput.-Aided Des. Integr. Circuits Syst.* **2020**, *39*, 747–751. [CrossRef]
38. Kaneko, M. A Novel Framework for Procedural Construction of Parallel Prefix Adders. In Proceedings of the 2019 IEEE International Symposium on Circuits and Systems (ISCAS), Sapporo, Japan, 26–29 May 2019; pp. 1–5. [CrossRef]
39. Ayala, C.L.; Takeuchi, N.; Yamanashi, Y.; Ortlepp, T.; Yoshikawa, N. Majority-Logic-Optimized Parallel Prefix Carry Look-Ahead Adder Families Using Adiabatic Quantum-Flux-Parametron Logic. *IEEE Trans. Appl. Supercond.* **2017**, *27*, 1–7. [CrossRef]

40. Levisse, A.; Giraud, B.; Noel, J.; Moreau, M.; Portal, J. RRAM Crossbar Arrays for Storage Class Memory Applications: Throughput and Density Considerations. In Proceedings of the 2018 Conference on Design of Circuits and Integrated Systems (DCIS), Lyon, France, 14–16 November 2018; pp. 1–6. [CrossRef]
41. Chang, Y.F.; Zhou, F.; Fowler, B.W.; Chen, Y.C.; Hsieh, C.C.; Guckert, L.; Swartzlander, E.E.; Lee, J.C. Memcomputing (Memristor + Computing) in Intrinsic SiO_x-Based Resistive Switching Memory: Arithmetic Operations for Logic Applications. *IEEE Trans. Electron Devices* **2017**, *64*, 2977–2983. [CrossRef]

Journal of
Low Power Electronics and Applications

MDPI

Article

A New Physical Design Flow for a Selective State Retention Based Approach

Joseph Rabinowicz [1] and Shlomo Greenberg [1,2,*]

[1] Department of Electrical and Computer Engineering, Ben Gurion University, Beer-Sheva 84105, Israel; rabinowi@post.bgu.ac.il
[2] Department of Electrical Engineering, Sami Shamoon College of Engineering, Beer-Sheva 84100, Israel
* Correspondence: shlomo.greenberg@gmail.com or shlomog@bgu.ac.il or shlomgr@sce.ac.il

Abstract: This research presents a novel approach for physical design implementation aimed for a System on Chip (SoC) based on Selective State Retention techniques. Leakage current has become a dominant factor in Very Large Scale Integration (VLSI) design. Power Gating (PG) techniques were first developed to mitigate these leakage currents, but they result in longer SoC wake-up periods due to loss of state. The common State Retention Power Gating (SRPG) approach was developed to overcome the PG technique's loss of state drawback. However, SRPG resulted in a costly expense of die area overhead due to the additional state retention logic required to keep the design state when power is gated. Moreover, the physical design implementation of SRPG presents additional wiring due to the extra power supply network and power-gating controls for the state retention logic. This results in increased implementation complexity for the physical design tools, and therefore increases runtime and limits the ability to handle large designs. Recently published works on Selective State Retention Power Gating (SSRPG) techniques allow reducing the total amount of retention logic and their leakage currents. Although the SSRPG approach mitigates the overhead area and power limitations of the conventional SRPG technique, still both SRPG and SSRPG approaches require a similar extra power grid network for the retention cells, and the effect of the selective approach on the complexity of the physical design has not been yet investigated. Therefore, this paper introduces further analysis of the physical design flow for the SSRPG design, which is required for optimal cell placement and power grid allocation. This significantly increases the potential routing area, which directly improves the convergence time of the Place and Route tools.

Keywords: physical design; power grid; power-gating; SRPG; selective SRPG; floorplanning; place and route

Citation: Rabinowicz, J.; Greenberg, S. A New Physical Design Flow for a Selective State Retention Based Approach. *J. Low Power Electron. Appl.* **2021**, *11*, 35. https://doi.org/10.3390/jlpea11030035

Academic Editor: Alex Serb

Received: 28 July 2021
Accepted: 9 September 2021
Published: 13 September 2021

1. Introduction

Leakage currents during standby mode become more significant in mobile devices as semiconductor processes continue to shrink [1]. These static leakage currents impact the battery standby time of low-power mobile devices when they are in an idle state. Therefore, to mitigate the static leakage currents, some Power-Gating (PG) techniques were developed [2–6]. Power-gating eliminates the static leakage but with no intention to retain the system state. As mobile devices are required to support many features and functions, resulting in a wide range of multitasking, a minimum delay for the state restoration of all active tasks is critical for user satisfaction [7]. Besides the additional delay, saving and restoring the system state presents additional dynamic power overhead that may not be acceptable for certain common applications.

Scan-based techniques, which are used for serially saving and restoring internal retention cells, also suffer from latency and energy overhead [8]. The State Retention Power Gating (SPRG) technique addresses the above-mentioned PG technique's limitations [9–13]. This technique uses unique retention cells to retain the flip-flops (FFs) values during power down (standby state). These cells have been widely adopted in standard library cells of major FAB vendors (such as TSMC). The SRPG approach aims to retain the systems state during standby, thus eliminating the disadvantages of the power-gating technique. However, common SRPG implementations require additional retention cells for all the FFs in the design resulting in significant area overhead. Moreover, these retention cells need to be connected to a dedicated power supply network and retention control signals. This additional wiring increases the area overhead and also complicates the physical design implementation in terms of tools runtime and the ability to handle large designs.

A more advanced approach, called Selective State Retention Power Gating (SSRPG), dramatically reduces the SRPG area overhead and further decreases the static power consumption. The main idea is to find a minimized set of FFs which are sufficient to retain the system state during standby. Chiang et al. [14] propose an empirical nonformal method for the selection of registers whose retention is unnecessary. Darbari et al. [15] present a formal approach based on symbolic simulation for implementing selective state retention. However, this method requires a formal representation of the entire design, which is not always available, and also no automated techniques are proposed. The two recently published SSRPG approaches introduced by [16,17] provide pure formal methods for automatic selecting of all the FF's, which require retention and are essential for a proper system recovery upon power-up. Experimental results show a significant reduction of about 80% of the retention cells area overhead. Recent SSRPG techniques can be efficiently applied to new modern SoC designs for automatic selection and formal validation of essential FFs requires retention. The current work is based on our previous formal SSRPG approach presented in [17], which utilizes formal verification methods and therefore can be easily implemented using the new proposed physical design flow.

Although the SSRPG approach mitigates the area and power overhead limitations of the conventional SRPG technique, still both SRPG and SSRPG approaches require a similar extra power supply network for the retention cells. The impact of the extra power supply when applying the selective approach has not yet been investigated. Therefore, further analysis of the physical design flow for SSRPG design is needed for optimal cell placement and power grid allocation. This may significantly increase the routing area, which in turn directly improves the convergence time of the place and route tools [18].

Furthermore, minimizing the number of retention FFs not only results in reducing the area overhead but also reduces the additional wiring required in SRPG. Although it is shown in [16] that a significant potential area reduction of about 9% of the chip area can be achieved, the added wiring required in SRPG is ignored. In SSRPG, the retention cells footprint can be simply deducted from the total cell area, but the wire-length deduction is not straightforward since it can only be obtained after completing the physical design flow. The wire-length overhead in the SRPG approach is derived from: (1) The connectivity of the retention cells to a new non-gated power supply network [19], and (2) the addition of retention control signals, which need to be connected to all FFs that are being preserved during standby by using retention cells [20]. This wiring overhead complicates the place and route physical design stages in SRPG. This work demonstrates the benefit of applying the SSRPG approach in a real physical design implementation concerning area, power saving, back-end runtime, and wire length.

Although some previous research works [16,17] try to estimate the area and power-saving factor results from applying the SRPG selective approaches, none of them validate it on real physical design implementation. Hence, one of the main objectives of this work is to quantify the real area and power saving factors while using SSRPG comparing to SRPG.

This work also demonstrates the benefit of applying a new, improved localized physical design flow using unique placement rules. The proposed localized improved flow yields significant power supply network area reduction in cases where selective state retention is used. It is shown that by applying these placement rules, metal layers that were originally used for power-supply distribution are freed up to be used for signal routing applied when connecting the different logic gates in the physical design during the routing stage, and therefore improving the routeability. This simplifies the implementation of selective state retention in the physical design flow and significantly reduces the tools' runtime.

Although the SSRPG approach [16,17] is not a new technique, the effect of the selective approach on the complexity of the physical design has not been yet investigated. Therefore, further analysis of the physical design flow for SSRPG design is needed for optimal cell placement and power grid allocation. This may significantly increase the potential routing area, which in turn directly improves the convergence time of the Place and Route tools. This paper aims at the physical implementation aspect to facilitate the complexity of the physical design suggesting a unique flow to efficiently address SoC design based on SSRPG. Moreover, this is the first work related to SSRPG implementation, which accurately quantifies the area, power, and tool runtime saving factors.

In this work, we provide a case study showing the accurate area, power, and tool runtime savings when comparing the physical design implementation of SSRPG to SRPG. Previous works provide area reduction estimations based on the percentage of FFs that does not require retention [16,17]. These area estimations suffer from inaccuracies since they do not take into account the additional wiring overhead required for connecting the retention cells to the non-gated power supply and power-gating controls. To quantify the selective state retention physical design flow benefits, a complete CMOS 28 nm physical design flow was carried out on a typical Double Data Rate (DDR) memory interface controller design.

This paper is organized as follows: Section 2 provides an improved physical design flow for an Application-Specific Integrated Circuit (ASIC) supporting state-retention. Section 3 describes the experiment and shows the comparison results for the four different physical design flows: no retention, full retention using SRPG, SSRPG without special placement rules, and an improved physical design flow for SSRPG. Finally, Section 4 summarizes the paper and states the conclusions.

2. An Improved SSRPG Physical Design Flow

We propose a new approach to the common SRPG technique, based on automatic classification of each of the design's FFs into one of two types: essential or non-essential. The flow begins with gathering the libraries and floor planning, followed by place and routing, and ends with verification of the physical design. Figure 1 depicts the five main stages of a typical physical design flow. Each stage is described in detail in the following section considering the specific additional requirements for state-retention. Two different physical SSRPG design flows are considered concerning the placement stage: distributed flow and improved localized SSRPG flow. Some unique placement rules are proposed for the implementation of the new localized SSRPG physical design approach.

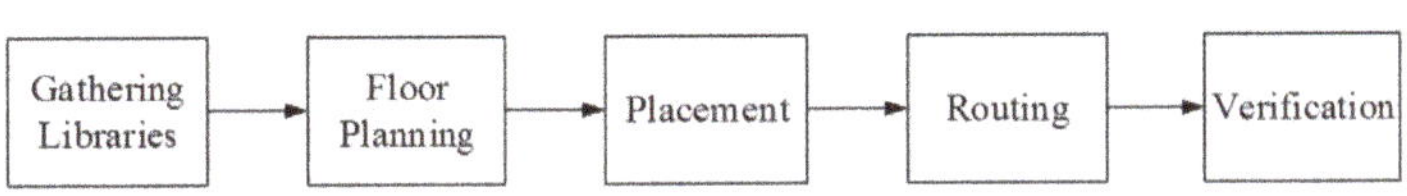

Figure 1. High-level stages of the physical design flow.

Although some physical implementation steps can be controlled by the common UPF and CPF industrial tools for power-aware content, those tools do not provide any specific placing rules except for limiting the logic cells placement to the appropriate power-domain (PDN).

2.1. Gathering Libraries

The libraries' physical design flow contains the list of basic cells and their attributes, such as physical layout abstractions, timing delay models, functional models, and transistor-level circuit descriptions [21].

To implement state retention, the libraries should contain special retention FFs. Such FFs are divided into different types that can be categorized by the two following criteria: (1) the transistors threshold voltages (low, high, or multi-threshold) (2) Using an additional latch (referred to as balloon latch) or rather than using the FF slave latch (in a common master-slave FF) for retention. Table 1 depicts the different types of retention FFs that are used in state retention approaches and their impact on low power, propagation delay, and physical design flow [13,22,23].

Table 1. Retention FFs types and tradeoffs.

FF Type	V_{TH}	Extra Latch	Low Power	Propagation Delay Impact	SoC Physical Design Flow Impact
1	Low	No	Week	Negligible	Need clock and reset gating during standby.
2	Low	Yes	Medium	Negligible	Additional area impact of the balloon latch.
3	High	No	Good	High	Need clock and reset gating during standby.
4	Multi	Yes	Good	Negligible	Extra balloon latch and extra power supply.

Retention FF's implemented with low threshold voltage transistors have less impact on the propagation delay since the low voltage threshold allows fast switching between off and on states. However, since the leakage increases exponentially when decreasing the threshold voltage, the efficiency of reducing the static leakage is limited for this type of FF. The static leakage is given by the following equation:

$$P_{leakage} = V_{dd} \cdot I_{leakage} = V_{dd} \cdot I_0 \cdot \exp\left\{ [(V_{GS} - V_{TH})/V_T] / [1 - \exp\left(\frac{-V_{DS}}{V_T}\right)] \right\} \tag{1}$$

where V_{TH} is the threshold voltage of the transistor, V_T is the thermal voltage, V_{GS} is the voltage between gate and source, and V_{DS} is the voltage between drain and source of a MOSFET transistor. Some improvement in static leakage reduction can be achieved by adding a specific balloon latch, as shown in Figure 2. This additional latch is designed to consume less power during standby since it does not affect the master-slave functional path and therefore supports higher frequencies compared to FFs that use the slave latch for retention.

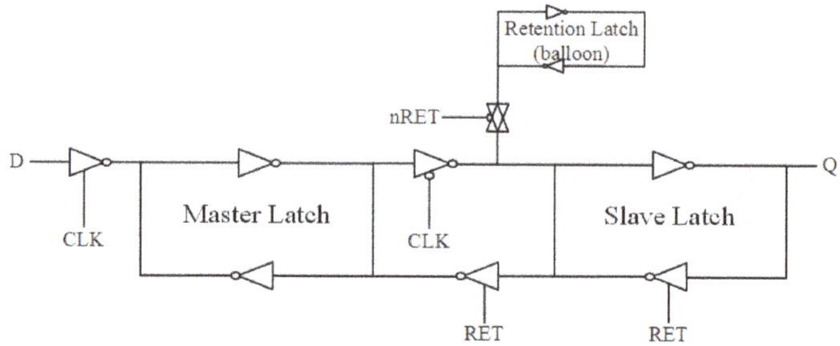

Figure 2. Retention FF implementation using a balloon latch.

Retention FFs that are implemented with high threshold voltage transistors, perform better with respect to static leakage reduction. A high voltage threshold leads to a better closure of the source/drain channels and thus preventing leakage currents when the transistor is in its off state. However, a high voltage threshold also impacts the propagation delay and therefore limits the clock frequency rates. Using both multi-voltage threshold transistors and an additional retention balloon latch allows better static leakage reduction and higher clock frequencies. However, this is at the expense of additional area overhead and extra external SoC power supply, which requires dedicated supply pads and balls, complicating the design [22]. Therefore, while choosing the physical design libraries in case of state retention, the SoC designer should consider the following factors and their tradeoffs: clock frequency, static leakage reduction, area overhead, and implementation complexity.

2.2. Floorplanning

A well-thought-out floor plan leads to a design with higher performance and optimum area [21]. In this stage, the physical designer determines the size of the macro instance, which includes the physical representation of the design. Additionally, the structure and placement of the power and ground strips referred to as power-supply networks are determined.

Some industrial SoCs may contain several power-gated domains and, therefore, many power switches to reduce IR drop [24]. This work aimed specifically at low power designs and referred to the hard macro level of implementation using only one or two power switches (as illustrated in Figure 3). To maintain minimum voltage drop and to prevent performance degradation, the power and ground strips should be as dense as possible. The following section refers to specific floorplanning adjustments required for state-retention-based designs. State-retention approaches require some modifications to the typical floorplan with respect to the power supply network. Specifically, two kinds of floorplan modifications are required: (1) adding an extra retention power supply network and (2) integration of dedicated sleep transistors for disconnecting the main power supply on standby. Figure 3 illustrates two power grids networks with a single power switch. The extra power grid uses a significant portion of the metal layers, which are actually needed for routing the logic gate connections (routeability) [13]. Although the strips of the extra power supply network are thinner compared to those of the main power supply, since there is no need to support full clock rate in standby, they should be spread over the entire macro instance.

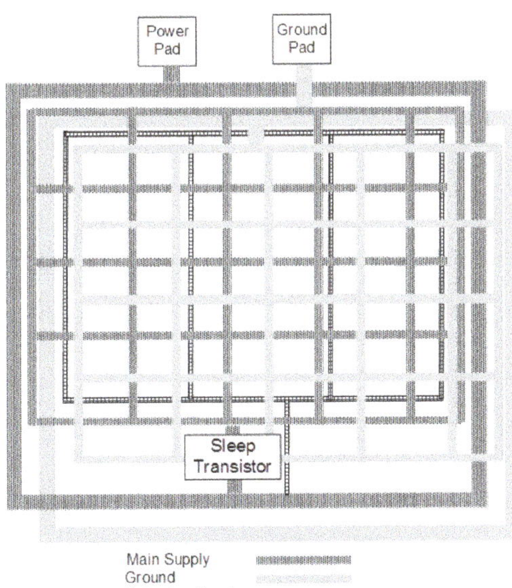

Figure 3. Power grids networks for State Retention-based SoC.

Any power gating implementation, including SRPG, requires a dedicated sleep transistor per gated power supply. The sleep transistors are based on high voltage threshold transistors and are responsible for disconnecting both the power supply source and the ground in standby, as shown in Figure 4. Unique SLEEP signals are used to control the sleep transistors and define two control modes: active and standby modes (SLEEP is driven to 1 during standby and 0 during active modes). The active mode utilizes the low voltage threshold transistors to operate at higher frequencies. In Standby mode, the SLEEP signals are activated to turn off the sleep transistors. Since the sleep transistors are based on high voltage threshold transistors, their static leakage is very small during standby. The size of the sleep transistor is critical in terms of performance, area, and leakage current [19]. While the sleep transistor should be large enough to drive sufficient current to meet frequency performance, it should not cause excessive leakage.

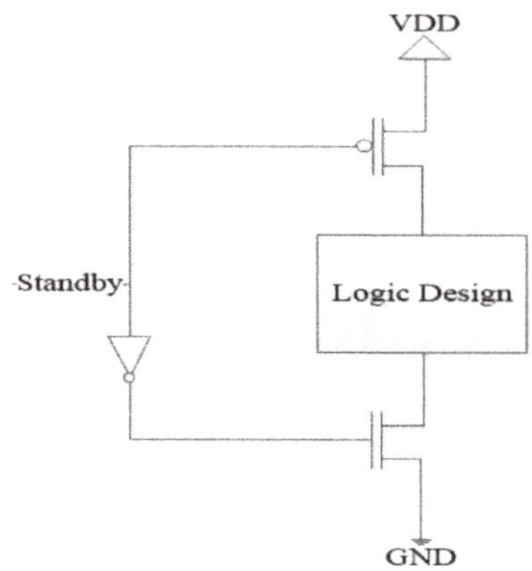

Figure 4. Sleep transistors.

2.3. Place and Route

The placement stage is responsible for placing the overall standard logic gates in a given macro instance and inserting buffer cells along with the clock and reset signal paths. Since the long wiring induces different propagation delays between different FFs, a clock balancing process is required. The buffer cells are used both for clock balancing and to support high fan-out and long wiring. This process of buffer insertion is commonly referred to as Clock Tree Synthesis (CTS) and has a significant impact on timing closure. In addition to the clock and reset signals, the CTS process is also applied to the retention FFs' control signals. This wiring and buffering overhead to support the additional retention control signals is significant in designs that include many sequential elements and might be similar to the overhead of the clock network [20]. Since the additional buffers should be connected to the retention power supply network, they have a significant impact on the routing to support the distributed retention controls signal paths. Power-supply network optimization is usually carried out after placement and before signal routing. The objective is to reserve more chip area for signal routing and, at the same time, maintain the performance of the power supply network. However, it is difficult to fully utilize the reserved chip-routing resource [25], especially in the case of a design that requires a dedicated power supply for the retention cells. Therefore, minimizing the area of the retention power supply network will lead a better routing utilization. The routeability in an SSRPG design can be further improved due to the small number of the required retention cells compared to SRPG. The routeability improvement can be achieved by making some appropriate adjustments both in the floorplan and the placement stages.

This work considers two different flows for SSRPG: the more straightforward distributed flow and a unique localized flow. In the distributed flow, the retention FFs are distributed all over the hard macro, while in the localized flow, the retention FFs are placed in a limited area using some placement constraints. Therefore, the region of the PDN of the always-on domain becomes smaller and requires less routing overhead. Furthermore, the proposed physical design flow is implemented within a hard macro level and applied to a specific functional design module. Therefore, since each hard macro commonly contains only one or two power domains, it is feasible to place all the retention FFs, connected to the always-on domain of the specific PDN, within a localized concentrated area.

We propose a unique physical design approach that is based on the assumption that the retention cells can be placed all together in a localized and relatively small area within the entire macro instance. This will lead to a reduced retention power supply network area. Figure 5 depicts placement results for two different physical design flows carried out on the proposed DDR controller design using the Cadence Encounter tool. Figure 5a shows the placement results for the distributed SSRPG flow in which the retention power grid (i.e., power supply network) is distributed throughout the entire macro instance area without any placement constraints as in the common SRPG flow. The figure depicts the spreading of the retention FFs. Figure 5b shows the placement results for the new proposed localized flow. It can be noticed that the retention FFs are now located together in a relatively small localized area.

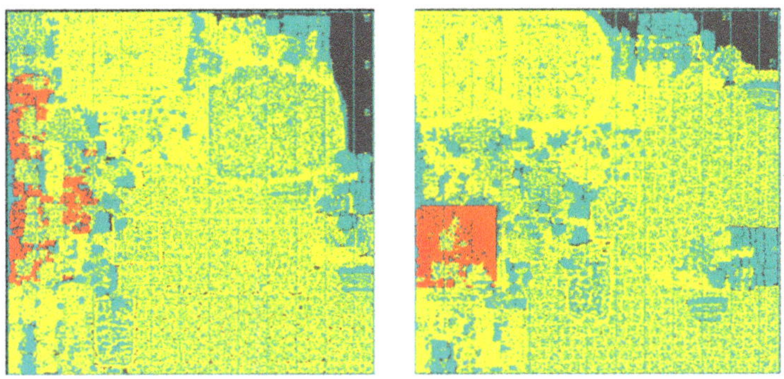

(a) (b)

Figure 5. (**a**) Placement of the distributed retention FFs (mark in red and spread mostly on the mid-left-hand side). (**b**) Placement for the proposed flow where retention FFs are placed in a localized area (red square on the mid-left-hand side).

Two modifications were applied to the localized physical design flow based on the distributed flow placement results and using the common SRPG flow. First, the power grid was limited to a specific and localized area in the floorplan stage. Then, some specific placement constraints were provided to the Encounter tool, forcing all retention cells to be placed in a limited minimized localized area within the retention power grid region. The results show that the retention cells and the relevant retention power grid were successfully placed in a minimized area enabling better routeability compared to the common approach. Since the extra power grid utilizes only a small part (about 1/16) of the metal layer used for the retention power supply network (Figure 5b), more metal area is freed up for routing. To further reduce wire-length and additional buffers, the external retention control input ports are also placed in the same selected area close to the retention power grid. Applying such constraints to the placement tool may result in timing violations since the interconnect length between FFs may significantly increase. However, since the number of retention cells in SSRPG is relatively small, and most of the retention FFs are not part of the data path, the timing violations are not critical [26]. In the next stage, the routing process is carried out. Routing is becoming more difficult, especially for state retention-based designs, like SRPG, since the design is getting more complex due to the additional retention cells and the required extra wiring. Therefore, SSRPG facilitates the routing process by significantly reducing the amount of routing and hence decreasing the route runtime.

2.4. Verification

The final stage of any physical design flow is verification. This stage focuses on functional testing and design manufacturability. A comprehensive design verification process consists of three categories: functional, timing, and physical. The functional verification includes logic simulations, formality checks, simulation randomization, in-circuit emulation, and hardware/software co-verification [27]. The timing closure is carried out using Static Timing Analysis (STA) to verify the timing of a digital design [28]. The physical verification checks the design layout against the specific process rules and includes Layout Versus Schematic (LVS) and Design Rule Check (DRC) [21]. In the case of state retention, some additional logic simulations scenarios should be considered. For example, entering standby and then restoring the design state upon power resumption and verifying the selection of the appropriate FF's which required retention.

3. Experiment and Results

In this section, we compare four different approaches in respect to the physical design flow: no retention, full retention using SRPG, SSRPG with no specific placement rules, and an improved SSRPG flow. All the flows were applied to a typical DDR controller design as a test case. The synthesis was carried out using the Cadence RTL compiler, and then a common full PD flow was applied using Cadence Encounter to each of the four approaches. One of the main purposes of this work was to quantify the efficiency of the selective approaches with respect to area and power saving. Additionally, this research compares the four different PD flows in respect to the ability of the tools to converge, tools runtime, total wiring length, static leakage, and area-saving factors. Figure 6 depicts the block diagram of the selected DDR controller design. The DDR controller contains about 62,000 FFs. The design contains a DDR control unit, a DDR PHY adaptor, and two ARM AXI bus interfaces. The control unit is used to configure the DDR controller and monitor the status registers. The DDR PHY interface is connected directly to the DDR PHY, while the AXI bus interfaces between the DDR PHY adaptor and the internal memories. The AXI bus is used to store and retrieve data to/from the internal memory using a First-in-First-out (FIFO) memory within the AXI interface. A clock generator is used to provide an accurate clock signal to the external DDR memory. The DDR controller has two different operating modes: consecutive and interleaving memory addressing. The DDR interleave mux selects the desired operating mode and supports data interleaving from two channels to one memory device, reducing the external memory access time. The chosen DDR controller is used in many common VLSI applications and is large enough to represent a typical macro instance. Moreover, the design has a significant amount of non-essential FFs and, therefore, can be efficiently implemented using the SSRPG flow. In addition, the working frequency of the DDR controller is relatively high (533 MHz) and makes the comparison qualify for high-frequency designs as well.

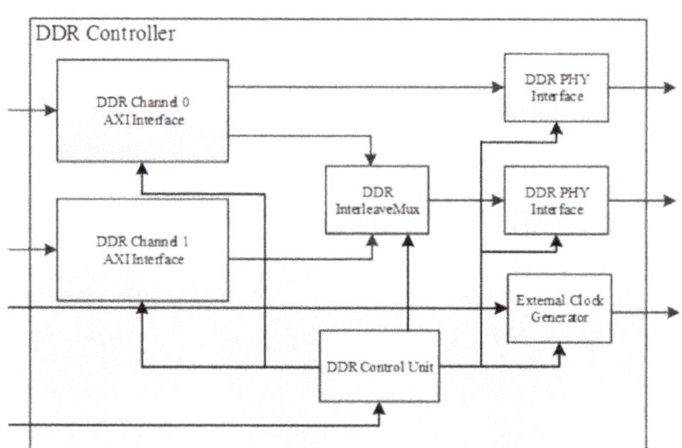

Figure 6. DDR Controller—block diagram.

3.1. Basic Synthesis

Physical Design Flow Implementation

The design was first synthesized using the Cadence RTL compiler (RC). The synthesis results provide the physical designer with the following data: (1) a standard library cell design representation referred to as netlist, (2) the total cell area estimation needed for floorplanning, and (3) critical timing paths that should be addressed in the synthesis stage. For timing closure, the clock frequencies and some specific timing constraints should be defined in the synthesis stage. In our test case, two frequencies were applied: 533 MHz for the AXI bus and DDR PHY interfaces and a lower frequency of 133 MHz for the control logic.

The delay constraints take into consideration 30% of the clock period for output ports and 70% for input ports. Some more delay adjustments were needed for certain ports according to specific timing issues. In order to extract the essential FFs for the DDR controller test case, we have used the SSRPG approach described in [16]. This approach is based on a gate-level analysis and suggests a fully automatic algorithm to classify the FFs in a typical design into two categories essential and non-essential FFs. Results show that only 2522 FFs (out of the total 61,944 FFs) were classified as essential FFs, and therefore only 4.1% of the FFs require retention cells. The netlist was updated accordingly with the additional retention cells.

3.2. Floorplanning

An important step in floor planning is to specify the appropriate area to place macros and standard cells. In general, the floorplan can be determined according to the dimensions of the total macro area, Utilization Factor (UF), and die area. The utilization factor is defined as follows [29].

$$\text{Utilization Factor} = \frac{Area\ of\ Standard\ cells}{Total\ Physical\ Design\ Area} \tag{2}$$

This means that a larger area of 1/UF multiplied by the standard cell area is allocated for the Encounter tool to place the standard cells and to permit enough routing resources for the cells' interconnections. Selection of the UF should both provide the Encounter tool with enough space to place the cells and route between them and still meet timing. As the UF decreases, the area to place cells increases, and therefore the Encounter tool has a better ability to successfully route the cells. The effects of choosing a Utilization Factor on total wire length, congestion, and DRC (Design Rule Constraints) violations have been explored (studied) in [21]. It was observed that a Utilization Factor of 0.5 to 0.7 is appropriate depending on the metal layers in which the Power and Ground planning is done.

The Cadence Encounter tool was used to determine the size of the macro instance for the chosen DDR Controller design. The total cell area (including FFs and logic gates) was extracted from the synthesis results for the four different physical designs. The utilization factor's selection should be considered a tradeoff between the motivation to minimize the macro instance area and the need to reduce the place and route complexity.

An initial recommended utilization factor of 0.7 was examined in the floor planning stage. Then a unique utilization factor was chosen for each of the four different proposed physical design flows according to congestion and DRC violations which directly affect the Encounter tool runtime.

For the no-retention physical design flow, the initial recommended utilization factor of 0.7 was found to be appropriate and did not have much effect on congestion, placement run time, and tool convergence compared to lower utilization factors. However, while applying this initial utilization factor for the SRPG and SSRPG physical design flows, the runtime was significantly higher (a factor of 5) compared to lower utilization factors.

Figure 7 shows the empiric place and route tool's runtime versus the utilization factor for various examined flows. The utilization factor (UF) is given in Equation (2). The available area for placing the cells increases as the UF factor decreases, and therefore the Encounter tool has a better ability to successfully route the cells.

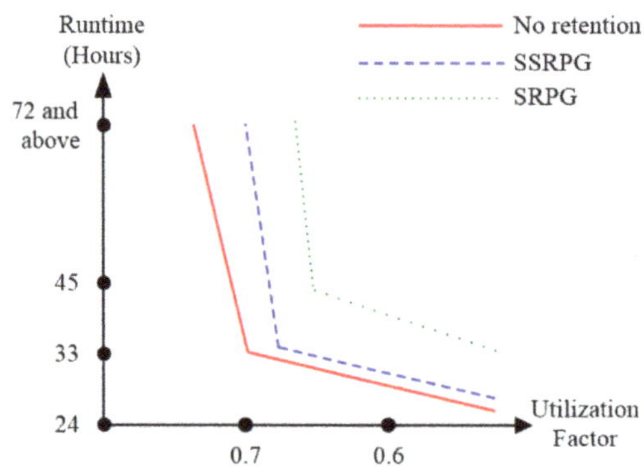

Figure 7. Place and Route runtime versus UF.

The effects of choosing a utilization factor on total wire length, congestion, and DRC (Design Rule Constraints) violations have been explored in [21]. The authors show that by using fewer number of metals to route between the standard cells spread across the core area (which is equivalent to the scenario of less available routing area), the tool has to do complex de-tour routing to avoid DRC violations. It was also observed that with fewer metals (a higher UF), the tool has fewer routing tracks to route between all the cells, introducing more congestion. Therefore, the number of available routing tracks available also decreases.

From Figure 7, we observe that the optimal UF factors are: 0.7, 0.65, and 0.67 for the no-retention, SRPG, and both SSRPG flows accordingly. Any attempt to increase those chosen utilization factors resulted in the divergence of the Encounter tool. In all our experiments, the convergence time limit was defined to be 72 h. The relatively lower UF factor achieved for the SRPG and SSRPG can be explained due to the additional extra power grid and its connections to the retention cells buffers required for the CTS process and the additional route connectivity. We observed that the UF for the SSRPG flow is higher than the UF obtained in the case of SRPG. This means that the SSRPG physical implementation required less area compared to SRPG.

As a part of the floor planning, certain physical elements, such as antenna and latch-up cells, were added to maintain the integrity of the macro instance [30]. Then, pin placement was done according to the SoC constraints. Finally, the appropriate power grid was defined according to the specific physical design flow. While in the case of no-retention flow, only one power grid is required and is spread out uniformly across the macro instance area, the SRPG and SSRPG flow require an extra power grid which should be connected to the additional retention cells.

Figure 8 shows a snapshot, taken from the floorplanning tool, of the two power grids required in SRPG and SSRPG. The common VDD grid is represented by the thick purple line wrapped by two thin red lines. The extra VDDG power grid is represented by two closely placed thin red lines. Since the VDDG supplies power only to the retention cells, it can be composed of fewer gridlines compared to VDD. It can be observed that the VDDG strips are less dense and are placed in a 1.8 µm interval once every second VDD strip. The distance between the VDD and VDDG grid lines was set to 0.125µm. These power grid configurations were validated using the Cadence encounter power analysis tool.

As discussed in Section 2.3, the power grid distribution in the localized SSRPG flow can be limited to a localized area in the floorplan. The exact flow used to determine the localized area in which the retention cells are located is described as follows. First, the floorplan with a uniform distributed power grid is used as an input to the placement stage. Then the results of this placement (location of the retention cells) are used to create a new floorplan in which the power grid is limited to a specific area. Finally, the retention control signals (RETN) which should be connected to all the retention cells, are placed close to this specific region to reduce routing.

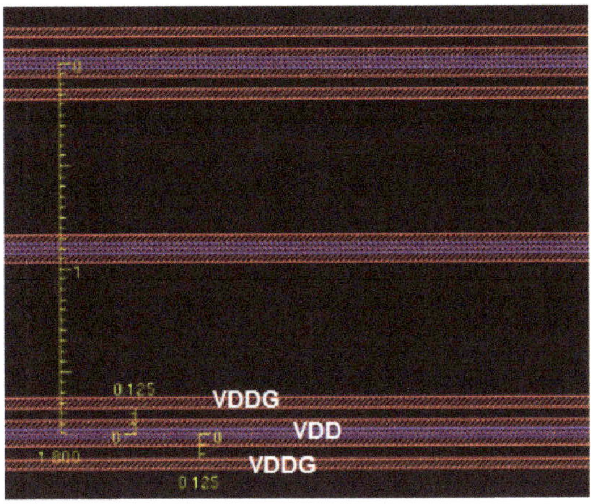

Figure 8. VDD and VDDG power grid floorplanning.

3.3. Placement and Routing

The placement stage was carried out the same way for the four physical design flows. The Cadence Encounter was used as the placement tool in order to meet timing and area constraints as derived from the floorplanning stage. The same clock tree methodology was used for the four examined flows using the CTS Cadence tool with the same timing constraints. In the case of SRPG and SSRPG flows, the additional RETN control signals used for retention purposes were also balanced in the clock tree process. The routing for the three-state retention flows also included the additional connections of the state-retention cells to the extra VDDG power grid.

3.4. Results

During the implementation of the four physical design flows DRC checks were carried out according to the 28 nm library requirements. The timing analysis implemented by the STA tool also included exhaustive signal integrity checks [28]. The difference in timing closure between all four physical design flows was less than 11 ps, which is less than 0.6% of the clock period. All flows were executed on a 64 bit Linux server (64 bit, 2.8 GHz with 64 GB RAM).

This section shows the comparison results for the four examined flows in terms of area, wire-length, static leakage, and runtime. First, we demonstrate the benefit of using the proposed improved SSRPG flow in terms of runtime. Then, we compare the proposed flow with the common SRPG and the no-retention flows. Table 2 depicts the comparison between the improved localized SSRPG flow, which uses the unique placement constraint rules, the common SRPG, and the distributed SSRPG physical design flows. It is shown that applying the extra placement rules, with regards to the selected retention FF's, improves the place and route Encounter tools' runtime by 11% compared to the distributed SSRPG and by 23% compared to the conventional SRPG flow. This is a considerable improvement compared to the runtime of the distributed flow, which does not apply any specific placement rules regarding the retention cells. The major improvement is achieved in the placement stage, in which the runtime is decreased by 29% compared to the distributed SSRPG flow. This is a significant result since the placement stage is an iterative stage due to the floorplan area estimation process. Moreover, the improved localized proposed flow outperforms the conventional SRPG by 63% in terms of placement runtime. The runtime for the routing stage is improved by 8% and 9% compared to the distributed SSRPG and SRPG, respectively. The runtime for the CTS stage is improved by 13% compared to the SRPG flow. Table 3 depicts the comparison between the four examined flows in terms of area, design density, number of library cells, wire-length, static leakage, and back-end tools runtime. As expected, the required area for SRPG implementation is 20% larger compared to the no-retention case. The implementation of the SSRPG approach results in a 16% area saving factor compared to SRPG. Moreover, almost no extra area is required for implementing the SSRPG flow compared to the no-retention case. While the wire length for SRPG is significantly larger compared to the no-retention flow, with about a 12% wiring increase, both SSRPG flows require only about 4% extra wiring compared to the no-retention case. This additional wiring overhead is required for connecting the retention cells to the non-gated power supply and power-gating controls. The increased wire-length induced by gathering all retention flip-FFs in a localized region is less than 1% compared to the distributed SSRPG.

Table 2. Place and Runtime Routing Comparison.

Run-Time (Hours)	No Retention	SRPG	Distributed SSRPG	Localized SSRPG
Placement	9.11	11.42	6.02	4.27
CTS	9.85	6.15	5.53	5.32
Routing	14.75	27.13	26.83	24.63
Total	33.71	44.7	38.38	34.22

Table 3. Physical design flow Comparison.

Physical Design Parameter	No Retention	SRPG	Distributed SSRPG	Localized SSRPG
Macro area (mm^2)	0.594	0.716	0.600	0.600
Design density (%)	72.2%	69.7%	72.3%	70.0%
Total library cells	315,837	318,052	313,679	309,369
Wire-length (m)	6.561	7.319	6.833	6.887
Static leakage (mW)	34.62	2.213	0.085	0.085
Backend Run-time (Hours)	33.72	44.7	38.38	34.22
Retained FFs	0	61,944	2522	2522

The increasing wiring can explain this since the retention FFs are associated along with other non-retention FFs. However, this wire-length is compensated due to the reduced distance between the retention cells to the always-on PDN and to the retention controls in the improved SSRPG flow. Table 3 shows that although the macro area is the same for both SSRPG flows, the design density (as measured by the Encounter Cadence tool) is reduced by 2.3% for the improved localized SSRPG compared to the distributed SSRPG. The lower density hints towards a lower crosstalk, though this still needs to be proved using bespoke benchmarks. Therefore, a better immune to crosstalk effects might be achieved using the localized PD approach. Spice simulations show that for both PD flows, the used gridlines meet the IR drop worst-case conditions (according to TSMC 28 nm library).

This can be explained due to the better routeability achieved by limiting the retention power grid to a specific localized region and therefore reducing the area occupied by both the always-on PDN and the retention control wiring. A significant improvement is also demonstrated for the static power leakage. Although SRPG reduces the static power leakage by 94% compared to the no-retention flow (whereas the supplies are always on), both SSRPG flows reduce the static power leakage by 99.7%. It is also important to notice that SSRPG outperforms the SRPG flow by 96% in terms of static leakage.

The efficiency of the improved SSRPG approach is expressed by the significant improvement in terms of back-end runtime. The required runtime for implementing the place and route stages is compared. While SRPG increases the runtime by a significant factor of 33%, the improved SSRPG flow can be implemented with a negligible overhead of only 3% compared to the non-retention flow. Moreover, the speed up comparing to the distributed SSRPG flow is about 11%. It should be noted that the improved SSRPG outperforms the distributed SSRPG in terms of back-end runtime in spite of the slightly increased wire length. This can be explained by the lower design density in the case of improved SSRPG due to the reduced buffers (as indicated by the total library cells) required to support the specific clock-tree for the retention controls compared to the distributed SSRPG flow.

4. Summary and Conclusions

This work presents a novel approach for SoC physical design implementation based on Selective State Retention techniques. The additional wiring required for the extra power grid network for the retention cells and power-gating controls for the state retention logic increases the complexity of the physical design and directly affects the tools' runtime and the ability to converge for large designs. Therefore, this work investigates the effect of the selective approach on the complexity of the physical design implementation and proposes a unique flow to efficiently address SoC design based on selective state retention techniques. We demonstrate a significant reduction of the metal area required for the extra power supply network using the proposed approach. This is done by applying some unique placement rules to the physical design implementation flow utilizing the selectivity feature. This results in optimal cell placement and power grid allocation, which significantly increase the potential routing area, directly improving the convergence time of the Place and Route tools. Furthermore, it is shown that reducing the extra power supply network area also leads to a significant reduction of the runtime required for the placement tools.

We also compare the SRPG and SSRPG physical design implementations in terms of power, area, wire-length, and physical design tools runtime and quantify the area and runtime saving factors result from selectivity. Experimental results show that implementing the SSRPG approach using the proposed physical design flow yields an area-saving factor of 16% compared to SRPG, which is in accordance with the previously estimated factor reported in recent publications. Furthermore, the static leakage is decreased by 96% compared to SRPG and is negligible compared to no retention. Tool complexity overhead was also reduced as such that the runtime overhead was negligible compared to the no retention physical design flow. Finally, by applying certain placement rules for the retention cells, the tool runtime for the improved SSRPG was further reduced by 11% compared to the common SSRPG and by 23% compared to SRPG.

The proposed improved localized SSRPG flow facilitates the complexity of the physical design implementation for retention-based design. This approach leads to both reducing the number of metal layers used for the always-on power distribution and therefore facilitates the signals routing, and reducing the wiring used for retention control signals as well as simplifying the isolation of the always-on domain from the power-gated domain. As a result, the runtime of the place and route tools is significantly reduced due to the wiring complexity reduction.

Moreover, to the best of our knowledge, this is the first work that demonstrates and quantifies the benefit of applying the SSRPG approach in real physical design implementation and demonstrating actual area, power, and tools runtime saving factor.

Author Contributions: Both authors contributed equally to this work. All authors have read and agreed to the published version of the manuscript.

Funding: This research received no external funding.

Institutional Review Board Statement: Not applicable.

Informed Consent Statement: Not applicable.

Data Availability Statement: Not applicable.

Conflicts of Interest: The authors declare no conflict of interest.

References

1. Austin, T.; Baauw, D.; Mudge, T.; Flautner, K.; Hu, J.S.; Irwin, M.J.; Kandemir, M.; Narayanan, V. Leakage current: Moore's law meets static power. *IEEE J. Comput.* **2003**, *36*, 68–75.
2. Horiguchi, M.; Sakata, T.; Itoh, K. Switched-source-impedance CMOS circuit for low standby sub-threshold current giga-scale LSI's, Solid-State Circuits. *IEEE J.* **1993**, *28*, 1131–1135.
3. Zhigang, H.; Buyuktosunoglu, A.; Srinivasan, V.; Zyuban, V.; Jacobson, H.; Bose, P. Micro architectural techniques for power gating of execution units. In Proceedings of the International Symposium on Low Power Electronics and Design, ISPLED, Newport Beach, CA, USA, 9–11 August 2004; pp. 32–37.
4. Henry, M.B. *Emerging Power-Gating Techniques for Low Power Digital Circuits*; Virginia Polytechnic Institute and State University: Blacksburg, VA, USA, 2011.
5. Weihan, W.; Ohta, Y.; Ishii, Y.; Usami, K.; Amano, H. Tradeoff analysis of fine-grained power gating methods for functional units in a CPU. In Proceedings of the Cool Chips XV (COOL Chips), Yokohama, Japan, 18–20 April 2012; pp. 1–3.
6. Henry, M.B.; Nazhandali, L. Design techniques for functional-unit power gating in the Ultra-Low-Voltage region. In Proceedings of the Design Automation Conference (ASP-DAC), Sydney, NSW, Australia, 30 January–2 February 2013; Association for Computing Machinery: New York, NY, USA, 2012; pp. 609–614.
7. Dasnurkar, S.; Datta, A.; Abu-Rahma, M.; Nguyen, H.; Villafana, M.; Rasouli, H.; Tamjidi, S.; Cai, M.; Sengupta, S.; Chidambaram, P.R.; et al. Experiments and analysis to characterize logic state retention limitations in 28 nm process node. In Proceedings of the IEEE 31st VLSI Test Symposium, VTS, Berkeley, CA, USA, 29 April–2 May 2013; pp. 1–6.
8. Henzler, S.; Nirschi, T.; Pacha, C.; Spindler, P.; Teichmann, P.; Fulde, M.; Fischer, J.; Eireiner, M.; Fischer, T.; Georgakos, J.; et al. Dynamic state-retention flip flop for fine-grained sleep-transistor scheme. In Proceedings of the European Solid-State Circuits Conference, ESSCIRC, Grenoble, France, 12–16 September 2005; pp. 145–148.
9. Shigematsu, S.; Mutoh, S.; Matsuya, Y.; Tanabe, Y.; Yamada, J. A 1-V high-speed MTCMOS circuit scheme for power-down application circuits. *IEEE J. Solid-State Circ.* **1997**, *32*, 861–869. [CrossRef]
10. Le-Coz, J.; Flatresse, P.; Clerc, S.; Belleville, M.; Valentian, A. 65 nm PD-SOI glitch-free Retention Flip-Flop for MTCMOS power switch applications. In Proceedings of the IEEE International Conference on IC Design & Technology, ICICDT, Kaohsiung, Taiwan, 2–4 May 2011; pp. 1–4.
11. Chul-Moon, J.; Kwan-Hee, J.; Eun-Sub, L.; Minh, V.H.; Kyeong-Sik, M. Zero-Sleep-Leakage Flip-Flop Circuit with Conditional-Storing Memristor Retention Latch. *IEEE Trans. Nanotechnol.* **2011**, *11*, 360–366.
12. Kyungho, R.; Jisu, K.; Jiwan, J.; Kim, J.P.; Kang, S.H.; Seong-Ook, J. A Magnetic Tunnel Junction Based Zero Standby Leakage Current Retention Flip-Flop. *IEEE Trans. Very Large Scale Integr. (VLSI) Syst.* **2011**, *20*, 2044–2053.
13. Jung-Hyun, P.; Heechai, K.; Dong-Hoon, J.; Kyungho, R.; Seong-Ook, J. Level-Converting Retention Flip-Flop for Reducing Standby Power in ZigBee SoCs. *IEEE Trans. Very Large Scale Integr. (VLSI) Syst.* **2014**, *23*, 413–421.
14. Ting-Wei Chiang, C.; Kai-Hui Chang, C.; Yen-Ting, L.; Jiang, J.-H.R. Scalable sequence-constrained retention register minimization in power gating design. In Proceedings of the Design Automation Conference, DAC, San Francisco, CA, USA, 7–11 June 2015; Association for Computing Machinery: New York, NY, USA; pp. 1–6.
15. Darbari, A.; Hashimi, B.M.A.; Flynn, D.; Biggs, J. Selective state retention design using symbolic simulation. In Proceedings of the 2009 Design, Automation and Test in Europe Conference and Exhibition, Nice, France, 20 April 2009; IEEE: Piscataway, NJ, USA; pp. 1644–1649.
16. Greenberg, S.; Rabinowicz, J.; Tsechanski, R.; Paperno, E. Selective State Retention Power Gating Based on Gate-Level Analysis. *IEEE Trans. Circ. Syst. I* **2013**, *61*, 1095–1104. [CrossRef]
17. Greenberg, S.; Rabinowicz, J.; Manor, E. Selective State Retention Power Gating Based on Formal Verification. *IEEE Trans. Circ. Syst. I* **2014**, *62*, 807–815. [CrossRef]
18. Wen-Hsiang, C.; Chao, M.C.-T.; Shi-Hao, C. Practical Routability-Driven Design Flow for Multilayer Power Networks Using Aluminum-Pad Layer. *IEEE Trans. Very Large Scale Integr. (VLSI) Syst.* **2013**, *22*, 1069–1081.
19. Hyung-Ock, K.; Youngsoo, S. Semicustom Design Methodology of Power Gated Circuits for Low Leakage Applications. *IEEE Trans. Circ. Syst. II Express Briefs* **2007**, *54*, 512–516.
20. Seomun, J.; Youngsoo, S. Design and Optimization of Power-Gated Circuits with Autonomous Data Retention. *IEEE Trans. Very Large Scale Integr. (VLSI) Syst.* **2009**, *19*, 227–236. [CrossRef]
21. Golshan, K. Physical Design Essentials. In *An ASIC Design Implementation Perspective*, 1st ed.; Springer: New York, NY, USA, 2007.
22. Flynn, D.; Aitken, R.; Gibbons, A.; Kaijian, S. Low Power Methodology Manual. In *For System-on-Chip Design*, 1st ed.; Springer: New York, NY, USA, 2007.
23. Mahmoodi-Meimand, H.; Roy, K. Data-retention flip-flops for power-down applications. In Proceedings of the 2004 IEEE International Symposium on Circuits and Systems, Vancouver, BC, Canada, 23–26 May 2004; pp. 1–4.
24. Henry, M.B.; Nazhandali, L. NEMS-Based Functional Unit Power-Gating: Design, Analysis, and Optimization. *IEEE Trans. Circ. Syst. I* **2013**, *60*, 290–302. [CrossRef]

25. Wang, K.; Marek-Sadowska, M. On-chip power-supply network optimization using multigrid-based technique. *IEEE Trans. Comput.-Aided Des. Integr. Circ. Syst.* **2005**, *24*, 407–417. [CrossRef]
26. Ye, T.T.; de Micheli, G. Data path placement with regularity. In *Proceedings of the ACM/IEEE Computer Aided Design, ICCAD*; San Jose, CA, USA, 5–9 November 2000, IEEE: Piscataway, NJ, USA; pp. 264–270.
27. Zhaohui, H.; Pierres, A.; Hu, S.; Chen, F.; Royannez, P.; Pek, S.E.; Ling, H.Y. Practical and efficient SOC verification flow by reusing IP testcase and testbench. In Proceedings of the SoC Design Conference, ISOCC, Jeju, Korea, 4–7 November 2012; pp. 175–178. Available online: http://www.isocc.org (accessed on 26 July 2021).
28. Bhasker, J.; Chadha, R. Static Timing Analysis for Nanometer Designs. In *A Practical Approach*, 1st ed.; Springer: New York, NY, USA, 2009.
29. Gunnala, V. Choosing Appropriate Utilization Factor and Metal Layer Numbers for an Efficient Floor Plan in VLSI Physical Design. *Int. J. Eng. Res. Appl. IJERA* **2012**, *2*, 456–462.
30. Voldman, S.H. *Latchup*, 1st ed.; Wiley: West Sussex, UK, 2007.

Journal of
Low Power Electronics and Applications

MDPI

Article

Energy-Efficient Non-Von Neumann Computing Architecture Supporting Multiple Computing Paradigms for Logic and Binarized Neural Networks

Tommaso Zanotti [1,*], Francesco Maria Puglisi [1] and Paolo Pavan [1]

Dipartimento di Ingegneria "Enzo Ferrari", Università di Modena e Reggio Emilia, Via P. Vivarelli 10/1, 41125 Modena, Italy; francescomaria.puglisi@unimore.it (F.M.P.); paolo.pavan@unimore.it (P.P.)
* Correspondence: tommaso.zanotti@unimore.it

Abstract: Different in-memory computing paradigms enabled by emerging non-volatile memory technologies are promising solutions for the development of ultra-low-power hardware for edge computing. Among these, SIMPLY, a smart logic-in-memory architecture, provides high reconfigurability and enables the in-memory computation of both logic operations and binarized neural networks (BNNs) inference. However, operation-specific hardware accelerators can result in better performance for a particular task, such as the analog computation of the multiply and accumulate operation for BNN inference, but lack reconfigurability. Nonetheless, a solution providing the flexibility of SIMPLY while also achieving the high performance of BNN-specific analog hardware accelerators is missing. In this work, we propose a novel in-memory architecture based on 1T1R crossbar arrays, which enables the coexistence on the same crossbar array of both SIMPLY computing paradigm and the analog acceleration of the multiply and accumulate operation for BNN inference. We also highlight the main design tradeoffs and opportunities enabled by different emerging non-volatile memory technologies. Finally, by using a physics-based Resistive Random Access Memory (RRAM) compact model calibrated on data from the literature, we show that the proposed architecture improves the energy delay product by $>10^3$ times when performing a BNN inference task with respect to a SIMPLY implementation.

Keywords: BNN; logic-in-memory; RRAM; SIMPLY

Citation: Zanotti, T.; Puglisi, F.M.; Pavan, P. Energy-Efficient Non-Von Neumann Computing Architecture Supporting Multiple Computing Paradigms for Logic and Binarized Neural Networks. *J. Low Power Electron. Appl.* **2021**, *11*, 29. https://doi.org/10.3390/jlpea11030029

Academic Editors: Alex Serb and Adnan Mehonic

Received: 31 May 2021
Accepted: 30 June 2021
Published: 6 July 2021

1. Introduction

The demand for more ubiquitous edge computing promoted by the rapidly growing volume of data exchanged over the communication network by devices for the Internet of Things (IoT) requires the development of more energy-efficient computing architectures [1,2]. Accordingly, several new computing paradigms [3–10] have been proposed, encouraging a departure from the traditional von Neumann architecture. All these new computing approaches aim at performing computation directly inside the memory by exploiting novel emerging non-volatile memory (ENVM) technologies, therefore bypassing the main performance bottleneck of traditional von Neumann architectures, i.e., the communication between the memory and the processing unit over a slow bus. While operation specific hardware accelerators can achieve very high performance when executing a specific task, the possibility to reconfigure the type of operations computed in-memory may benefit resource-constrained devices for edge computing applications [11]. Among in-memory computing paradigms providing reconfigurability [4,12–16], architectures based on resistive memory devices and the material implication (IMPLY) logic are a promising solution [13]. Also, a smart IMPLY (SIMPLY) [17] architecture was proposed for solving the reliability issues of conventional IMPLY-based architectures demonstrating high reliability and high energy efficiency when implementing logic operations. Recently, a binarized neural network (BNN) [18] implementation based on the SIMPLY architecture was proposed [19,20] and shown to improve energy efficiency with respect to conventional embedded system implementations. Nevertheless, specific BNN hardware accelerators based on resistive memory technologies [21–25] which accelerate in analog the multiply and accumulate (MAC) operation can achieve higher performance if properly designed, lacking, however, reconfigurability options. Thus, a solution enabling the coexistence of both computing approaches on the same resources would enable the development of reconfigurable ultra-low-power edge computing hardware, but such a solution is still missing.

In this work, we design a new in-memory computing architecture enabling the coexistence on the same 1T1R crossbar array of both the SIMPLY logic-in-memory paradigm and the analog acceleration of the multiply and accumulate operation for BNN inference applications. We analyze the design tradeoffs of the proposed architecture and indicate the opportunities and limitations introduced using different emerging non-volatile memory technologies. Finally, exploiting a physics-based Resistive Random Access Memory (RRAM) compact model calibrated on a TiN/HfOx/AlOx/Pt RRAM technology from the literature [26], we benchmark with respect to a SIMPLY implementation the performance improvements for an inference task on a BNN provided by the novel architecture.

2. Results

2.1. Logic-in-Memory and the SIMPLY Architecture

2.1.1. Material Implication Logic

The IMPLY logic is based on two logic operations, i.e., the IMPLY, the truth table of which is shown in Figure 1c, and the FALSE which always results in a logic zero. These two operations can be implemented with RRAM devices and the circuit shown in Figure 1a, which comprises a resistor (i.e., R_G in Figure 1a), a control logic and analog tri-state buffers to deliver appropriate voltages to RRAM devices. Since the IMPLY and the FALSE form a complete logic group, all logic operations can be implemented with a sequence of these two operations [27]. In this framework, logic bits are mapped to RRAM devices resistances, and a logic 0 and a logic 1 are encoded into a high-resistive state (HRS) or low-resistive state (LRS), respectively. When performing computations RRAMs act at the same time both as the inputs and the outputs of computation. In particular, to perform an IMPLY operation between two bits (i.e., P and Q in Figure 1b,c), the control logic simultaneously drives the top electrodes of the two input RRAM devices with two voltage pulses with amplitudes V_{COND} and V_{SET} (see Figure 1b) on the two devices, respectively. By determining an appropriate value for V_{SET} and V_{COND} voltages, which must satisfy all the different requirements for each input combinations reported in Figure 1c, the state of the device receiving V_{COND} (i.e., P in Figure 1b,c) never changes while the state of the other device (i.e., Q in Figure 1b,c) changes according to the IMPLY truth table (see Figure 1c, where Q′ is the state of Q after the IMPLY operation execution). The FALSE operation is executed by applying a negative voltage pulse with amplitude V_{FALSE} to a single device to reset it into a HRS (see Figure 1d). However, this IMPLY scheme is affected by several reliability challenges, such as logic state degradation and small tolerance to voltage variations, which hinder its implementation [28,29].

2.1.2. SIMPLY

A solution to the reliability issues affecting the conventional IMPLY architecture is the SIMPLY architecture [17]. In SIMPLY, the computation of the IMPLY operation is split into two steps, i.e., a read step and a conditional write step. As shown in the IMPLY truth table (see Figure 1c), the state of Q, which is the device storing the result of the IMPLY operation, changes only when both P and Q (i.e., the inputs of the IMPLY operation) are in HRS. This condition can be detected by applying two simultaneous small read voltage pulses with amplitude V_{READ} to P and Q and comparing the voltage across R_G (V_N) with a threshold (V_{TH}) using a comparator, as shown in Figure 2a. In fact, V_N is lower when both inputs are zero than in all the other cases (see Figure 2b), providing a sufficient read margin (RM) for the comparator. The output of the comparator is fed to a control logic which then pulses

V_{SET} on Q only when necessary. By using a sufficiently low V_{READ} voltage the problem of logic state degradation is effectively solved [17]. Also, the drivers in the peripheral circuitry of the array can be simplified, as V_{COND} is no longer required. In addition, the high V_{SET} voltage pulse is applied only in the first case of the truth table, while in the other three cases the main energy consumption is due to the small V_{READ} pulses and the comparator. As described in previous works [19,20], the latter can be implemented with the voltage sense amplifier (VSA) in Figure 2d, which is fast and energy efficient (i.e., the VSA implemented with a 45 nm technology from [30], and a V_{DD} of 2V dissipated just 8 fJ per comparison on average). Therefore, SIMPLY considerably improves the energy per IMPLY operation in three out of four cases of the truth table compared to the conventional IMPLY architecture [17,19].

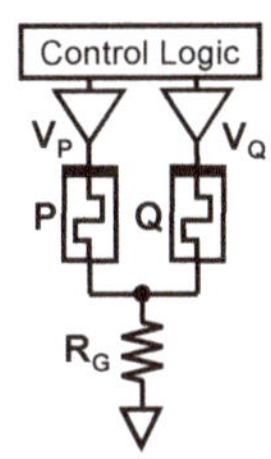

(a)

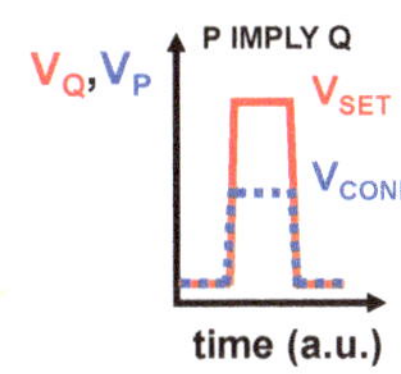

(b)

Input Combination		Output (Q′)	Requirements
P=0	Q=0	1	High V_{SET}, Low V_{COND}
P=0	Q=1	1	Low $\lvert V_{COND}-V_{SET}\rvert$
P=1	Q=0	0	Low $\lvert V_{SET}-V_{COND}\rvert$
P=1	Q=1	1	-

(c)

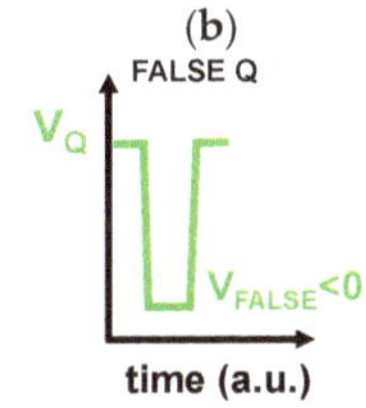

(d)

Figure 1. (**a**) Circuit implementing the elementary IMPLY logic gate. (**b**) Driving voltage scheme implementing the P IMPLY Q operation. (**c**) IMPLY operation truth table highlighting the contrasting requirements on V_{SET} and V_{COND} for a reliable gate functionality. Q′ represents the state of Q after the IMPLY operation execution. (**d**) Driving voltage scheme implementing the FALSE Q operation.

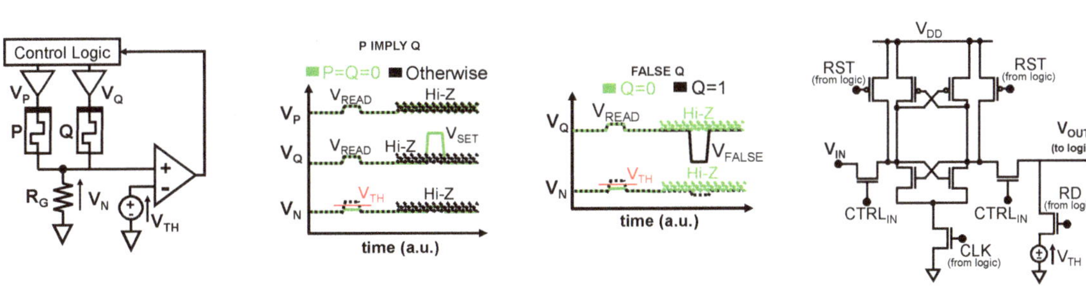

Figure 2. (**a**) Circuit implementation of the elementary IMPLY gate in the SIMPLY framework. (**b**) Driving voltage scheme used to implement the P IMPLY Q operation. The control logic pulses V_{SET} on Q only when the comparator detects P = Q = 0 (green lines) while the drivers are kept in high impedance (Hi-Z) in all other cases (dashed black lines). (**c**) Driving voltage scheme used to implement the FALSE Q operation in the SIMPLY framework. The comparator detects when Q = 1 (black lines) and pulses V_{FALSE} accordingly. (**d**) Voltage sense amplifier implemented and simulated with a 45nm technology [30]. All FETs have minimum size (i.e., L = 50 nm W = 90 nm).

To further improve the energy efficiency, the same approach can be used for the FALSE operation [19,20]. When a device is in HRS the high V_{FALSE} voltage results in unnecessary energy dissipation. This can be prevented by first reading the state of the device and then applying the V_{FALSE} only when the device is in LRS (see Figure 2c). The effectiveness in reducing the energy per operation depends on the employed RRAM technology [19]. In fact, the achieved energy reduction with RRAM technology characterized by very high HRS is limited, while it is relevant for technologies with relatively low HRS.

The RM is the most important reliability metric, and enough RM must be ensured even in presence of resistive state variability and random telegraph noise (RTN). While a higher RM can be achieved by slightly increasing V_{READ} (see Figure 3a), a higher RM is also obtained by using the optimal value for R_G (see Figure 3b) that is determined using equation (1) from [31], where $R_{HRS,MAX}$ and $R_{HRS,MIN}$ are the $\pm 3\sigma$ values of the R_{HRS} distribution, while $R_{LRS,MAX}$ is the $+3\sigma$ value of the R_{LRS} distribution.

$$R_G = \sqrt{\frac{1}{\frac{1}{R_{HRS_{MAX}}} + \frac{1}{R_{LRS_{MAX}}}} \cdot R_{\frac{HRS_{MIN}}{2}}}, \tag{1}$$

The SIMPLY framework can be implemented also on crossbar arrays [19]. Specifically, to implement SIMPLY on the 1T1R crossbar array the architecture shown in Figure 4 is needed. Additional field effect transistor (FET) devices in the array periphery are used to connect adjacent rows of the crossbar to perform IMPLY operations between devices on the same column but different rows, and to select specific rows. Also, the degree of parallelism in the SIMPLY architecture can be increased by adding more VSAs in the array periphery, as shown in Figure 4. Using multiple VSAs enables the realization of single instruction multiple data (SIMD) architectures, in which the IMPLY and FALSE operations can be performed in parallel on data stored in different rows but aligned on the same columns. For instance, to perform IMPLY operations in parallel, two columns are driven with the read voltage, then only the FETs implementing R_G and those enabling the selected rows are enabled, therefore routing each active row to the specific VSA. The control logic receives in input the results of all the comparisons and activates only the rows where an RRAM device should be switched during the device SET step. As shown in previous works [19,31], the use of SIMPLY-based SIMD architectures results in high computing efficiency and throughput.

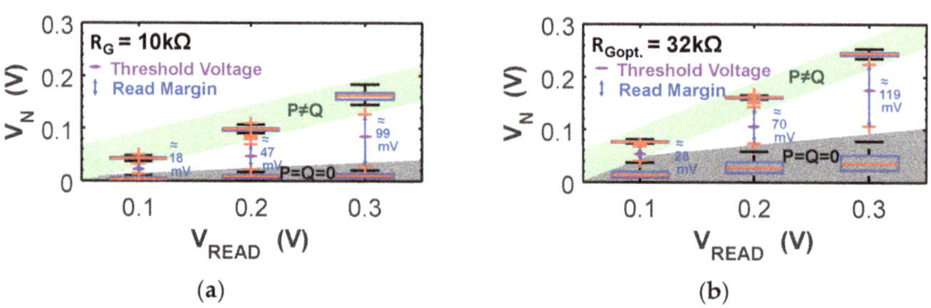

Figure 3. (**a**) Distribution of the comparator input voltage (V_N) for increasing V_{READ} considering the TiN/HfO_x/AlO_x/Pt RRAM devices from [26], when considering a suboptimal R_G in (**a**) and the optimal R_{Gopt} in (**b**), which maximize the read margin (RM). The distributions for P = Q = 0 (grey bands) and P $\neq$ Q (green bands) are reported together with the read margins (RM—blue arrows) and associated threshold voltages (V_{TH}—violet line) for the comparator. The effects of cycle-to-cycle, device-to-device variability, and random telegraph noise (RTN) are considered by repeating the simulations 50. The extreme points of the distributions are indicated with black whiskers, and outliers due to RTN with red crosses.

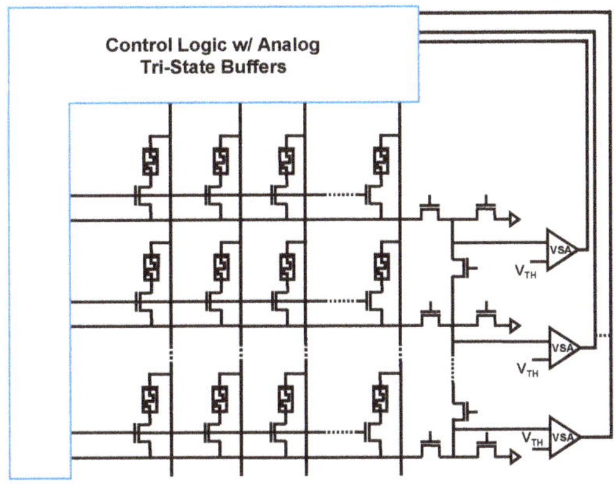

Figure 4. SIMPLY implementation on a 1T1R crossbar array. FET devices are used to implement R_G, select specific rows, and to connect adjacent columns.

2.2. Binarized Neural Networks (BNNs) Hardware Accelerator Architectures

2.2.1. Binarized Neural Networks with SIMPLY

BNNs are an effective solution that enables the implementation of neural networks at a lower computational cost, while retaining sufficiently high accuracies [18], as compared to full-precision neural networks. In fact, by using 1-bit weights and activations only logic operations are required to perform an inference task. Thanks to this simplification, previous works [19,20] showed how BNN inference can be implemented in the SIMPLY computing framework. The network parameters are trained offline using for example the DoReFa-Net algorithm [32] (see Section 4.2) and directly mapped to the resistance of RRAM devices in the crossbar array, as shown in Figure 5b where a single crossbar row is reported. In BNNs, the multiply and accumulate (MAC) can be implemented with bitwise XNOR operations between each neuron weights and input activations, the accumulation with the popcount operation, and each neuron activation by performing a comparison with a trained threshold. Finally, the output class is predicted by using the hardmax function, which selects the class corresponding to the neuron with the highest output activation. By enabling the realization of SIMD architectures, the resulting SIMPLY implementation exploits the intrinsic parallelism in BNN computations, to efficiently compute in parallel the operations in each neural network layer. Also, thanks to its reconfigurability, SIMPLY enables the possibility to easily implement different neural networks topologies. However, the high degree of reconfigurability comes at the price of a high number of computing steps, which limits the latency performance. Specifically, among the different operations, the computation of MAC operations is the main limitation in SIMPLY-based BNN implementations, and accounts for almost all the computing steps of a network layer when considering a layer with 1000 input activations, as shown in Figure 5a. While each bitwise XNOR requires nine computing steps (see Figure 5c), the number per accumulation operations for the implemented accumulator rapidly rises as the number of inputs to a network layer increases. In fact, to implement the accumulator, a chain of half-adders (HAs) is used where the first HA is fed its current output and the bit to be accumulated, while the following HAs are fed their current output and the carry-out from the previous HA in the chain [19,20], as shown in Figure 5e. Each HA requires 13 computing steps to accumulate a single bit (see Figure 5d). Thus, the whole accumulation operation is computed in $\sum^{m}{}_{i=1} 13i{\cdot}2^{(i-1)}$. This is because each HA is activated only after a number of input bits equal to two to the power of their respective bit position has been accumulated, since the carry-out bits from the preceding HA stage is necessarily zero when fewer bits have been accumulated. Therefore, the latency for computing the accumulation operation rises exponentially when the size of a neural network layer increases, thus suggesting that BNN SIMPLY implementations are more suitable for small neural network implementations, while the implementation of larger networks would require more efficient MAC execution.

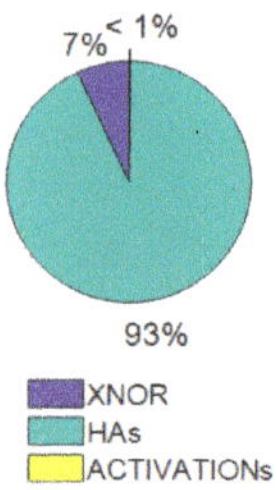

(a)

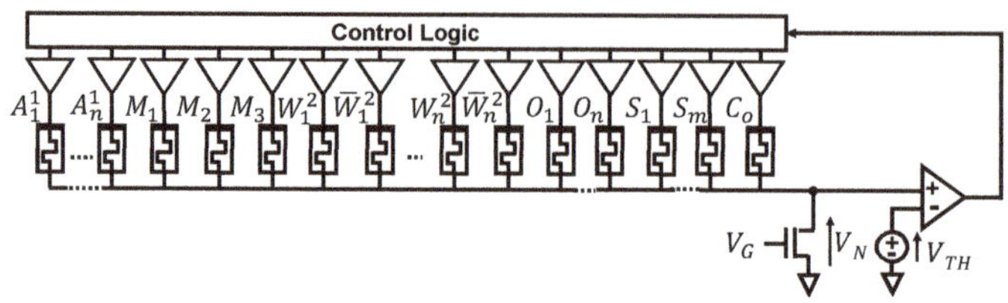

(b)

XNOR(A, W)

Step Numb.	Operation
1-3	$FALSE\ (O, M_1, M_2)$
4	$IMPLY\ (A, M_1)$
5	$IMPLY\ (M_1, M_2)$
6	$IMPLY\ (W, M_1)$
7	$IMPLY\ (\overline{W}, M_2)$
8	$IMPLY\ (M_1, O)$
9	$IMPLY\ (M_2, O)$

(c)

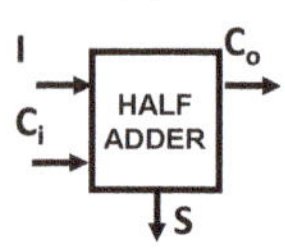

1-bit Half Adder

Step Numb.	Operation
1-4	$FALSE\ (C_0, M_1, M_2, M_3)$
5	$IMPLY\ (I, M_1)$
6	$IMPLY\ (C_i, M_2)$
7	$IMPLY\ (C_i, M_1)$
8	$IMPLY\ (M_2, I)$
9	$IMPLY\ (M_1, M_3)$
10	$IMPLY\ (I, M_3)$
11	$FALSE\ (S)$
12	$IMPLY\ (M_3, S)$
13	$IMPLY\ (M_1, C_o)$

(d)

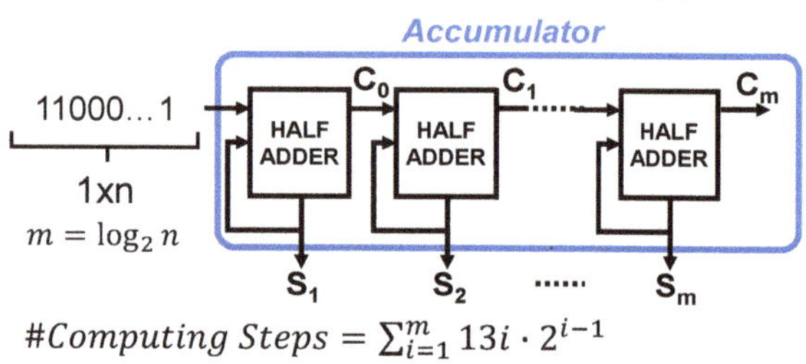

(e)

Figure 5. (**a**) Breakdown of the percentage number of computing steps performed in a binarized neural network (BNN) layer with 1000 input activations. (**b**) Example of SIMPLY implementation of the multiply and accumulate (MAC) operation for a single neuron. Devices storing the neural network weights (W), their complement ($\overline{W}$), the results of the bitwise XNOR (O), the result of the accumulation (S), the carry-out (C_0), and supporting intermediate computations (M_1, M_2, M_3) are reported. (**c**) Sequence of IMPLY and FALSE operations implementing a two input XNOR [19]. (**d**) SIMPLY-based half-adder (HA) implementation. (**e**) SIMPLY-based accumulator operation implementing the popcount operation.

2.2.2. Binarized Neural Networks with Analog Vector Matrix Multiplication

The computation in the analog domain of the vector matrix multiplication with resistive memory devices has been shown to be a promising solution for accelerating in hardware the execution of deep neural networks. In this framework, the weights of the neural network are mapped into the analog non-volatile resistance of RRAM devices of a crossbar array [3], while the input activations to a neuron are mapped to voltage pulses with amplitude or duration proportional to the input value. By applying such input activations to the rows of the crossbar and by providing a virtual ground to the end of each crossbar column, the current flowing in each crossbar column is linearly proportional to the result of the vector matrix multiplication, that is computed in a single step thanks to Ohm's and Kirchhoff's current laws. However, such architecture presents some challenges due to resistive memory devices non-idealities such as the resistive state variability, which limits the number of bits that can be reliably encoded into a single RRAM device. Furthermore, providing a virtual ground at each crossbar column comes at the expense of a large chip area occupied by the peripheral circuitry due to the need of operational amplifiers [33] and analog to digital converters (ADCs) which limit the area efficiency especially when many rows are read in parallel. Thus, the need for the virtual ground becomes the main bottleneck for such architecture, introducing a tradeoff between crossbar density and latency (i.e., the same operational amplifier and ADC pair can be shared among multiple crossbar columns; however, reducing the maximum throughput). A more robust solution to RRAM variability, which is also more efficient in terms of chip area occupancy, are BNNs. In fact, binary weights can be reliably stored in a crossbar array using a pair of RRAM devices in the same column, as shown in Figure 6b. To store a +1 weight, the two RRAM devices are programmed into an LRS/HRS configuration, while to store a -1 the opposite configuration is used (see Figure 6b). When computing the binary vector matrix multiplication, which is equivalent to the combined bitwise XNOR and popcount operations, the two FETs in series with the RRAMs representing a single weight are driven with complementary signals that encode the +1/-1 input activations so that only one FET is active at time. As a result, the current flowing through each crossbar column is proportional to the sum of all the positive results of the products between the input activations and each neuron's weights. Using this approach, the operational amplifier and ADC can be replaced by a much more compact voltage sense amplifiers (VSAs) circuit [21,22], implementing the architecture shown in Figure 6a, thus reducing the required chip area and improving the throughput and the energy efficiency. Instead of voltage sensing, the same approach could also be implemented by using current mode sense amplifiers [34], however resulting in lower energy efficiency [35]. Nevertheless, when using a sense amplifier, no virtual ground is available at the end of the crossbar columns. Thus, a FET implementing a pull-up or pull-down resistor must be used, thereby realizing a voltage divider between the equivalent parallel resistance of the active 1T1R devices in a column and the pull-up or pull-down resistor. Due to the use of a voltage divider, the linear relation between the number of active devices and the input to the VSA is lost. Nevertheless, He et al. [21] showed that the method is robust and that the output activation can be reliably determined by comparing the voltage from the voltage divider with an appropriate threshold using the VSA, retaining high inference accuracy also when process variations are considered. However, the number of devices that can be reliably read in parallel to compute the MAC operation is limited and strictly dependent on R_{HRS}, R_{LRS}, the pull-down resistance (R_{PD}) and the VSA threshold voltage. As shown in Figure 7a, considering the case with 15 devices read in parallel during each MAC, increasing R_{PD} changes the required VSA threshold voltage. Ideally, a linear relation between V_N and the number of positive products is desirable. However, to achieve such linearity very low R_{PD} values should be used, resulting in a considerable reduction of the dynamic range at the input of the comparator thus increasing the probability of errors due to the effect of resistive state variability. On the other hand, a too high R_{PD} value causes the input voltage (i.e., V_N) to the VSA to rapidly saturate to V_{READ}. While, considering a fixed V_{TH} and lowering R_{PD} increases the number of devices that can be read in parallel during each MAC operation, as shown in Figure 7b. However, too low R_{PD} values would require large FET devices and the effect of line parasitic resistances may affect the circuit reliability making the circuit more susceptible to noise. Thus, in this framework, MAC operations are split into multiple computing steps using the input split method [21,36], so that partial MAC operations are computed using the maximum parallelism enabled by the designed architecture. These partial results need to be accumulated and the result of the accumulation is compared to a trained threshold to produce the neuron output activation.

Compared to the SIMPLY implementation of the BNN MAC operation, this approach is considerably faster, as it requires fewer computing steps, and more energy efficient since no RRAM device is switched during computations. However efficient, this approach is specialized for BNNs and do not provide the reconfigurability of the kind of operations computed in-memory enabled by SIMPLY. Thus, the crossbar array can only be used for BNN inferencing and storage applications.

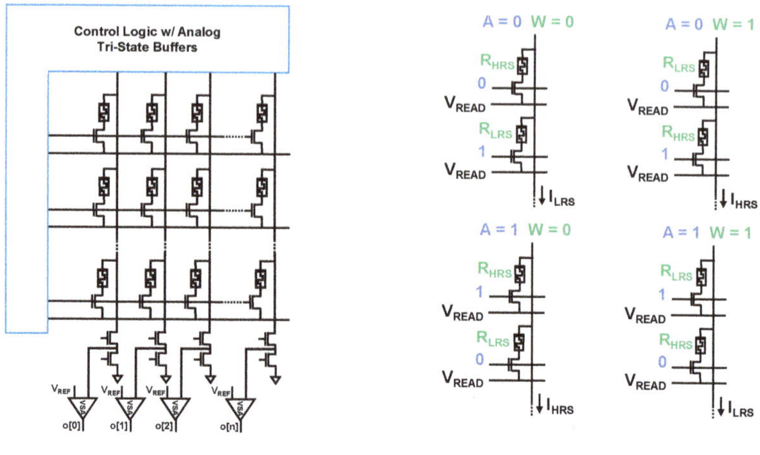

Figure 6. (**a**) Example of an in-memory computing architecture based on a 1T1R crossbar array enabling the analog BNN vector matrix multiplication acceleration using voltage sense amplifiers (VSAs). (**b**) Implemented binary multiplication between the input activation and the neuron weight. A pair of 1T1R devices with complementary resistive states is used to map the neuron weights. The input activation is realized with two complementary signals driving the two selector transistors of each weight.

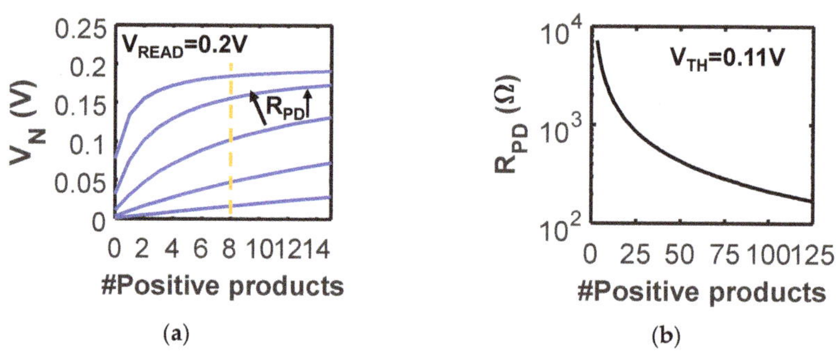

Figure 7. (**a**) Qualitative trends of the voltage at the input of the comparator for different R_{PD} values at increasing number of +1 products results with a V_{READ} = 0.2V. The comparator commute when the #positive products greater or equal than 8, thus the voltage threshold, the trend and slope change with R_{PD}. (**b**) Optimal R_{PD} values at increasing number of devices read in parallel for a fixed threshold voltage V_{TH} (i.e., the same used for SIMPLY). Increasing the computation parallelism requires lowering R_{PD}, thus leading to a tradeoff between area (i.e., lower R_{PD} require a larger FET area) and parallelism. In all cases, the nominal R_{HRS} and R_{LRS} for a TiN/HfOx/AlOx/Pt RRAM technology from the literature [26], are considered.

2.3. Merging SIMPLY and BNN Analog Vector Matrix Multiplication Accelerator

As discussed in the previous sections, both the SIMPLY computing paradigm and the BNN analog vector matrix multiplication (AVMM) accelerator are promising solution for IoT and edge computing devices and applications, as they provide considerable energy savings when computing different kinds of operations. While using multiple crossbar arrays specialized for different applications would be a solution to implement both computing paradigms on the same chip, it would result in an inefficient exploitation of the already scarce resources available to low-power devices. A better solution would be introducing the possibility to reconfigure the available resources to implement both computing paradigms on the same crossbar, provided that the additional flexibility must not determine the need for a much more complex, large, and inefficient peripheral circuitry. As it can be noted from Figures 4 and 6a, the circuit architecture used to implement both computing paradigms are indeed similar, relying on some FET devices used to implement R_G in the SIMPLY paradigm and R_{PD} in the BNN AVMM acceleration, VSAs and corresponding voltage thresholds. However, when considering the same 1T1R crossbar array, there are some differences between the two architectures and in their respective control signals management. Specifically, when performing a read step in the SIMPLY computing paradigm the select line corresponding to the row where the devices are located is activated, the read voltages are applied to the crossbar columns and the output is read from the appropriate crossbar row by means of the VSA and of a threshold. This holds true both when performing an IMPLY between devices located in the same row and when the devices are located in the same column.

On the other hand, to execute a MAC operation in the analog BNN AVMM accelerator the select lines encode the neurons input activations, and each select line must be shared among the neuron in the same neural network layer. Thus, read voltages need to be applied to the crossbar rows while the voltages encoding the result of the MAC operations are read out from the crossbar columns using VSAs with appropriate thresholds.

Therefore, to merge the two approaches the peripheral circuitry comprising the VSAs and pull-down resistance needs to be repeated both at the columns and the rows of the crossbar with limited additional complexity, resulting in the architecture shown in Figure 8. While the need of additional VSAs increases the chip area, it enables the coexistence of the two in-memory computing paradigms on the same crossbar. In addition, it improves the SIMPLY architecture by increasing achievable parallelism when performing operations on devices on the same columns but different rows, thus accelerating the copy of data between the rows of the crossbar array. In fact, only one IMPLY operation between devices on the same column using the SIMPLY architecture in Figure 4 can be executed in one computing step due to the lack of SA at the crossbar columns. Instead, the addition of SAs at each crossbar column enables the parallel execution of IMPLY operation on multiple columns, by applying V_{READ} to the crossbar rows and comparing the V_N voltage with the appropriate threshold at each column. Since IMPLY operations can be performed both on devices on the same row or column by applying V_{READ} at the crossbar columns and rows, respectively, the selector transistor in series with each RRAM device is subject to different source-bulk voltages. Nevertheless, since V_{READ} is small, the influence of the body effect can be minimized by driving these transistors with sufficiently high gate voltages. Also, using the same VSA threshold voltage for the two computing paradigms translates to different optimal R_G and R_{PD} values, requiring appropriate control of the gate voltage of the FET devices implementing such resistances. In fact, R_{PD} is much lower than R_G since more devices are read in parallel compared to SIMPLY. A too high R_{PD} would let the input of the VSA saturate at V_{READ} with only a few active devices in LRS, thus hindering the correct circuit operation.

A specific advantage of the proposed architecture is the possibility to exploit both the SIMPLY and BNN AVMM computing paradigms on the same crossbar array, which is particularly useful for some applications. For instance, when implementing a complete BNN exploiting the AVMM in-memory acceleration the MAC operations are computed in multiple steps, using the input split strategy [21]. Thus, the computation of logic operations is required to determine each neuron output activation, and consist in the accumulation of the intermediate MAC results and a comparison with a trained threshold. While as discussed in Section 2.2.1 the cost for performing accumulations with SIMPLY rapidly increases with the number of bits to be added, the use of the BNN AVMM for computing intermediate MAC results drastically reduces the number of bits that needs to be accumulated with SIMPLY. Thus, while intermediate computations could also be executed on task-specific CMOS digital circuits, merging the two computing approaches enables achieving high performance by exploiting the intrinsic high degree of parallelism in the computation, without requiring additional circuits complexity. Also, this approach is particularly advantageous for large neural network implementations that require storing the network parameters over multiple chips thus incurring in the inter-chip communication penalty that can exceed the RRAM programming time and energy. Overall, the proposed architecture is an extremely flexible solution for ultra low-power applications.

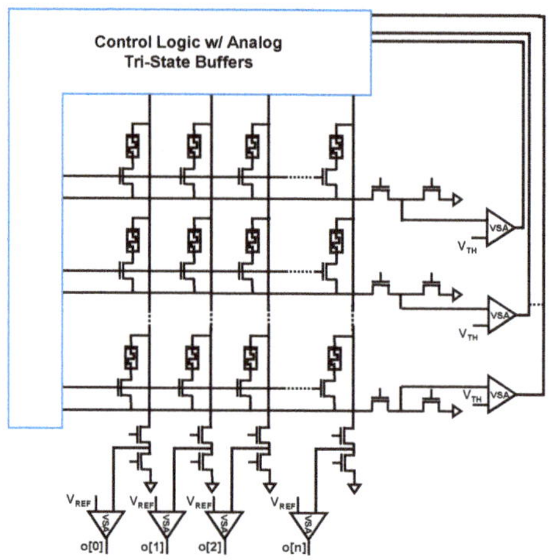

Figure 8. Proposed architecture, enabling the coexistence of the SIMPLY and BNN analog vector matrix multiplication computing paradigms on the same 1T1R crossbar array.

2.4. Circuit Design Tradeoffs for Performance and Reliability

The circuit design of the proposed architecture is directly connected to the specific requirements of possible use case applications [3,10], which may require the use of different resistive memory technologies to meet specific requirements. Thus, the correct selection of the most appropriate resistive memory technology becomes very important and is governed by the existing design tradeoffs that are aimed at providing low operation energy, fast speed, high integration density, and high reliability, as discussed in this section.

Specifically, to minimize the energy consumption different approaches are possible. The most effective solution is to employ resistive memory technologies with low current compliance (I_C), and therefore higher R_{LRS}. Firstly, the use of lower I_C leads to lower energy dissipation when programming a device, thus tackling the main energy limitation associated to the SIMPLY paradigm, largely improving the energy efficiency. Secondly, higher R_{LRS} values also lower the energy required for each parallel read both when computing an IMPLY in the SIMPLY paradigm and when implementing the BNN AVMM. Furthermore, this strategy results in additional advantages on the overall area consumption and speed. By lowering I_C, the required size of the FET devices used as selector devices and in the array periphery is reduced, thus reducing the chip area. Also, using higher R_{LRS} increases the parallelism of the BNN AVMM implementation, since more rows can be read in parallel when using the same R_{PD} resistance. However, cycle-to-cycle (C2C) and device-to-device (D2D) variability is inversely proportional to I_C [31,37], thus too low I_C values may affect the circuit reliability depending on the memory technology employed. The energy efficiency is also improved by reducing V_{READ}, which in turns reduces the energy consumption during the read operations performed both in the SIMPLY and in the BNN AVMM computing paradigms. Also, in this case the reliability issue may arise, since too low read voltages would reduce the RM and the SNR at the input of the VSAs.

Limiting the overall chip area is indeed very important to reduce the fabrication costs. Thus, higher crossbar array densities are beneficial. To this end, two main technology features are prominently relevant, namely the memory cell feature size and the possibility to implement dense 3D structures. While the use of a selector transistor effectively solves the sneak-path problem, it increases the chip area, due to the larger feature size of the 1T1R device (i.e., $8F^2$) with respect to a passive 1R device [38] (i.e., $4F^2$), and requires a higher number of control signals and interconnections. Other passive selector devices could introduce the required high non-linearity to solve the sneak-path issue while retaining the $4F^2$ [39] device feature size and could be also used with the proposed architecture by changing the driving voltage scheme. Also, the equivalent number of memory devices per chip area can be increased by fabricating 3D array structures. These can be implemented by stacking horizontal crossbars arrays, and even more efficiently by realizing a 3D vertical structure that would lead to the highest densities and costs reduction [40].

Also, crossbar line parasitic effects, such as line resistance and coupling capacitance, influence the maximum attainable computing speed. In fact, these effects together with the resistance of RRAM devices introduce propagation delays that grow as the size of the crossbar arrays is increased [41], therefore introducing a tradeoff between computing speed and maximum array size.

Finally, to ensure high circuit reliability, in addition to providing a sufficient RM at the input of the VSA, memory technologies with high endurance and retention should be preferred. In particular, endurance is a key parameter for the SIMPLY paradigm. In fact, while the analog BNN VMM only requires reading the state of RRAM devices, the SIMPLY principle of operations relies on the conditional programming of RRAM devices to perform computation. Long retention, on the other hand, is required to prevent periodic memory refresh cycles that would degrade the architecture's efficiency.

3. Discussion

While SIMPLY was shown in previous works to be an effective solution for the in-memory computation of logic operations (e.g., XNOR [19], full adders [31]), its effectiveness for the computation of the set of logic operations required to implement a BNN inference task was limited due to the large number of computing steps required to implement the multiply and accumulate (MAC) operation. In fact, as previously mentioned, the number of computing steps needed for a MAC operation grows exponentially with the number of inputs to a BNN layer [19] making SIMPLY suitable only for smaller networks. As shown in Table 1, the energy reduction (i.e., $\approx$400 times lower) with respect to a conventional embedded system implementation reported in a previous work [19] is much larger than the latency improvement (i.e., $\approx$26 times lower), which would further reduce as the network size increases. By enabling the coexistence on the same crossbar array of both the SIMPLY and AVMM computing paradigms at the cost of a limited complexity increase, the proposed in-memory computing architecture achieves substantial performance improvements when considering a BNN inference task, while retaining the hardware reconfigurability feature that is required by edge computing devices and applications. This is clearly reported in Table 1, where we show, considering the ideal case both for SIMPLY and the proposed architectures where all the network parameters and computing devices are stored in a single crossbar, that also in the worst-case the proposed architecture drastically reduces the latency and energy consumption compared to the SIMPLY architecture when computing the same inference task, achieving an energy delay product (EDP) improvement larger than 10^3.

Table 1. Benchmark of the performance of the proposed architecture on a classification task of black and white 20 × 20 pixels images from the MNIST dataset performed with a shallow multilayer perceptron neural network with 1 hidden layer of 1000 neurons and 10 output neurons.

Implementation [1]	Average Energy	Latency	Average EDP	EDP Improvement
Embedded system [42]	5.37 mJ	17.35 ms	9.3×10^{-5} Js	1
SIMPLY parallel [1,2] [19]	11.4 µJ	663 µs	7.6×10^{-9} Js	1.2×10^4
SIMPLY parallel [1] **w** $R_{G,Opt}$	78.9 µJ	663 µs	5.2×10^{-9} Js	1.8×10^4
This work [1] **w** $R_{G,Opt}$	231 nJ	31.6 µs	7.3×10^{-12} Js	1.3×10^7

[1] Estimates were determined considering the RRAM technology from [26], and the ideal case where all the network parameters can be stored in a single crossbar. Only the worst-case estimates for RRAM variability are reported. The energy estimates do not include the decoder and driver energy overhead. [2] In [19], a suboptimal R_G = 10 kΩ was used.

As discussed in Section 2.4, the performance, reliability, and target application for the proposed computing architecture strongly depend on the resistive memory technology employed. While the device endurance does not impact on the AVMM implementation, it is a very important discriminant for SIMPLY. Applications performing intensive computations require very high endurance (i.e., $>10^{14}$). Thus, for this application spin-transfer torque magnetic RAM (STT-MRAM) devices are more suitable candidate, thanks to their high retention (>10 years), endurance ($>10^{14}$) and switching speed (~ns) [43]. However, STT-MRAM devices have usually small tunnel magneto resistance (TMR) which leads to a very small memory window that can affect the circuit reliability if not address properly, especially when implementing the AVMM. For instance, Gao et al. [44] showed that STT-MRAM can indeed be used to accelerate the AVMM for BNNs, however at the cost of additional in-hardware calibration steps and more complex peripheral circuitry, which include operational amplifiers to implement the virtual ground, thus limiting the throughput, energy efficiency, and chip density. Conversely, devices characterized by lower endurance are better suited for applications requiring less frequent burst operations, such as smart sensors. For instance, to provide a reliable device operation over a 10-year period with a memory technology providing a 10^8 endurance would limit the computing speed to 20 inferences per minute without introducing mitigation strategies, such as periodically changing devices used for computations. Currently, several emerging non-volatile memory (NVM) technologies were shown to achieve endurance $>10^8$. Among these technologies, phase change memory (PCM) devices are the most mature and offer long retention, high endurance ($>10^{12}$) [45–47], but require higher switching currents compared to other ENVMs, therefore limiting the integration density. Also, ferroelectric tunnel junction (FTJ) devices are a promising candidate for the development of ultra-low-power in-memory computing architectures thanks to low programming energy and fast speed. However, high endurance, retention, and scalability still need to be fully demonstrated [45,47]. At the state of the art, RRAM technologies provide the best overall characteristics. Depending on the used stack of materials, RRAMs can achieve endurance up to 10^{10} [48], long retention, large memory window (≈10), and can be used to realize vertical 3D structures similar to flash memory technology, leading to ultra-dense arrays. However, two main technology-related challenges remain to be solved. Specifically, C2C and D2D variability lead to random resistance distributions which spread when lowering I_C [31,37] thus introducing a tradeoff between energy efficiency, reliability, and throughput when performing the AVMM. Also, to achieve ultra-dense vertical 3D arrays while preventing the sneak path issue, a particular research focus must be directed to the development of compatible selector devices with a strongly non-linear conduction behavior [39,49].

4. Materials and Methods

4.1. Circuit Simulations

4.1.1. RRAM Physics-Based Compact Model

The performance of the proposed architecture was estimated by means of circuit simulation performed on Cadence Virtuoso® software, using the RRAM physics-based compact model from [50] available on nanoHub, that was calibrated on a TiN/HfOx/AlOx/Pt RRAM technology from [26] that is programmed with an I_C of 100 μA. A sketch of the compact model is reported in Figure 9a,b. While other general-purpose memristors [51–54] and physics-based RRAM compact models [55–60] exist in the literature, the used RRAM physics-based compact model is particularly suited for estimating the circuit performance and reliability, as it includes all the relevant RRAM devices' characteristics and non-idealities (e.g., dynamic temperature modelling, resistive state variability, and RTN) which only a few other physics-based compact models [56,57] consider, as discussed in [28]. Specifically, the compact model approximates the device resistance as the sum of a conductive filament (CF) and a dielectric barrier component (see Figure 9a,b). Differential equations model the field-activated and temperature-accelerated bond breaking during set, and the field-driven oxygen ions' drift and recombination during reset, thus reproducing the dielectric barrier thickness dynamics. Thermal effects are also modeled with differential equation, leading to accurate results also when ultra-fast pulses are considered. As shown in Figure 9c,d, the compact model well reproduces with a single set of parameters both the DC IV and the response to fast reset pulses. Additionally, the compact model includes all the RRAM non-idealities that are relevant to accurately estimate circuit performance and reliability, specifically RTN and variability [28,50]. The complete list of calibrated parameters is available in [20].

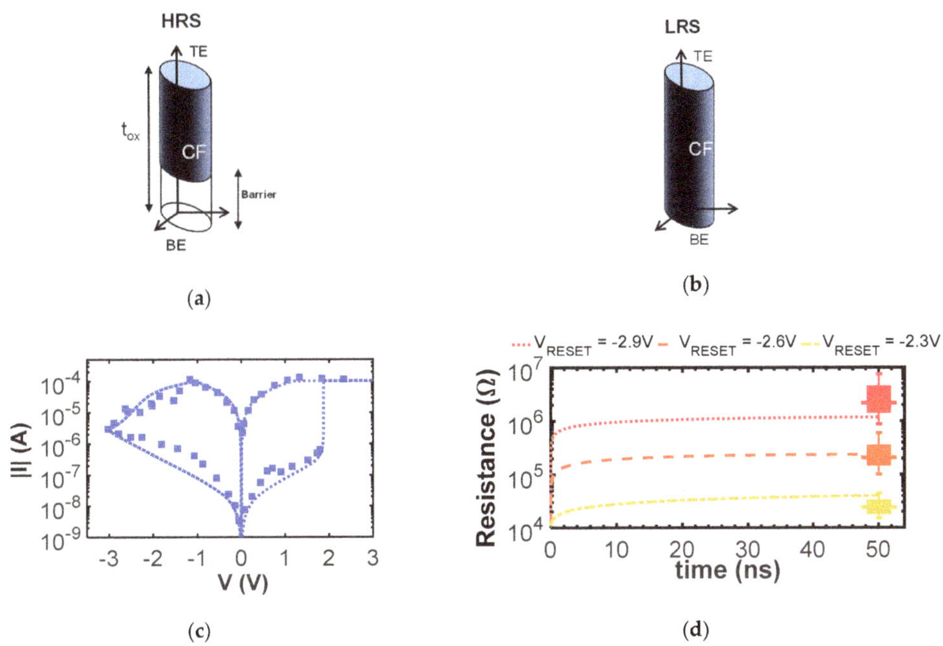

Figure 9. Sketch of an RRAM device in (**a**) high-resistive state (HRS) and in (**b**) low-resistive state (LRS) as represented in the compact model. (**c**) Experimental (square symbols) and simulated (dotted line) IV characteristic of the RRAM technology from [26]. (**d**) Experimental (boxes) and simulated (lines) response to 50 ns reset voltage pulses at different reset voltages (V_{RESET}) data from [26].

4.1.2. SIMPLY Simulations

The performance of the proposed architecture when operating as SIMPLY were estimated considering 1ns read and write pulses which result in a 4 ns execution time of a single IMPLY or FALSE operation. The R_G resistors were simulated using planar NMOS devices in a 45 nm technology [30] with a channel width of 250 nm and a channel length of 50 nm. Also, the energy contribution of the SA is considered by simulating the circuit shown in Figure 4 in the same 45 nm technology, which results in an average energy consumption of 8 fJ when operated with a 2 V V_{DD} over a temperature range from 0 °C to 85 °C, as reported in [20]. The device SET and RESET operations are achieved using 1 ns voltage pulses with amplitudes 3 V and –2.9 V, respectively. The SET and RESET amplitudes were determined using the physics-based compact model to ensure that a memory window larger than 10 is achieved also for very short voltage pulses, as discussed in [19]. The read margin (RM) and performance for both IMPLY and FALSE operations reported in Tables 2 and 3 were estimated including the effect of variability and RTN during the read operation by repeating the simulations (i.e., 50 trials) and reseeding the random sources. Additional information regarding SIMPLY circuit simulations and the list of variability and RTN model parameters are available in [20,50], respectively.

Table 2. Performance estimates of the IMPLY operation implemented on the SIMPLY architecture using $R_{G,opt}$.

Input Configuration		Energy [1] (min-avg-max)
0	0	139 – 429 – 509 fJ
0	1	6.18 – 6.183 – 6.185 fJ
1	0	6.18 – 6.183 – 6.185 fJ
1	1	6.184 – 6.184 – 6.185 fJ

[1] Device-to-device (D2D) and cycle-to-cycle (C2C) variability are included by repeating the circuit simulations with different seed for the random noise sources (50 trials).

Table 3. Performance estimates of the FALSE operation implemented on the SIMPLY architecture using $R_{G,opt}$.

Input Configuration	Energy [1] (min-avg-max)
0	9.6 – 11.2 – 12 fJ
1	100 – 145 – 190 fJ

[1] D2D and C2C variability are included by repeating the circuit simulations with different seed for the random noise sources (50 trials).

4.2. Implemented Neural Network

To benchmark the performance of the proposed architecture against the SIMPLY implementation, the same BNN from [19,42] was implemented. The network consists of a single hidden layer with 1000 neurons, and classify the digits 0–9 from the MNIST handwritten digits dataset [61]. The 20 × 20 pixels images are binarized to black and white images before training. The training was performed on 9500 images using the DoReFa-Net algorithm [32] considering one bit for weights and activations and 32 bits for the gradients. 2500 and 2000 images were used for validation and testing, respectively. The trained network achieves an accuracy of 91.4% [19].

4.3. BNN Performance Estimates

The trained network parameters were mapped to RRAM devices' resistance values including the effect of resistive state variability. The performance of the proposed architecture was estimated on an inference task by means of circuit simulation, where the AVMM is used to compute intermediate results of the MAC operations while SIMPLY is used to accumulate intermediate results and to compute each layer output activations. Intermediate MAC operations are needed to preserve the same NMOS size for implementing both R_{PD} and R_G by just adjusting V_{GS} (i.e., V_{GS} is 1.48 V and 2.9 V when operating the crossbar array as SIMPLY or BNN AVMM accelerator, respectively). With the considered RRAM technology, a maximum of 15 crossbar rows can be reliably read in parallel during the

AVMM. Thus, as an example, 27 (i.e., 400/15) computing steps are needed to compute all the intermediate MAC operations in the first layer. After each parallel read operation, the intermediate MAC results are stored in RRAM devices in the crossbar array and accumulated using the SIMPLY accumulator implementation shown in Figure 5e [20]. The results of the accumulations are compared with a threshold to produce each layer output activations using the SIMPLY comparator implementation from [19], which requires $9 \cdot m + \frac{m(m+1)}{2}$ where m is the number of compared bits. Finally, the output layer computes the predicted class using the hardmax SIMPLY implementation reported in [19] which accounts for 1457 computing steps on the proposed architecture and determines the predicted class as the class with the highest activation value. Thus, a total of 7902 computing steps are required for each inference, resulting in a 31.6 μs inference latency when 1 ns voltage pulses are used, as reported in Table 1. The worst-case energy for an inference task reported in Table 1 is estimated by running the neural network on the complete test set and considering only the worst-case energy for each IMPLY, SET, and FALSE operations for each specific input combination. The VSA energy contribution is included when performing both SIMPLY and BNN MAC operations. By considering the worst-case energy for each SIMPLY operation, which is the main contribution to the overall energy consumption, the energy assessments are indeed slightly overestimated, and roughly account for additional energy contributions possibly introduced by the peripheral circuitry. Nevertheless, even when increasing by 20% the energy consumption to account for the decoders and drivers considering the power breakdown reported by He at al. [21], the results (see Table 1) underline the remarkable energy efficiency in comparison with conventional embedded system implementations.

5. Conclusions

In this work, we proposed a novel in-memory computing architecture that enables the coexistence on the same crossbar array of the SIMPLY logic-in-memory computing approach and of the BNN AVMM. Design tradeoffs and requirements for circuit performance and reliability were analyzed in depth. The performance of the proposed architecture on an inference task were benchmarked against a pure SIMPLY implementation by means of circuit simulations enabled by a calibrated RRAM physics-based compact model. The results show that the proposed approach drastically improves the EDP by a factor $>10^3$, indicating that the proposed architecture is a viable solution for the realization of reconfigurable ultra-low-power hardware accelerators for edge computing applications.

Author Contributions: Conceptualization, T.Z. and F.M.P.; methodology, T.Z.; software, T.Z., F.M.P. and P.P.; validation, T.Z.; writing—original draft preparation, T.Z.; writing—review and editing, T.Z., F.M.P. and P.P.; supervision, F.M.P. and P.P. All authors have read and agreed to the published version of the manuscript.

Funding: This research received no external funding.

Data Availability Statement: The data presented in this study are available within the article.

Conflicts of Interest: The authors declare no conflict of interests.

References

1. Zhang, W.; Gao, B.; Tang, J.; Yao, P.; Yu, S.; Chang, M.-F.; Yoo, H.-J.; Qian, H.; Wu, H. Neuro-Inspired Computing Chips. *Nat. Electron.* **2020**, *3*, 371–382. [CrossRef]
2. Deng, S.; Zhao, H.; Fang, W.; Yin, J.; Dustdar, S.; Zomaya, A.Y. Edge Intelligence: The Confluence of Edge Computing and Artificial Intelligence. *IEEE Internet Things J.* **2020**, *7*, 7457–7469. [CrossRef]
3. Pedretti, G.; Ielmini, D. In-Memory Computing with Resistive Memory Circuits: Status and Outlook. *Electronics* **2021**, *10*, 1063. [CrossRef]
4. Kvatinsky, S.; Belousov, D.; Liman, S.; Satat, G.; Wald, N.; Friedman, E.G.; Kolodny, A.; Weiser, U.C. MAGIC—Memristor-Aided Logic. *IEEE Trans. Circuits Syst. II: Express Briefs* **2014**, *61*, 895–899. [CrossRef]
5. Ziegler, T.; Waser, R.; Wouters, D.J.; Menzel, S. In-Memory Binary Vector–Matrix Multiplication Based on Complementary Resistive Switches. *Adv. Intell. Syst.* **2020**, *2*, 2000134. [CrossRef]

6. Kingra, S.K.; Parmar, V.; Chang, C.-C.; Hudec, B.; Hou, T.-H.; Suri, M. SLIM: Simultaneous Logic-in-Memory Computing Exploiting Bilayer Analog OxRAM Devices. *Sci. Rep.* **2020**, *10*. [CrossRef]
7. Pei, J.; Deng, L.; Song, S.; Zhao, M.; Zhang, Y.; Wu, S.; Wang, G.; Zou, Z.; Wu, Z.; He, W.; et al. Towards Artificial General Intelligence with Hybrid Tianjic Chip Architecture. *Nature* **2019**, *572*, 106–111. [CrossRef] [PubMed]
8. Xiao, T.P.; Bennett, C.H.; Feinberg, B.; Agarwal, S.; Marinella, M.J. Analog Architectures for Neural Network Acceleration Based on Non-Volatile Memory. *Appl. Phys. Rev.* **2020**, *7*, 031301. [CrossRef]
9. Saxena, V. Neuromorphic Computing: From Devices to Integrated Circuits. *J. Vac. Sci. Technol. B* **2021**, *39*, 010801. [CrossRef]
10. Berggren, K.; Xia, Q.; Likharev, K.K.; Strukov, D.B.; Jiang, H.; Mikolajick, T.; Querlioz, D.; Salinga, M.; Erickson, J.R.; Pi, S.; et al. Roadmap on Emerging Hardware and Technology for Machine Learning. *Nanotechnology* **2020**, *32*, 012002. [CrossRef]
11. Benoit, P.; Dalmasso, L.; Patrigeon, G.; Gil, T.; Bruguier, F.; Torres, L. Edge-Computing Perspectives with Reconfigurable Hardware. In Proceedings of the 2019 14th International Symposium on Reconfigurable Communication-centric Systems-on-Chip (ReCoSoC); York, UK, 1–3 July 2019; pp. 51–58.
12. Yu, J.; Du Nguyen, H.A.; Abu Lebdeh, M.; Taouil, M.; Hamdioui, S. Enhanced Scouting Logic: A Robust Memristive Logic Design Scheme. In Proceedings of the 2019 IEEE/ACM International Symposium on Nanoscale Architectures (NANOARCH), Qingdao, China, 17−19 July 2019; pp. 1–6.
13. Borghetti, J.; Snider, G.S.; Kuekes, P.J.; Yang, J.J.; Stewart, D.R.; Williams, R.S. 'Memristive' Switches Enable 'Stateful' Logic Operations via Material Implication. *Nature* **2010**, *464*, 873–876. [CrossRef]
14. Siemon, A.; Menzel, S.; Waser, R.; Linn, E. A Complementary Resistive Switch-Based Crossbar Array Adder. *IEEE J. Emerg. Sel. Top. Circuits Syst.* **2015**, *5*, 64–74. [CrossRef]
15. Siemon, A.; Drabinski, R.; Schultis, M.J.; Hu, X.; Linn, E.; Heittmann, A.; Waser, R.; Querlioz, D.; Menzel, S.; Friedman, J.S. Stateful Three-Input Logic with Memristive Switches. *Sci. Rep.* **2019**, *9*, 1–13. [CrossRef] [PubMed]
16. Hu, S.-Y.; Li, Y.; Cheng, L.; Wang, Z.-R.; Chang, T.-C.; Sze, S.M.; Miao, X. Reconfigurable Boolean Logic in Memristive Crossbar: The Principle and Implementation. *IEEE Electron Device Lett.* **2019**, *40*, 200–203. [CrossRef]
17. Puglisi, F.M.; Zanotti, T.; Pavan, P. SIMPLY: Design of a RRAM-Based Smart Logic-in-Memory Architecture Using RRAM Compact Model. In Proceedings of the ESSDERC 2019—49th European Solid-State Device Research Conference (ESSDERC), Krakow, Poland, 23−26 September 2019; pp. 130–133.
18. Courbariaux, M.; Hubara, I.; Soudry, D.; El-Yaniv, R.; Bengio, Y. Binarized Neural Networks: Training Deep Neural Networks with Weights and Activations Constrained To+ 1 or-1. *arXiv* **2016**, arXiv:1602.02830.
19. Zanotti, T.; Puglisi, F.M.; Pavan, P. Reliability and Performance Analysis of Logic-in-Memory Based Binarized Neural Networks. *IEEE Trans. Device Mater. Reliab.* **2021**, 1. [CrossRef]
20. Zanotti, T.; Puglisi, F.M.; Pavan, P. Reconfigurable Smart In-Memory Computing Platform Supporting Logic and Binarized Neural Networks for Low-Power Edge Devices. *IEEE J. Emerg. Sel. Top. Circuits Syst.* **2020**, 1. [CrossRef]
21. He, W.; Yin, S.; Kim, Y.; Sun, X.; Kim, J.-J.; Yu, S.; Seo, J.-S. 2-Bit-Per-Cell RRAM-Based In-Memory Computing for Area-/Energy-Efficient Deep Learning. *IEEE Solid State Circuits Lett.* **2020**, *3*, 194–197. [CrossRef]
22. Sun, X.; Peng, X.; Chen, P.; Liu, R.; Seo, J.; Yu, S. Fully Parallel RRAM Synaptic Array for Implementing Binary Neural Network with (+1, −1) Weights and (+1, 0) Neurons. In Proceedings of the 2018 23rd Asia and South Pacific Design Automation Conference (ASP-DAC), Jeju, Korea, 22−25 January 2018; pp. 574–579.
23. Vieira, J.; Giacomin, E.; Qureshi, Y.; Zapater, M.; Tang, X.; Kvatinsky, S.; Atienza, D.; Gaillardon, P.-E. A Product Engine for Energy-Efficient Execution of Binary Neural Networks Using Resistive Memories. In Proceedings of the 2019 IFIP/IEEE 27th International Conference on Very Large Scale Integration (VLSI-SoC), Cuzco, Peru, 6−9 October 2019; pp. 160–165.
24. Yi, W.; Kim, Y.; Kim, J.-J. Effect of Device Variation on Mapping Binary Neural Network to Memristor Crossbar Array. In Proceedings of the 2019 Design, Automation Test in Europe Conference Exhibition (DATE), Florence, Italy, 25−29 March 2019; pp. 320–323.
25. Qin, Y.-F.; Kuang, R.; Huang, X.-D.; Li, Y.; Chen, J.; Miao, X.-S. Design of High Robustness BNN Inference Accelerator Based on Binary Memristors. *IEEE Trans. Electron Devices* **2020**, *67*, 3435–3441. [CrossRef]
26. Yu, S.; Wu, Y.; Chai, Y.; Provine, J.; Wong, H.-S.P. Characterization of Switching Parameters and Multilevel Capability in HfOx/AlOx Bi-Layer RRAM Devices. In Proceedings of the 2011 International Symposium on VLSI Technology, Systems and Applications, Hsinchu, Taiwan, 25–27 April 2011; pp. 1–2.
27. Lehtonen, E.; Poikonen, J.H.; Laiho, M. Two Memristors Suffice to Compute All Boolean Functions. *Electron. Lett.* **2010**, *46*, 239–240. [CrossRef]
28. Zanotti, T.; Puglisi, F.M.; Pavan, P. Reliability-Aware Design Strategies for Stateful Logic-in-Memory Architectures. *IEEE Trans. Device Mater. Reliab.* **2020**, *20*, 278–285. [CrossRef]
29. Kvatinsky, S.; Satat, G.; Wald, N.; Friedman, E.G.; Kolodny, A.; Weiser, U.C. Memristor-Based Material Implication (IMPLY) Logic: Design Principles and Methodologies. *IEEE Trans. Very Large Scale Integr. (VLSI) Syst.* **2014**, *22*, 2054–2066. [CrossRef]
30. Stine, J.E.; Castellanos, I.; Wood, M.; Henson, J.; Love, F.; Davis, W.R.; Franzon, P.D.; Bucher, M.; Basavarajaiah, S.; Oh, J.; et al. FreePDK: An Open-Source Variation-Aware Design Kit. In Proceedings of the 2007 IEEE International Conference on Microelectronic Systems Education (MSE'07), San Diego, CA, USA, 3−4 June 2007; pp. 173–174.
31. Zanotti, T.; Zambelli, C.; Puglisi, F.M.; Milo, V.; Pérez, E.; Mahadevaiah, M.K.; Ossorio, O.G.; Wenger, C.; Pavan, P.; Olivo, P.; et al. Reliability of Logic-in-Memory Circuits in Resistive Memory Arrays. *IEEE Trans. Electron Devices* **2020**, *67*, 4611–4615. [CrossRef]

32. Zhou, S.; Wu, Y.; Ni, Z.; Zhou, X.; Wen, H.; Zou, Y. DoReFa-Net: Training Low Bitwidth Convolutional Neural Networks with Low Bitwidth Gradients. *arXiv* **2018**, arXiv:1606.06160.
33. Krestinskaya, O.; Otaniyozov, O.; James, A.P. Binarized Neural Network with Stochastic Memristors. In Proceedings of the 2019 IEEE International Conference on Artificial Intelligence Circuits and Systems (AICAS), Hsinchu, Taiwan, 18–20 March 2019; pp. 274–275.
34. Chen, W.-H.; Dou, C.; Li, K.-X.; Lin, W.-Y.; Li, P.-Y.; Huang, J.-H.; Wang, J.-H.; Wei, W.-C.; Xue, C.-X.; Chiu, Y.-C.; et al. CMOS-Integrated Memristive Non-Volatile Computing-in-Memory for AI Edge Processors. *Nat. Electron* **2019**, *2*, 420–428. [CrossRef]
35. Wan, W.; Kubendran, R.; Gao, B.; Joshi, S.; Raina, P.; Wu, H.; Cauwenberghs, G.; Wong, H.S.P. A Voltage-Mode Sensing Scheme with Differential-Row Weight Mapping for Energy-Efficient RRAM-Based In-Memory Computing. In Proceedings of the 2020 IEEE Symposium on VLSI Technology, Honolulu, HI, USA, 16–19 June 2020; pp. 1–2.
36. Yin, S.; Kim, Y.; Han, X.; Barnaby, H.; Yu, S.; Luo, Y.; He, W.; Sun, X.; Kim, J.-J.; Seo, J. Monolithically Integrated RRAM- and CMOS-Based In-Memory Computing Optimizations for Efficient Deep Learning. *IEEE Micro.* **2019**, *39*, 54–63. [CrossRef]
37. Grossi, A.; Nowak, E.; Zambelli, C.; Pellissier, C.; Bernasconi, S.; Cibrario, G.; El Hajjam, K.; Crochemore, R.; Nodin, J.F.; Olivo, P.; et al. Fundamental Variability Limits of Filament-Based RRAM. In Proceedings of the 2016 IEEE International Electron Devices Meeting (IEDM), San Francisco, CA, USA, 3–7 December 2016.
38. Mahmoodi, M.R.; Vincent, A.F.; Nili, H.; Strukov, D.B. Intrinsic Bounds for Computing Precision in Memristor-Based Vector-by-Matrix Multipliers. *IEEE Trans. Nanotechnol.* **2020**, *19*, 429–435. [CrossRef]
39. Xia, Q.; Yang, J.J. Memristive Crossbar Arrays for Brain-Inspired Computing. *Nat. Mater.* **2019**, *18*, 309–323. [CrossRef]
40. Yu, M.; Cai, Y.; Wang, Z.; Fang, Y.; Liu, Y.; Yu, Z.; Pan, Y.; Zhang, Z.; Tan, J.; Yang, X.; et al. Novel Vertical 3D Structure of TaO_x-Based RRAM with Self-Localized Switching Region by Sidewall Electrode Oxidation. *Sci. Rep.* **2016**, *6*, 21020. [CrossRef]
41. Fouda, M.E.; Eltawil, A.M.; Kurdahi, F. Modeling and Analysis of Passive Switching Crossbar Arrays. *IEEE Trans. Circuits Syst. I: Regul. Pap.* **2018**, *65*, 270–282. [CrossRef]
42. McDanel, B.; Teerapittayanon, S.; Kung, H.T. Embedded Binarized Neural Networks. In Proceedings of the 2017 International Conference on Embedded Wireless Systems and Networks, Uppsala, Sweden, 20–22 February 2017; pp. 168–173.
43. Kim, C.-H.; Lim, S.; Woo, S.Y.; Kang, W.-M.; Seo, Y.-T.; Lee, S.-T.; Lee, S.; Kwon, D.; Oh, S.; Noh, Y.; et al. Emerging Memory Technologies for Neuromorphic Computing. *Nanotechnology* **2019**, *30*, 032001. [CrossRef]
44. Gao, S.; Chen, B.; Qu, Y.; Zhao, Y. MRAM Acceleration Core for Vector Matrix Multiplication and XNOR-Binarized Neural Network Inference. In Proceedings of the 2020 International Symposium on VLSI Technology, Systems and Applications (VLSI-TSA), Hsinchu, Taiwan, 10–13 August 2020; pp. 153–154.
45. Slesazeck, S.; Mikolajick, T. Nanoscale Resistive Switching Memory Devices: A Review. *Nanotechnology* **2019**, *30*, 352003. [CrossRef] [PubMed]
46. Ielmini, D.; Wong, H.-S.P. In-Memory Computing with Resistive Switching Devices. *Nat. Electron.* **2018**, *1*, 333–343. [CrossRef]
47. Chen, A. A Review of Emerging Non-Volatile Memory (NVM) Technologies and Applications. *Solid State Electron.* **2016**, *125*, 25–38. [CrossRef]
48. Nail, C.; Molas, G.; Blaise, P.; Piccolboni, G.; Sklenard, B.; Cagli, C.; Bernard, M.; Roule, A.; Azzaz, M.; Vianello, E.; et al. Understanding RRAM Endurance, Retention and Window Margin Trade-off Using Experimental Results and Simulations. In Proceedings of the 2016 IEEE International Electron Devices Meeting (IEDM), San Francisco, CA, USA, 3–7 December 2016.
49. Shi, L.; Zheng, G.; Tian, B.; Dkhil, B.; Duan, C. Research Progress on Solutions to the Sneak Path Issue in Memristor Crossbar Arrays. *Nanoscale Adv.* **2020**, *2*, 1811–1827. [CrossRef]
50. Puglisi, F.M.; Zanotti, T.; Pavan, P. Unimore Resistive Random Access Memory (RRAM) Verilog-A Model. *nanoHUB* **2019**. [CrossRef]
51. Yakopcic, C.; Taha, T.M.; Subramanyam, G.; Pino, R.E.; Rogers, S. A Memristor Device Model. *IEEE Electron Device Lett.* **2011**, *32*, 1436–1438. [CrossRef]
52. Kvatinsky, S.; Friedman, E.G.; Kolodny, A.; Weiser, U.C. TEAM: ThrEshold Adaptive Memristor Model. *IEEE Trans. Circuits Syst. I: Regul. Pap.* **2013**, *60*, 211–221. [CrossRef]
53. Kvatinsky, S.; Ramadan, M.; Friedman, E.G.; Kolodny, A. VTEAM: A General Model for Voltage-Controlled Memristors. *IEEE Trans. Circuits Syst. II: Express Briefs* **2015**, *62*, 786–790. [CrossRef]
54. Messaris, I.; Serb, A.; Stathopoulos, S.; Khiat, A.; Nikolaidis, S.; Prodromakis, T. A Data-Driven Verilog-A ReRAM Model. *IEEE Trans. Comput-Aided Des. Integr. Circuits Syst.* **2018**, *37*, 3151–3162. [CrossRef]
55. La Torre, C.; Zurhelle, A.F.; Breuer, T.; Waser, R.; Menzel, S. Compact Modeling of Complementary Switching in Oxide-Based ReRAM Devices. *IEEE Trans. Electron Devices* **2019**, *66*, 1268–1275. [CrossRef]
56. Wiefels, S.; Bengel, C.; Kopperberg, N.; Zhang, K.; Waser, R.; Menzel, S. HRS Instability in Oxide-Based Bipolar Resistive Switching Cells. *IEEE Trans. Electron Devices* **2020**, *67*, 4208–4215. [CrossRef]
57. González-Cordero, G.; González, M.B.; Campabadal, F.; Jiménez-Molinos, F.; Roldán, J.B. A Physically Based SPICE Model for RRAMs Including RTN. In Proceedings of the 2020 XXXV Conference on Design of Circuits and Integrated Systems (DCIS), Segovia, Spain, 18–20 November 2020; pp. 1–6.

58. Yu, S.; Gao, B.; Fang, Z.; Yu, H.; Kang, J.; Wong, H.-P. A Neuromorphic Visual System Using RRAM Synaptic Devices with Sub-PJ Energy and Tolerance to Variability: Experimental Characterization and Large-Scale Modeling. In Proceedings of the 2012 International Electron Devices Meeting, San Francisco, CA, USA, 10–13 December 2012.
59. Jiang, Z.; Yu, S.; Wu, Y.; Engel, J.H.; Guan, X.; Wong, H.-P. Verilog-A Compact Model for Oxide-Based Resistive Random Access Memory (RRAM). In Proceedings of the 2014 International Conference on Simulation of Semiconductor Processes and Devices (SISPAD), Yokohama, Japan, 9–11 September 2014; pp. 41–44.
60. Li, H.; Jiang, Z.; Huang, P.; Wu, Y.; Chen, H.-; Gao, B.; Liu, X.Y.; Kang, J.F.; Wong, H.-P. Variation-Aware, Reliability-Emphasized Design and Optimization of RRAM Using SPICE Model. In Proceedings of the 2015 Design, Automation Test in Europe Conference Exhibition (DATE), Grenoble, France, 9–13 March 2015; pp. 1425–1430.
61. Lecun, Y.; Bottou, L.; Bengio, Y.; Haffner, P. Gradient-Based Learning Applied to Document Recognition. *Proc. IEEE* **1998**, *86*, 2278–2324. [CrossRef]

Journal of
Low Power Electronics and Applications

MDPI

Article

Minimization of the Line Resistance Impact on Memdiode-Based Simulations of Multilayer Perceptron Arrays Applied to Pattern Recognition

Fernando Leonel Aguirre [1,2,3,*], Nicolás M. Gomez [4], Sebastián Matías Pazos [1,2], Félix Palumbo [1,2], Jordi Suñé [3] and Enrique Miranda [3,*]

1 Unidad de Investigación y Desarrollo de las Ingenierías (UIDI), Facultad Regional Buenos Aires, Universidad Tecnológica Nacional (UTN-FRBA), Buenos Aires C1179AAQ, Argentina; spazos@frba.utn.edu.ar (S.M.P.); felix.palumbo@conicet.gov.ar (F.P.)
2 Consejo Nacional de Investigaciones Científicas y Técnicas (CONICET), Buenos Aires C1425FQB, Argentina
3 Departament d'Enginyeria Electrònica, Universitat Autònoma de Barcelona (UAB), 08193 Cerdanyola del Vallès, Spain; jordi.sune@uab.cat
4 Departamento de Ingeniería Electrónica, Facultad Regional Buenos Aires, Universidad Tecnológica Nacional (UTN-FRBA), Buenos Aires C1179AAQ, Argentina; nigomez@est.frba.utn.edu.ar
* Correspondence: aguirref@ieee.org (F.L.A.); enrique.miranda@uab.cat (E.M.)

Abstract: In this paper, we extend the application of the Quasi-Static Memdiode model to the realistic SPICE simulation of memristor-based single (SLPs) and multilayer perceptrons (MLPs) intended for large dataset pattern recognition. By considering ex-situ training and the classification of the handwritten characters of the MNIST database, we evaluate the degradation of the inference accuracy due to the interconnection resistances for MLPs involving up to three hidden neural layers. Two approaches to reduce the impact of the line resistance are considered and implemented in our simulations, they are the inclusion of an iterative calibration algorithm and the partitioning of the synaptic layers into smaller blocks. The obtained results indicate that MLPs are more sensitive to the line resistance effect than SLPs and that partitioning is the most effective way to minimize the impact of high line resistance values.

Keywords: RRAM; resistive-switching; cross-point; memory; memristor; neuromorphic; pattern recognition; multilayer perceptron

Citation: Aguirre, F.L.; Gomez, N.M.; Pazos, S.M.; Palumbo, F.; Suñé, J.; Miranda, E. Minimization of the Line Resistance Impact on Memdiode-Based Simulations of Multilayer Perceptron Arrays Applied to Pattern Recognition. *J. Low Power Electron. Appl.* **2021**, *11*, 9. https://doi.org/10.3390/jlpea11010009

Academic Editor: Alex Serb

Received: 4 January 2021
Accepted: 2 February 2021
Published: 5 February 2021

1. Introduction

In-memory-computation [1] has been recently proposed as an alternative approach to overcome the inherent bottleneck that limits the performance improvement of traditional Von-Neuman architectures, while also allowing significant energy saving. The key elements to enable the further maturing of this technology are the memory cells, which are required to be nonvolatile (*nonvolatile memory*, NVM) and to operate at low power [2]. Resistive memories (RRAM) [1] were found to meet these requirements as well as allowing dense memory integration (up to $4F^2$, F being the feature size of the technology node [3]) via architectures such as memristor cross-bar arrays (MCA see Figure 1a). In particular, MCAs are of great interest for the development of hardware-based deep neural networks (DNN, Figure 1b) as they are suitable for implementing the matrix-vector-multiplication (MVM) method necessary to perform operations and propagate signals through the neural layers [2] with reduced power consumption. Such applications have been extensively studied in previous works [4–8] considering various MCA architectures as well as different memristor models. For instance, Li et al. reported in [9] the case of character classification using an MCA-based multilayer perceptron (MLP) of 64 × 54 × 10 neurons with a single layer of hidden neurons.

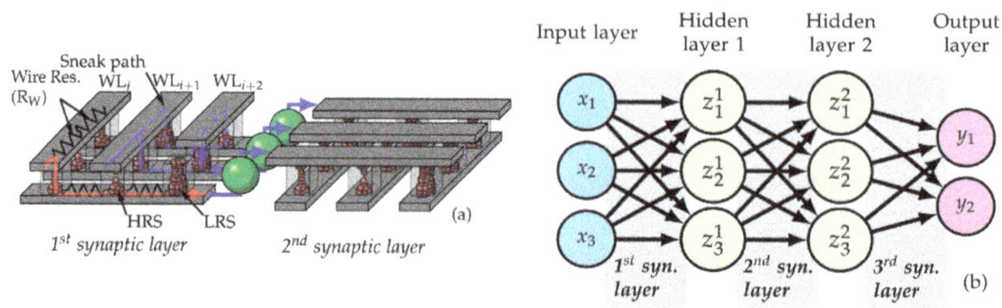

Figure 1. (**a**) Sketch of the MCA structure. Red and blue arrows show currents from the top (word lines—WL) to the bottom lines (bit lines—BL). Different resistance states are represented (high (HRS) to low (LRS) resistance states). The dashed blue line depicts the sneak path problem. The parasitic R_L is indicated for WL_i and BL_i. Two MCAs are depicted, representing two layers of synapses. (**b**) Sketch of a DNN with two hidden layers.

However, despite these promising studies, the development of in-memory computation is still hindered by the many practical limitations faced by MCAs, such as the line or wire resistances (R_L), the limited resistance window of the devices (R_{ON} and R_{OFF}) as well as the inherent features associated with the integration of memristors in an MCA such as the so-called sneakpath problem (see Figure 1a). While the former are mainly a consequence of the R_L increase as the fabrication technology node scales down [8,10] and which in combination with a reduced resistance window or low R_{ON} causes a significant voltage drop across the MCA lines, the latter refers to the non-negligible current flowing through the unselected devices. This causes errors in the read and write processes [10]. Although hardware-based techniques were proposed to address these challenges, they are in general both time, power and cost demanding [9]. Instead, software solutions [4–8,10–14] allow a more systematic study and thus different approaches have been proposed. Among them, SPICE simulation appears to be the most suitable approach as it allows studying the full system, i.e., the MCA and the control electronics necessary to operate the network. However, this approach is normally constrained to the limitations of the memristor model considered and to the size of the memristor-based MLP given the high computational requirements [15,16].

In this regard, the results presented by Aguirre et al. in [17] represent a step forward in the realistic circuital modeling of MCA-based single-layer perceptrons (SLP) involving thousands of devices intended for the classifications of large pattern datasets. A key element in that study is the Quasi-Static Memdiode Model (QMM), a memristor model originally proposed by Miranda in [18,19], that provides high simulation accuracy at reduced computational cost. The closed-form expression for the transport equation, i.e., the current-voltage (*I-V*) curve (continuous and differentiable) and the recursive nature of the state variable computation, makes the QMM suitable for dealing with arbitrary input signals (continuous or discontinuous, differentiable or nondifferentiable). This is a significant advantage when compared to other widely explored memristor models such as the general phenomenological models (Yakopcic [20], TEAM [21], VTEAM [22], Eshraghian [23], etc.) that although capable of successfully fitting experimental data, rely on various internal equations or artificial window functions (commonly used for modeling the SET/RESET transitions) in the memory equation (ME, a first order differential equation that links the current flowing through or the voltage applied to the structure with its internal memory state) that can seriously affect the model's convergence [24,25]. Nevertheless, the extension of the results obtained for the SLP test structures to more practical implementations such as MLPs considering the aforementioned line parasitics is still to be addressed.

It is worth pointing out that other memristor device nonidealities threaten the performance of MCA DNNs and are currently the focus of intense research: nonlinearity in the *I-V* characteristics [26], retention failures [27–29], nonuniformity [17,30], Device-to-Device (D2D) and Cycle-to-Cycle (C2C) variability are some of the most representative challenges.

However, nonlinearity factors well below 10 have been obtained by optimizing the device fabrication process [31,32] and have also been addressed through specific training [33,34] and voltage mapping [26] methodologies. Additionally, the use of devices with a higher R_{OFF}/R_{ON} ratio has been shown to reduce the impact of the D2D variability [17]. Moreover, for the specific case of on-line training, nonlinear weight update [35–37] is another relevant source of inaccuracy. In this regard, it has been shown that activation function engineering and threshold weight update schemes effectively suppress training noise [36]. Particularly, the write–verify approach, as the one described in [17,38], allows to mitigate the impact of this effect while also providing robustness against C2C and D2D variability [39]. Line resistance (R_L) is another nonideal factor that worsens as the technology scales down [8,10]. Therefore, the realistic simulation and optimization of DNNs considering the impact of line resistance is of utmost importance to enable robust implementation of neuromorphic circuits independently of the technology node and RRAM device optimization.

In this paper, we demonstrate the applicability of the QMM to SPICE simulations of MCA-based MLPs and evaluate the inference accuracy degradation as a function of R_L. Ex-situ training is considered and the classification of the grayscale images from the MNIST dataset [40] is assumed for benchmarking purposes. The simulation workflow presented in [17] was modified so as to account for multiple synaptic layers and hidden neural layers. In order to minimize the impact of R_L, two approaches were evaluated, they are the divisions of each synaptic layer into smaller partitions and the inclusion of a calibration procedure that compensates the effects associated with R_L. The rest of this paper is organized as follows: the fundamentals (*I-V* and ME characteristics) of the QMM are presented in Section 2. Section 3 explains the MCA-based MLP training and simulation procedures, including the MCA partitioning and R_L-dependent calibration. In Section 4, the obtained simulation results in terms of the aforementioned features are discussed. Finally, in Section 5, the general conclusions of this paper are presented. To the best of the authors knowledge, the study of MLPs including the parasitic effects by means of SPICE simulations and considering a realistic memristor model has not been published before.

2. Quasi-Static Memdiode Model

The resistive switching (RS) mechanism is the fundamental phenomenon behind RRAM devices. In the particular cases of CBRAMs and OxRAMs, RS relies on the displacement of metal ions/oxygen vacancies within the dielectric film in a metal–insulator–metal (MIM) structure originated from the application of an external electrical stimulus, current or voltage [41–44]. Such migration of ions causes the alternate completion and destruction of a conductive filament (CF) spanning across the insulating film. For a ruptured CF, the device is in the high resistance state (HRS), often characterized by an exponential *I-V* relationship, while the completion of the CF leads to the low resistance state (LRS), which often exhibits a linear *I-V* curve [45,46]. In between these two extreme situations, the modulation of the CF transport properties renders intermediate states by voltage-controlled redox reactions. From the modeling viewpoint, the compact model originally proposed by Miranda in [18] and later extended by Patterson et al. in [19] is able to describe the major and minor *I-V* loops and the gradual transitions in bipolar resistive switches. This is accomplished, as shown in the inset of Figure 2a, by considering a nonlinear transport equation based on two identical opposite-biased diodes in series with a resistor. The *I-V* relationship resembles a diode with memory and that is why this device was termed memdiode. Notice that the antiparallel connected diodes allow the bidirectional current flow through the memdiode device, as for both positive and negative polarities there will be a forward biased diode. For the sake of completeness, the QMM is succinctly reviewed in the next paragraphs.

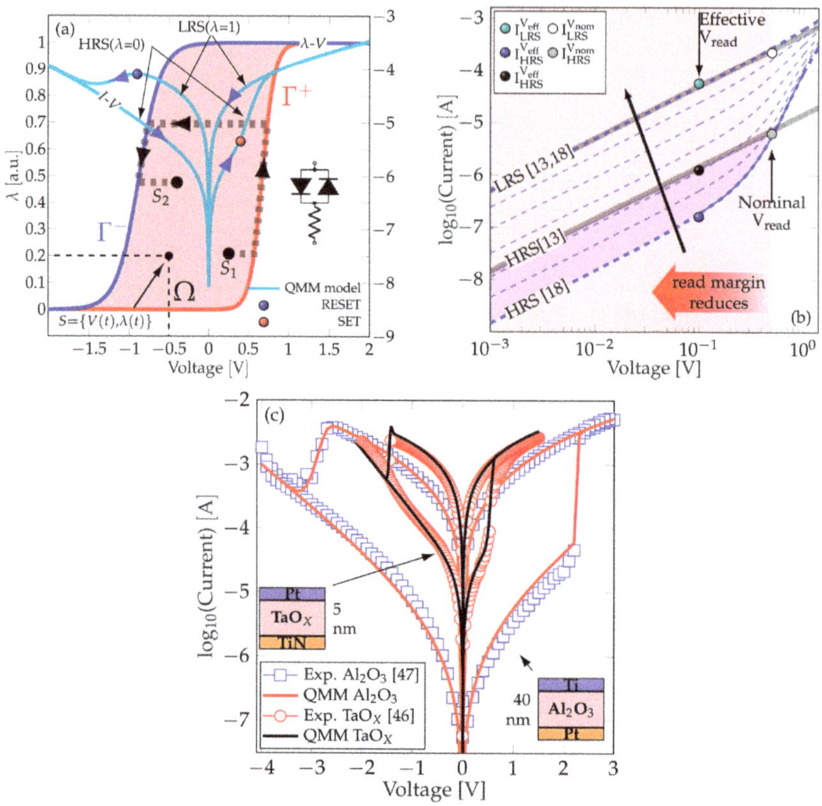

Figure 2. (**a**) Hysteron model with logistic ridge functions Γ^+ (Equation (3)) and Γ^- (Equation (4)). Ω is the space of feasible states S. The red thick faded line superimposed to the hysteron model indicates the trajectory of the state variable λ inside Ω from an initial S_1 to a final S_2 state. The inset in the right shows the equivalent circuit model for the current equation (Equation (1)) including the series resistance. The diodes are driven by the memory state of the device and one diode is activated at a time. Typical *I-V* characteristic for a memdiode obtained via simulation of the proposed model are superimposed. Current evolution is indicated by the blue arrows. (**b**) *I-V* characteristics of the memdiode showing the exponential (HRS) to lineal (LRS) transition by varying λ. The red shaded region indicates the possible voltages applied to the device as the read margin reduces. I_{HRS} and I_{LRS} currents are pinpointed at nominal V_{read} with the grey and white circle markers, respectively. Overestimation of I_{HRS} may occur when considering a linear model for the HRS regime and lower effective V_{read} voltages as indicated by the cyan, blue and black ball markers. (**c**) Experimental *I-V* loops of different materials reported in the literature fitted with the QMM model: Al_2O_3 [47] and TaO_X [46].

Physically, the memdiode is associated with a potential barrier that controls the electron flow in the CF. The conduction properties of this nonlinear device change according to the variation of this barrier. Due to the uncertainty in the area of the CF, instead of the potential barrier height, the diode current amplitude is used as the reference variable. Following Chua's memristive device theory, the proposed model comprises two equations, one for the electron transport and a second equation for the memory state of the device (ME), which is controlled by a hysteresis operator. The equation for the *I-V* characteristic of a memdiode is given by the expression:

$$I = sgn(V)\left\{\frac{W\left(\alpha R I_0(\lambda) e^{\alpha(abs(V)+RI_0(\lambda))}\right)}{\alpha R} - I_0(\lambda)\right\} \quad (1)$$

where $I_0(\lambda) = I_{min}(1-\lambda) + I_{max}\lambda$ is the diode current amplitude, α a fitting constant, and R a series resistance. Equation (1) is the solution of a diode with series resistance and W is the Lambert function. I_{min} and I_{max} are the minimum and maximum values of the current amplitude, respectively. $abs(V)$ is the absolute value of the applied bias and $sgn()$ the sign function. As I_0 increases in Equation (1), the *I-V* curve changes its shape from exponential to linear through a continuum of states as experimentally observed for this kind of device. λ is a control parameter that runs between 0 (HRS) and 1 (LRS) and is given by the recursive operator (Equation (2)):

$$\lambda(V) = min\left\{\Gamma^-(V), max\left[\lambda\left(\overleftarrow{V}\right), \Gamma^+(V)\right]\right\} \tag{2}$$

where $min()$ and $max()$ are the minimum and maximum functions, respectively, and $\overleftarrow{V}$ is the voltage a timestep before V. The positive and negative ridge functions in Equation (2), $\Gamma^+(V)$ and $\Gamma^-(V)$ represent the transitions from HRS to LRS (SET) and vice versa (RESET) and can be physically linked to the completion and destruction of the CF [45,46], respectively. They are defined by Equations (3) and (4)

$$\Gamma^+(V) = \left\{1 + e^{-\eta^+(V-V^+)}\right\}^{-1} \tag{3}$$

$$\Gamma^-(V) = \left\{1 + e^{-\eta^-(V-V^-)}\right\}^{-1} \tag{4}$$

where η^+ and η^- are the transition rates and V^+ and V^- the threshold voltages for SET and RESET, respectively. $\lambda(V)$ defines the so-called logistic hysteron or memory map of the device and keeps track of the history of the device as a function of the applied voltage (see Figure 2a). λ calculated from Equation (2) yields the transition from HRS to LRS and vice versa through a change in the properties of the diodes depicted in the inset of Figure 2a. The combination of Equations (1) and (2) results in a *I-V* loop such as that superimposed to the hysteron loop in Figure 2a, which starts in HRS ($\lambda = 0$) and evolves as indicated by the blue arrows. The name quasi-static comes from the fact that the characteristic time of the ions/vacancies responsible of the switching phenomenon is assumed to be infinite for a state within the hysteron structure. This implies that for a state located inside the hysteron loop no changes occur in the conduction characteristics, unless it reaches the ridge functions $\Gamma^+(V)$ or $\Gamma^-(V)$. The QMM can be transformed into a dynamic model by incorporating the time module described in [19].

Figure 2b shows the HRS (exponential) to LRS (linear) transition, altogether with some intermediate states (solid blue lines). Note that the memdiode model can successfully describe both HRS and LRS curves by solely changing a single parameter in the transport equation. As λ is swept from 10^{-7} to 1, I_0 in Equation (1) varies between I_{min} and I_{max}, causing the *I-V* curve to gradually change its shape from linear-exponential (HRS regime) to linear (LRS regime). This is a consequence of the potential drop in the series resistance which linearizes the transport equation. In a neuromorphic application such as the one discussed in this paper, the intermediate conductance states are achieved by means of a Write–Verify iterative loop approach. In such method, pulses of incremental amplitude are applied to the devices (Write) until the required conductance is reached (Verify) [17,38]. If the target conductance is exceeded, then increasing pulses with the opposite polarity are applied in a similar fashion to gradually reduce the conductance value (within an error margin). This writing methodology implies a transition as the one depicted in Figure 2a by the red-thick faded line, where the incremental pulses cause the system to evolve from the initial state S_1 up to the final state S_2 following Γ^+. If the conductance target is exceeded, then the system moves down along Γ^- by the application of voltage pulses with the appropriate polarity. Another relevant feature of the proposed model is that it can be described by a simple SPICE script as shown in [17]. Finally, the accuracy of the model is reported in Figure 2c by fitting experimental data extracted from different published works. In particular, results obtained for Al_2O_3 [47] and TaO_x [46] structures at room temperature under DC voltage sweeps are presented. In summary, the proposed QMM not only provides a simple SPICE-compatible implementation for the resistive memory devices but also a versatile one, as it can accurately fit the major and minor *I-V* loops measured in a wide variety of RRAM devices

3. MCA-Based MLP Modeling and R_L Calibration

Based on the procedure previously reported in [17] to create and simulate realistic circuital MCA-based SLPs intended for large dataset pattern recognition tasks, a novel procedure is derived here to account for a more practical case such as the MLP. For simplicity, ex-situ (off-line) supervised learning will remain as the training method of choice. To evaluate the MLP performance, the recognition of patterns from the MNIST [40] database (see Figure 3a,b) will be considered. Besides the extension to MLP classifiers, this modified workflow also involves an iterative calibration algorithm intended to minimize the R_L-induced degradation of the inference accuracy. The chart depicted in Figure 3c summarizes the workflow. The tasks can be split into three parts: the first one comprises a set of MATLAB subroutines for creating, training, and writing the SPICE netlist for an ideal feed-forward MLP. The second part creates an idealized fully linear model of the MCA-based artificial neural networks (ANNs) in Python to calibrate the synaptic weights obtained during the training to account for the parasitic line resistances (the details can be seen in Figure 3d). Last but not least, the third part relates to the SPICE simulation of the proposed circuit during the inference phase.

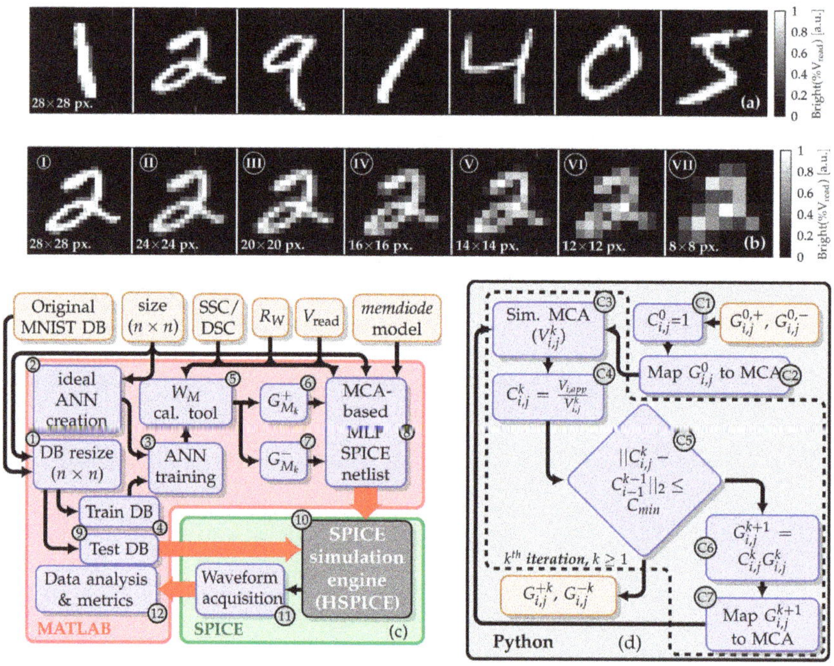

Figure 3. (**a**) Samples of the MNIST database considered in this article. In all cases images are represented in 28 × 28 px. Pixel brightness (or intensity) is codified in 256 levels ranging from 0 (fully OFF, black) to 1 (fully ON, white). (**b**) Readability loss as the resolution decreases from 28 × 28 px (case I) to 8 × 8 (case VII). (**c**) Flowchart diagram for the simulation procedure. Starting with the image size specification, R_L, V_{read}, and connection scheme, the routine creates the dataset, trains the MLP, translates it into an MCA, performs the simulations and processes the results. (**d**) Flowchart diagram of the calibration method to minimize the impact of R_L. It is included in block 5 from (**c**).

3.1. Simulation Flow

Regarding the MATLAB-implemented part of the procedure, the first step consists in creating the image ($n \times n$ pixels) database. This includes rescaling each image of the original database (item (1) in the flowchart shown in Figure 3c). The MNIST (Modified National Institute of Standards and Technology) is a large database of handwritten digits from 0 to 9 commonly used for training and testing image processing systems including ANNs in the field of machine learning. This database contains 60,000 training images and 10,000 testing images, both in grayscale and with a 28 × 28 pixels resolution [40]. A few examples of these images can be seen in Figure 3a where the x and y axes stand for the pixel index. Pixel brightness is codified into 256 gray levels between 0 (fully OFF, black) and 1 (fully ON, white). Resizing to different resolutions can be seen in Figure 3b.

Then, a software-based SLP or MLP with n^2 inputs, 10 outputs and a number N of hidden neural layers (each of them comprising m_i neurons) is created (2) and trained (3) using the previously rescaled database of training images (4). The MLP (or SLP) is ex-situ trained considering the scaled conjugate gradient (SCG) [48] as the training algorithm, as proposed in [17]. Further details concerning the training function are beyond the scope of this work, as we focus on the MCA-based implementation of the MLP. This produces $N + 1$ weight matrices $W_{M_k} \in \mathbb{R}$, with k ∈ {1, 2, ... , N + 1} (5) (for instance for two hidden layers with m_1 and m_2 neurons each, three weight matrices W_{M_1}, W_{M_2} and W_{M_3} are obtained, with sizes $n^2 \times m_1$, $m_1 \times m_2$ and $m_2 \times 10$, respectively). To allow rendering both the positive and negative elements of W_{M_k} with the always positive conductance of the MCA, each synaptic weight is implemented using two memdiodes as suggested in [49–51] resulting in two MCAs per synaptic layer. Thereby, each W_{M_K} matrix is split into two matrices $W^+_{M_k}$ and $W^-_{M_k}$ as:

$$w^+_{M_{k_{i,j}}} = \begin{cases} w_{M_{k_{i,j}}}, w_{M_{i,j}} > 0 \\ 0, w_{M_{i,j}} \leq 0 \end{cases} \tag{5}$$

$$w^-_{M_{k_{i,j}}} = \begin{cases} -w_{M_{k_{i,j}}}, w_{M_{i,j}} < 0 \\ 0, w_{M_{i,j}} \geq 0 \end{cases} \tag{6}$$

each of them containing only positive weights, so that $W_{M_K} = W^+_{M_k} - W^-_{M_k}$. In the next step, the conductance matrices $G^+_{M_k}$ and $G^-_{M_k}$ ((6) and(7)) to be mapped onto the MCAs are calculated by the linear transformation [52]:

$$G^{+,-}_{M} = \frac{G_{max} - G_{min}}{max\{W_{M_k}\} - min\{W_{M_k}\}} W^{+,-}_{M_k} + \left[G_{max} - \frac{(G_{max} - G_{min}) max\{W_{M_k}\}}{max\{W_{M_k}\} - min\{W_{M_k}\}} \right] \tag{7}$$

where $[G_{min}, G_{max}]$ is a selected conductance range for a linear computation in matrix-vector calculations. For simplicity, we consider $G_{max} = G_{LRS} = \frac{1}{R_{ON}}$ and $G_{min} = G_{HRS} = \frac{1}{R_{OFF}}$, where $max\{W_{M_K}\}$ and $min\{W_{M_K}\}$ are the maximum and minimum synaptic weight values in the software obtained W_{M_K}. In this way, the synaptic weights in the $W^+_{M_k}$ and $W^-_{M_k}$ matrices are converted to conductance values within the range $[G_{HRS}, G_{LRS}]$.

The subsequent subroutines generate the circuit netlist for the dual-$n^2 x\ m_i$, $m_i\ x\ m_{i+1}$, ..., $m_N \times 10$ memdiode MCA-based MLP (8), adding the parasitic wire resistance, connection scheme, and control logic necessary to perform the inference phase. As reported in [17], a single MCA is not efficient for implementing large matrices. Given that both R_L and R_{ON}/R_{OFF} are normally defined by the selected fabrication node and RS mechanism, respectively, a widely accepted [51,53] alternative design consists of dividing the large matrices into smaller partitions, whose reduced size improves the voltage effectively delivered to the memristive cell. Figure 4a shows the simplified circuit schematic of the partitioned MCA and the interconnections required to realize the complete matrix-vector multiplication (MVM) in the 1st synaptic layer. Exploding the integrability of the MCA with CMOS circuitry, vertical interconnects used to connect the outputs of the vertical MCA partitions may be placed under the partitioned structure, as well as the analogue sensing electronics, allowing the partitioned MCA to maintain a similar area consumption than the original nonpartitioned case [51]. The vertical interconnects are grounded through the sensing circuit to absorb the currents within the same vertical wire.

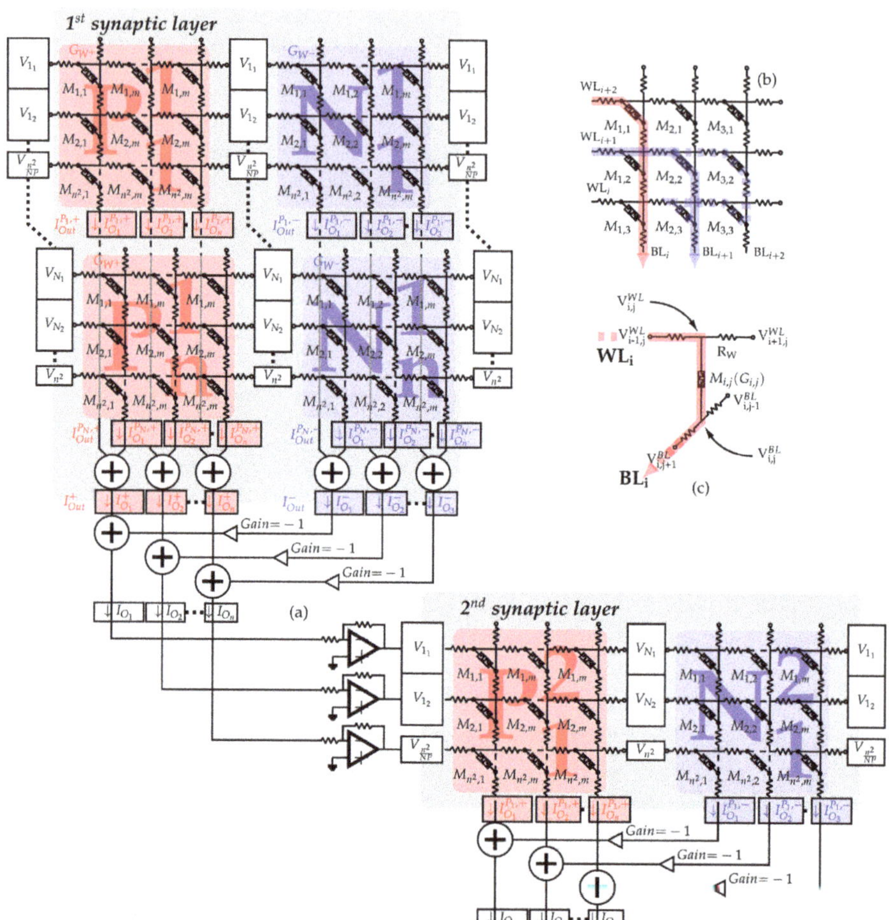

Figure 4. (**a**) Simplified equivalent circuit schematic for a partitioned MCA-based MLP. Each MCA in the 1st synaptic layer is subdivided into N identically sized partitions to minimize the parasitic voltage drops. Partial output current vectors are indicated in the output of each partition. (**b**) Equivalent circuit schematic for an MCA based SLP. Red and blue arrows exemplify the electron flow through the memdiodes connecting the top (word lines—WL) and bottom lines (bit lines—BL). The dashed blue line depicts the so-called sneakpath problem. (**c**) Individual RRAM cell with the associated R_L resistors.

Each memdiode in the MCAs is set to the corresponding conductance value from the $G^+_{M_k}$ and $G^-_{M_k}$ matrices by adjusting the control parameter λ. The required value of λ is obtained by solving Equation (1), $I = g^{+(-)}_{k_{1,j}} V$, $g^{+(-)}_{k_{1,j}}$ being each of the elements of $G^+_{M_k}$ ($G^-_{M_k}$). As in this work we focus on the artificial synapses modeling using the memdiode model, hidden neurons in the kth hidden neural layer connecting the two adjacent layers of synapses $k-1$ and $k+1$ are implemented in terms of a behavioral SPICE model. The model for each neuron involves a trans-impedance amplifier (TIA) that translates the output current in the associated bitline on the $i-1$ synaptic layer to a voltage which is fed to a nonlinear activation function and then propagated to the corresponding wordline in the $i+1$ synaptic layer. In this paper, we consider a log-sigmoidal $(1/(1+e^{-x}))$ activation function, though a tan-sigmoidal activation function could be used as well. The input stimulus in each synaptic layer is delivered following a dual side connection (DSC) scheme, as shown in the simplified equivalent circuit in Figure 4a. Despite the increased peripheral circuitry complexity, this scheme improves the voltage delivery to each synapse [10] by connecting the two wordline terminals to the same input stimuli. The input stimuli of the 1st synaptic layer is obtained by unrolling each of the rescaled grayscale $n \times n$ images of the test database (9) into an equivalent $n^2 \times 1$ vector and scaling it by a voltage V_{read}. V_{read} is chosen such as to prevent altering the memdiode states during the inference simulation. In this way, during the inference process each of the test images is presented to the MLP as a vector of analogue voltages in the range $[0, V_{read}]$. Once the circuit netlist has been generated, it is passed to the HSPICE simulator (10) which evaluates the voltage and current distributions in the MCA-based MLP circuit while it processes and classifies the input images (11) and then passes the resulting waveforms back to the MATLAB routine for metrics extraction (12).

3.2. Calibration Tool Workflow

In an ideal case scenario, that is with R_L negligible, the output current for a column (or bitline) in a MCA of the kth synaptic layer of a MLP or SLP is given by Equation (8), where the $g_{k_{i,j}}^{+(-)}$ elements are the junction conductances along one bitline and $V_{k_{i,app}}$ the wordline voltages. Note that the voltage applied to each artificial synapse (conductances) is independent of the device location within the MCA provided that no voltage drops occur in the interconnection lines. Instead, in a real case scenario, the voltage applied across each junction depends on the device location in the MCA partition as indicated by Equation (9). This happens because a significant IR-drop occurs in the line resistances. Consequently, the voltage applied across each junction is always lower than the applied wordline voltage, and so it is the resulting output current ${I^{real}}_{k_j}$

$$I_{ideal}^{jk+(-)} = g_{k_{1,j}}^{+(-)} V_{k_{1,app}} + g_{k_{2,j}}^{+(-)} V_{k_{2,app}} + g_{k_{3,j}}^{+(-)} V_{k_{3,app}} + \ldots + g_{k_{N,j}}^{+(-)} V_{k_{1,app}} \tag{8}$$

$$I_{real}^{jk+(-)} = g_{k_{1,j}}^{+(-)} V_{k_{1,j}} + g_{k_{2,j}}^{+(-)} V_{k_{2,j}} + g_{k_{3,j}}^{+(-)} V_{k_{3,j}} + \ldots + g_{k_{N,j}}^{+(-)} V_{k_{1,j}} \tag{9}$$

An interesting approach to compensate for the smaller currents was presented by Lee et al. in [13]. In their study, the authors propose to increase the conductance level of each individual memory cell proportionally to the voltage reduction. Let us then consider a calibration factor $c_{k_{i,j}}^{+(-)} = \frac{V_{k_{i,app}}}{V_{k_{i,j}}} \geq 1$ for each element in the MCA. Then the compensated conductance of each memristive device is calculated as $g_{k_{i,j}}^{\prime +(-)} = g_{k_{i,j}}^{+(-)} c_{k_{i,j}}^{+(-)}$. Since the calibrated conductances ($g_{k_{i,j}}^{\prime +(-)}$) are higher than the previous ones ($g_{k_{i,j}}^{+(-)}$), the overall current increases and consequently so does the IR-drops along the word and bitlines. Thereby, this method implies multiple iterations until convergence is reached.

To speed-up the iterative calibration process, a parametric fully linear model of the MCA-based MLP was developed. In this scenario each memristor is represented as a resistor of fixed value and the overall MCA model (see Figure 4b) is expressed in terms of a system of coupled equations arising from considering the Current Kirchhoff Law at each junction of the MCA (see Figure 4c). The details of such modeling approach first considered in [10] are included in Appendix A. This method avoids calculating the required values of λ for each memdiode in each iteration, which significantly reduces the calibration time, especially for large networks. The simulation code was implemented in Python, taking advantage of the object oriented programming characteristic of such language. In this context, each MCA in a partitioned multilayer perceptron can be easily created as an instance of a unique class that describes the properties and behavior of a single MCA.

The details of the iterative calibration process (block (5) in the flowchart of Figure 3c) are illustrated in Figure 3d and Algorithm 1. First, the synaptic weight matrices W_{M_K}, delivered from the training process (block (4) in the flowchart of Figure 3c) are mapped onto each MCA of the complete MLP (block (C2) in the flowchart of Figure 3d). The input stimuli feed to each MCA during calibration consists of a vector of analogue voltages obtained from averaging the brightness of each pixel from the images of the training set (C3). By solving the system of coupled equations, the effective voltage delivered to each memristor is calculated and used to compute the required calibration factor $c_{k_{i,j}}^{+(-)} = \frac{V_{k_{i,app}}}{V_{k_{i,j}}}$ (C4). Then, the absolute distances to the values calculated in the previous loop are compared against a predefined target, which represents a termination criterion for the process (C5). If the distance to the target exceeds the criterion, the conductance matrices are calculated as $g'^{+(-)}_{k_{i,j}} = g^{+(-)}_{k_{i,j}} c^{+(-)}_{k_{i,j}}$ (C6) and remapped onto the MCA object (C7) and the voltages at the nodes recalculated (C3). The iterative loop from (C3)–(C7) is then repeated until the termination criterion is met. The results of this iterative calibration process are the $2k$ matrices of calibrated conductance values $G^{+}_{M_k}$, $G^{-}_{M_k}$, (blocks (6) and (7) in the flowchart of Figure 3c).

Algorithm 1: Iterative calibration algorithm

```
Input: GMk+(-)(i,j)
Output: GMk+(-)calibrated(i,j)
Define cal_vector as the average brightness of each pixel
finish_calibration==false
while finish_calibration==false do
  | finish_calibration=true
  | get_WL_voltages(cal_vector)
  | for i in row_numbers do
  |   | for j in column_numbers do
  |   |   | prev_Cij=Cij
  |   |   | Cij=WL_voltage[I,j]/V_app[i]
  |   |   | if abs(Cij-prev_Cij)>criterion_value then
  |   |   |   | Finish_calibration==false
  |   |   |   | GMk+(-)(i,j)=GMk+(-)(i,j)*Cij
  |   |   | else
  |   | end
  | end
end
GMk+(-)calibrated(i,j)= GMk+(-)(i,j)
```

4. Simulation Results and Discussion

The line resistance between adjacent cells can be calculated as $R_L = \rho \cdot L/(W \cdot T)$, where L and W are the wire length between adjacent cells and wire width, respectively. For simplicity, they are taken equal to the feature size F. T is the metal thickness which is assumed >10 nm. In this context, R_L ranges from 1 to 10 Ω, as the resistivity of conventional metal wires (ρ) ranges from 10^{-8} to 10^{-7} Ω·m. Thereby, R_L can be estimated to be ≈4.53, 2.97, and 1.55 Ω for the 16, 22, and 32 nm technology nodes, respectively [13]. However, in Cu-wires there is a non-negligible size-dependent resistivity for technology nodes below the 10 nm limit, caused by the surface and grain boundary scattering as the mean free path of electrons becomes comparable to the wire dimensions. According to the Fuchs–Sondheimer (FS) and the Mayadas–Shatzkes (MS) models [53], R_L for highly scaled nodes can be as large as ≈100 kΩ. Considering the QMM model, the influence of the line resistance is evaluated in the following Sections 4.1 and 4.2, assuming a different number of hidden layers and two alternative approaches to minimize the parasitic voltage drop, respectively.

4.1. Influence of the Number of Hidden Layers

Unlike the case of SLP, where the network size (in terms of the number of devices) is fixed by the pattern features and possible output classes, in case of MLP, the introduction of hidden neural layers results in multiple possible networks for the classification of a given pattern dataset [54]. In this regard, it is known that as the number of hidden layers increases, so does the overall network accuracy. Nevertheless, when considering a realistic memristor-based implementation as done in this paper, there is a degradation of the signals propagated across each synaptic layer due to the line resistance in combination with the sneakpath effect. Consequently, the hidden neurons are prone to propagate erroneous signals, thus threatening the accuracy of the MLP. To shed light on this issue, five MLPs comprising different numbers of hidden layers and neurons per layer were simulated. The obtained results are summarized in Table 1, considering for all cases the MNIST images resized to 8 × 8 px, dual-side-connection, no partitioning of the MCA used for each synaptic layer, V_{read} = 300 mV and R_L swept from 100 mΩ to 1 kΩ. The synaptic connections are modeled with the QMM SPICE subcircuit described in [17] and considering the following set of parameters: I_{min} = 85 nA, I_{max} = 52 µA, α_{min} = 4.5, α_{max} = 2.5, R_{min} = R_{min} = 110 Ω, and β = 0.5. This combination of values renders (at the nominal V_{read}) resistances $R_{OFF} \approx$ 577 kΩ and $R_{ON} \approx$ 7.5 kΩ (approx. an R_{OFF}/R_{ON} ratio in the order of 100).

Table 1. Structures of the MLPs considered in the simulations of this section. In all cases the MNIST images resized to 8 × 8 px are considered as the input pattern.

Hidden Layers	Code	Network Structure	Number of Memristive Sys.	Accuracy at $R_L \to 0$ Ω	Accuracy (Soft. Case)
0	SLP	64 × 10	1280 sys.	89.6%	91.14%
1	MLP-2a	64 × 54 × 10	7992 sys.	92.3%	95.95%
	MLP-2b	64 × 100 × 10	14,800 sys.	92.7%	96.89%
2	MLP-3a	64 × 54 × 34 × 10	11,263 sys.	95.2%	96.30%
	MLP-3b	64 × 100 × 50 × 10	23,800 sys.	96%	96.92%
3	MLP-4	64 × 54 × 34 × 24 × 10	12,696 sys.	94.3%	95.81%

The simulation results are graphically reported in Figure 5, where the inference accuracy as function of R_L is shown normalized against the inference accuracy for $R_L \to 0$ Ω. A central point to highlight here is the notorious increase of the MLP sensitivity to R_L when compared against the reference SLP, regardless of the number of hidden layers and neurons per layer. This could be explained by taking into account the larger size of the synaptic layers involved for the MLPs cases (the MCAs used for the SLP has a maximum size of 64 × 10 while for the MLP-#a it increases up to 64 × 54). The use of larger MCAs with no partitions to implement the synaptic connections degrades the effective voltage delivered to the synapses located away from the driving ports of the MCA. The ratio between the effective voltage delivered to each synapse and the nominal applied voltage is known as the read margin, and it has been shown in [17] that for a given value of R_L, it decreases as the size of the MCA grows, directly degrading the inference accuracy. This interpretation is also supported by the results obtained for the set of MLPs named MLP-#b. For these simulations, the R_L sensitivity further increases as it could be expected given the bigger size of the largest MCA involved in the network (64 × 100 for the set MLP-#b against 64 × 54 for MLP-#a). It is also worth noting that both the MLP-#a and MLP-#b sets follow unique decreasing trends with R_L regardless of the number of layers. Thereby the increase in the number of hidden layers does not significantly compromise the R_L sensitivity but allows a non-negligible increase in the inference accuracy, as shown in the inset of Figure 5. Instead, the number of neurons per layer causes a sensible increase of the inference accuracy degradation caused by R_L, as it implies changes in the MCAs used to implement the MLP.

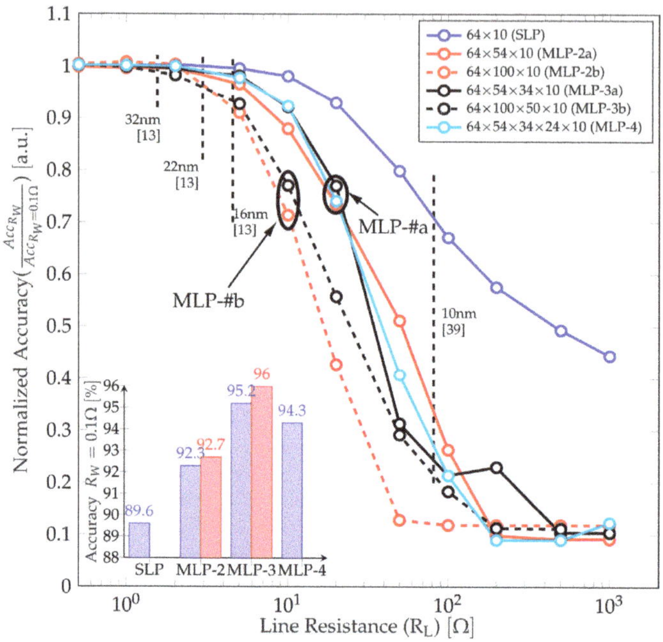

Figure 5. Inference accuracy vs. R_L, normalized against the inference accuracy obtained for $R_L \to 0\ \Omega$. Two different sets of MLP (#a and #b, see Table 1 for details) are considered as well as a SLP for comparison purposes. The inset in the lower left shows the inference accuracy at $R_L \to 0\ \Omega$ for the different cases considered. Note that the R_L dependency of the inference accuracy is determined by the size of the largest MLP layer, and it presents a very shallow dependence on the number of hidden layers. In fact, the increase in the number of hidden layers allows boosting the inference accuracy as R_L decreases without compromising the MLP sensitivity to R_L variations.

4.2. Techniques to Minimize the Impact of the Line Resistance (R_L)

As mentioned in the previous subsection, the voltage drop occurring across the parasitic line resistances imposes a serious limitation to the number of neurons that can be included in each neural layer without causing a major reduction of the inference accuracy. Methods to minimize this problem are thereby mandatory to allow rendering MLPs capable of dealing with large input patterns. For instance, in [55], Truong et al. proposed a circuit to compensate the voltage drop across the interconnections. Although capable of improving the inference metrics, the proposed method implies a significant circuit overhead and might be not suitable for networks involving a large number of neurons. Therefore, the search for alternative solutions requiring lesser additional circuitry is encouraged. Two of them were discussed in Sections 3.1 and 3.2, namely the MCA partitioning and the iterative calibration of the synaptic weights. Although tested in [13,17], their applicability in MLP has not yet been addressed considering realistic electrical simulations. Thus, in this section the capability of such techniques to mitigate the line resistance impact on MLP is studied based on the framework defined in Section 3.1 and using the same values for the QMM as in Section 4.1. Only one hidden layer is considered as it was shown in Section 4.1 that the number of layers does not significantly alter the R_L dependency. Instead two different MNIST representation sizes are considered: 8 × 8 px. (64 × 54 × 10 as reported in [9]) and 14 × 14 px. (196 × 20 × 10 as reported in [38]) to account for the MCA size dependency.

For comparison purposes we also report the case of pattern classification with SLPs (of sizes 64 × 10 and 196 × 10).

Let us first consider the nonpartitioned (NP = 1), uncalibrated cases (blue empty markers). As it can be seen in Figure 6, in all cases (MLP and SLP for 8 × 8 px. and 14 × 14 px. images) the inference accuracy approaches the ideal case as R_L tends to zero. Nonetheless, when considering the 14 × 14 px. images (Figure 6b,d) a higher accuracy degradation is observed, as expected for the use of a larger MCA as the first synaptic layer. This can be seen as a left-shift of the accuracy vs. R_L curves when the image size is increased and occurs both for the SLP (see the displacement of the trend from Figure 6a,b) and the MLP (Figure 6c,d) cases. Note that for the 14 × 14 px. images there is a significant accuracy loss even for low values of the line resistance (see Figure 6d for instance, where a value of R_L of approx. 5 Ω obtained for a feature size of 16 nm causes the inference accuracy to drop from approx. 96% to 73%). It is also worth mentioning that the steeper decrease of the inference accuracy vs. R_L observed in MLPs vs. SLP trained to classify the 8 × 8 px. images in Figure 5 is also present for the 14 × 14 px. images (see Figure 6b,d).

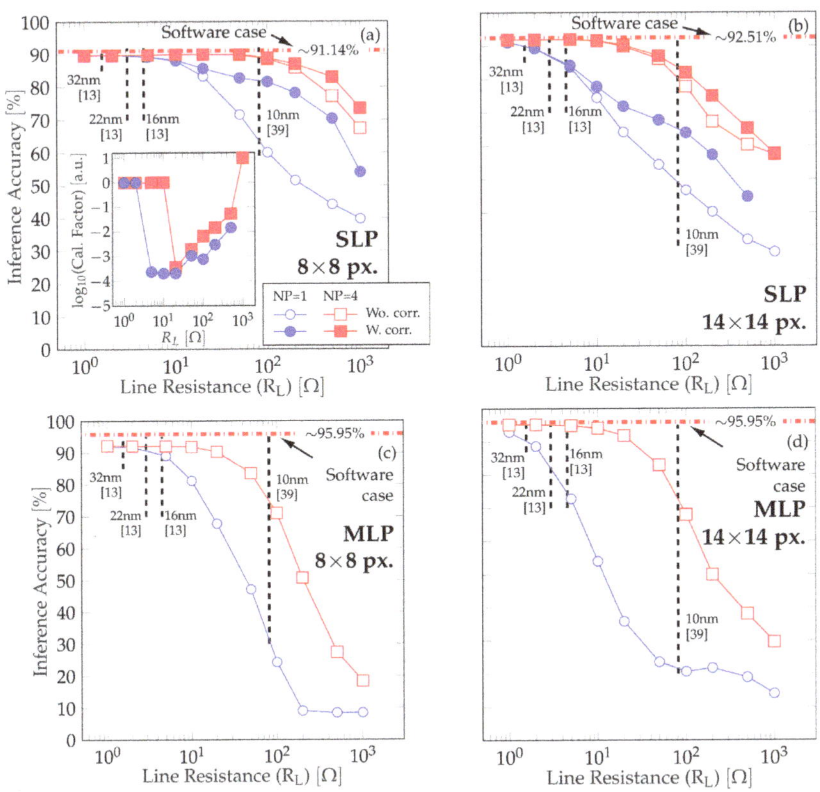

Figure 6. Inference accuracy plotted against the line resistance (R_L) for four different scenarios: SLPs considering (**a**) 8 × 8 px. images (the SLP structure is 64 × 10) and (**b**) 14 × 14 px. (196 × 10) and MLP considering (**c**) 8 × 8 px. images (64 × 54 × 10) and (**d**) 14 × 14 px. images (196 × 20 × 10). The ideal results considering software implementation of the same networks is added in each subfigure by a red dash-dotted line. For (**a**,**b**) both the calibration and partitioned implementations are superimposed for comparison. Instead for (**c**,**d**) only the partitioned scenario is shown as no relevant improvement was found by the calibration procedure. The inset in (**a**) shows the optimal calibration factor as function of R_L.

To improve the metrics discussed in the previous paragraph, the post-training iterative calibration of the synaptic weights is first performed on the nonpartitioned SLP (filled blue lines). This process has two different outcomes: on one hand when considering the SLP case, a clear improvement of up to approx. 30% for the 8 × 8 px images (Figure 6a) can be observed for R_L values approaching 100 Ω in highly scaled fabrication nodes [53]. Beyond this limit, the capability of the calibration method to reduce the voltage drop in the interconnections is not enough and thereby the accuracy improvement becomes smaller. A very similar behavior is shown in Figure 6b for the 14 × 14 px images. However, given the larger size of the MCAs involved, the improvement is smaller (not bigger than 20%). As the target calibration factor passed to the calibration routine is defined by the user, in this paper we performed an iterative loop to automatically determine the calibration factor that allows maximizing the inference accuracy. The resulting factors are plotted against the inference accuracy in the inset of Figure 6a for the 8 × 8 px. images. Note that for low R_L values, the calibration factor plays no role as no calibration is indeed required (the parasitic voltage drop due to the line resistance is negligible). Then the factor is tightened and progressively relaxed as the line resistance increases, as if the calibration factor is too exigent the calibration cannot yield a real accuracy improvement.

When addressing the case of the MLPs with different sizes, it was found that the calibration process produces a marginal improvement, resulting in identical inference vs. R_L trends as in the noncalibrated cases (and thereby not plotted in Figure 6c,d as they would coincide with the noncalibrated trends). Instead, the use of partitioned schemes for the realization of the complete MCA-based synaptic layers is shown to be efficient both for SLPs and MLPs. For instance, when the 64 × 10 SLP shown in Figure 6a is partitioned into four blocks of 16 × 10 the inference accuracy notably increases (note the empty red markers). The same effect is observed for the 196 × 10 MCA (partitioned into four blocks of 49 × 10) from which the results presented in Figure 6b were extracted. Furthermore, the inference accuracy of the partitioned SLP can also be improved by using the calibration algorithm (filled red markers in Figure 6a,b). For the MLP, the enhancement achieved with the partitioning is seen as a right shift in the accuracy vs. R_L trends. Note that in these cases, the first layer in the 64 × 54 × 10 MLP (Figure 6c) was implemented with 12 blocks of 16 × 18 and the second layer with three blocks of 18 × 10 and for the 196 × 20 × 10 MLP (Figure 6d), the first layer was implemented using four partitions of 49 × 20 and the second layer was not partitioned (20 × 10).

5. Conclusions

In this paper we extended the use of the Quasi-static Memdiode Model (QMM) previously proven for single-layer perceptrons (SLPs) to the SPICE modeling and simulation of multilayer perceptrons (MLPs) intended for large dataset pattern recognition. The versatility and reduced computational cost of this model allow performing electrical simulations without losing accuracy. The inference performance was tested considering the MNIST dataset of grey-scale handwritten digits, rescaled to different resolutions to test MLPs of different sizes. Two aspects were analyzed: the impact of the MLP structure (number of layers and neurons per layer) on the inference accuracy and alternative techniques to mitigate the impact of the line resistance. Concerning the first point, it was found that the number of hidden layers does not cause major variations in the line resistance dependence of the inference accuracy. Instead, it is the size of the largest synaptic layer what acts as a bottleneck, severely limiting the overall accuracy. Thereby the addition of memristive-based synaptic layers helps improving the accuracy without inducing further R_L-related degradation. Concerning the second point, the use of partitioned schemes was shown to provide the best performance results both in SLP and MLP when compared to the calibration technique. In fact, the calibration technique resulted in no gain in terms of accuracy when applied to MLP networks. This should be taken into account by circuit designers.

Author Contributions: Conceptualization, F.L.A. and E.M.; methodology, F.L.A., N.M.G. and E.M.; software, F.L.A., N.M.G. and E.M.; validation, F.L.A., N.M.G., S.M.P., F.P., J.S. E.M.; formal analysis, F.L.A., E.M.; investigation, F.L.A., N.M.G., S.M.P., F.P., J.S. E.M.; resources, F.P, J.S and E.M.; data curation, F.L.A.; writing—original draft preparation, F.L.A., N.M.G. and E.M.; writing—review and editing, F.L.A., N.M.G., S.M.P., F.P., J.S. and E.M.; visualization, F.A, E.M.; supervision, E.M.; project administration, F.P., J.S. and E.M; funding acquisition, F.P., J.S. and E.M. All authors have read and agreed to the published version of the manuscript.

Funding: This research has been funded by both Argentinean and European institutions. Argentine funding was provided by MINCyT (Contracts PICT2013/1210, PICT 2016/0579 and PME 2015-0196), CONICET (Project PIP-11220130100077CO) and UTN.BA (Projects PID-UTN EIUTIBA4395TC3, CCUTIBA4764TC, MATUNBA4936, CCUTNBA5182 and CCUTNBA6615). E.M. and J. S. acknowledge the support from TEC2017-84321-C4-4-R and WAKeMeUP 783176 projects, cofunded by grants from the Spanish Ministerio de Ciencia e Innovación and the ECSEL-EU Joint Undertaking.

Conflicts of Interest: The authors declare no conflict of interest.

Appendix A

Each MCA is described using the equivalent circuit schematic represented in Figure 4b, by considering the 2nd Kirchhoff's law on the terminals of each memristive device (see Figure 4c), which will have the form of any of the Equations (A1)–(A6) depending on the device location within the MCA. G_L is the line conductance ($1/R_L$), $G^{BL}_{i,in} = G^{WL}_{i,in}$ are the wordline and bitline access conductances (resistance in the BL/WL terminals) ($G^{WL}_{i,in} = 1/R^{WL}_{i,in}$ and $G^{BL}_{i,in} = 1/R^{BL}_{i,in}$), $V^{WL}_{i,app}$ are the applied voltages in the WL terminals, corresponding to the MNIST images and $V^{BL}_{j,app}$ are grounded through a sensing resistor. Node voltages in the WLs are indicated as $V^{WL}_{i,j}$ and, in the same way, $V^{BL}_{i,j}$ refers to node voltages in the BLs. Six different equations arise as they account for the elements located at the BL/WL terminals (Equations (A2), (A3), (A5) and A6) or somewhere in between (Equations (A1) and (A4)).

$$(WL,(i,j)) : G_L\left(V^{WL}_{i,j} - V^{WL}_{i,i-1}\right) - G_{i,j}\left(V^{BL}_{i,j} - V^{WL}_{i,i}\right) - G_L\left(V^{WL}_{i,j+1} - V^{WL}_{i,j}\right) = 0 \quad \text{(A1)}$$

$$(WL,j=1) : G^{WL}_{i,in}\left(V^{WL}_{i,1} - V^{WL}_{i,app}\right) - G_{i,j}\left(V^{BL}_{i,1} - V^{WL}_{i,1}\right) - G_L\left(V^{WL}_{i,2} - V^{WL}_{i,1}\right) = 0 \quad \text{(A2)}$$

$$(WL,j=n) : G_L\left(V^{WL}_{i,n} - V^{WL}_{i,n-1}\right) - G_{i,n}\left(V^{BL}_{i,n} - V^{WL}_{i,n}\right) = 0 \quad \text{(A3)}$$

$$(BL,(i,j)) : G_L\left(V^{BL}_{i+1,j} - V^{BL}_{i,i}\right) - G_{i,j}\left(V^{BL}_{i,j} - V^{WL}_{i,j}\right) - G_L\left(V^{BL}_{i,j} - V^{BL}_{i-1,j}\right) = 0 \quad \text{(A4)}$$

$$(BL,\ i=m) : G^{BL}_{in,j}\left(V^{WL}_{i,j} - V^{WL}_{i,i-1}\right) - G_{i,j}\left(V^{BL}_{i,j} - V^{WL}_{i,i}\right) - G_L\left(V^{BL}_{i,j+1} - V^{BL}_{i,j}\right) = 0 \quad \text{(A5)}$$

$$(BL,i=1) : G_L\left(V^{BL}_{2,j} - V^{BL}_{1,j}\right) - G_{i,j}\left(V^{BL}_{1,j} - V^{WL}_{1,j}\right) = 0 \quad \text{(A6)}$$

This results in a system of 2 mn coupled equations, with 2mn unknown voltages corresponding to the WL ($V_{WL} = \left[V^{WL}_{1,1}, V^{WL}_{1,2}, \ldots, V^{WL}_{1,n}, V^{WL}_{2,1}, \ldots, V^{WL}_{n,m}\right]^T$) and BL ($V_{BL} = \left[V^{BL}_{1,1}, V^{BL}_{1,2}, \ldots, V^{BL}_{1,n}, V^{BL}_{2,1}, \ldots, V^{BL}_{n,m}\right]^T$) voltages. By defining the column vectors E_{WL} and E_{BL} as $\left[G^{WL}_{1,in}V^{WL}_{1,in}, 0, \ldots, G^{WL}_{2,in}V^{WL}_{2,in}, 0, \ldots, G^{WL}_{m,in}V^{WL}_{m,in}\right]$ and $\left[G^{BL}_{1,in}V^{BL}_{1,in}, 0, \ldots, G^{BL}_{2,in}V^{BL}_{2,in}, 0, \ldots, G^{BL}_{n,in}V^{BL}_{n,in}\right]$ respectively, Equations (A1)–(A6) can be represented following a matrix formulation as in Equation (A7):

$$\begin{bmatrix} A & B \\ C & D \end{bmatrix} \begin{bmatrix} V_{WL} \\ V_{BL} \end{bmatrix} = \begin{bmatrix} E_{WL} \\ E_{BL} \end{bmatrix} \quad \text{(A7)}$$

where all A, B, C, and D matrix are $m \times n$. Further details regarding the structure of these matrices can be found in [10]. Then the output of the $m \times n$ MCA is a row vector of $1 \times n$ currents, defined as $I_{Out} = V^{BL}_{n,j}G_L$, with $1 \le j \le m$, obtained by solving Equation (A7). It should be noted that the system of coupled equations presented in the matrix formulation of Equation (A7) allow representing both the case of the input voltage being applied from one single side or from both sides of WLs.

References

1. Li, C.; Belkin, D.; Li, Y.; Yan, P.; Hu, M.; Ge, N.; Jiang, H.; Montgomery, E.; Lin, P.; Wang, Z.; et al. In-Memory Computing with Memristor Arrays. In Proceedings of the 2018 IEEE International Memory Workshop (IMW), Kyoto, Japan, 13–16 May 2018; pp. 1–4.
2. Upadhyay, N.K.; Joshi, S.; Yang, J.J. Synaptic electronics and neuromorphic computing. *Sci. China Inf. Sci.* **2016**, *59*, 1–26. [CrossRef]
3. Sasago, Y.; Kinoshita, M.; Morikawa, T.; Kurotsuchi, K.; Hanzawa, S.; Mine, T.; Shima, A.; Fujisaki, Y.; Kume, H.; Moriya, H.; et al. Cross-Point Phase Change Memory with $4F^2$ Cell Size Driven by Low-Contact-Resistivity Poly-Si Diode. In Proceedings of the Digest of Technical Papers-Symposium on VLSI Technology, Kyoto, Japan, 16–18 June 2009; pp. 24–25.
4. Truong, S.N.; Ham, S.-J.; Min, K.-S. Neuromorphic crossbar circuit with nanoscale filamentary-switching binary memristors for speech recognition. *Nanoscale Res. Lett.* **2014**, *9*, 629. [CrossRef] [PubMed]
5. Truong, S.N.; Min, K.-S. New Memristor-Based Crossbar Array Architecture with 50-% Area Reduction and 48-% Power Saving for Matrix-Vector Multiplication of Analog Neuromorphic Computing. *J. Semicond. Technol. Sci.* **2014**, *14*, 356–363. [CrossRef]
6. Truong, S.N.; Shin, S.; Byeon, S.-D.; Song, J.; Min, K.-S. New Twin Crossbar Architecture of Binary Memristors for Low-Power Image Recognition with Discrete Cosine Transform. *IEEE Trans. Nanotechnol.* **2015**, *14*, 1104–1111. [CrossRef]
7. Hu, M.; Li, H.; Chen, Y.; Wu, Q.; Rose, G.S.; Linderman, R.W. Memristor Crossbar-Based Neuromorphic Computing System: A Case Study. *IEEE Trans. Neural Netw. Learn. Syst.* **2014**, *25*, 1864–1878. [CrossRef]
8. Liu, B.; Li, H.; Chen, Y.; Li, X.; Huang, T.; Wu, Q.; Barnell, M. Reduction and IR-drop compensations techniques for reliable neuromorphic computing systems. In Proceedings of the 2014 IEEE/ACM International Conference on Computer-Aided Design (ICCAD), San Jose, CA, USA, 2–6 November 2014; pp. 63–70. [CrossRef]
9. Li, C.; Belkin, D.; Li, Y.; Yan, P.; Hu, M.; Ge, N.; Jiang, H.; Montgomery, E.; Lin, P.; Wang, Z.; et al. Efficient and self-adaptive in-situ learning in multilayer memristor neural networks. *Nat. Commun.* **2018**, *9*, 1–8. [CrossRef]
10. Chen, A. A Comprehensive Crossbar Array Model with Solutions for Line Resistance and Nonlinear Device Characteristics. *IEEE Trans. Electron Devices* **2013**, *60*, 1318–1326. [CrossRef]
11. Park, S.; Kim, H.; Choo, M.; Noh, J.; Sheri, A.; Jung, S.; Seo, K.; Park, J.; Kim, S.; Lee, W.; et al. RRAM-based synapse for neuromorphic system with pattern recognition function. In Proceedings of the 2012 International Electron Devices Meeting, San Francisco, CL, USA, 10–13 December 2012; Institute of Electrical and Electronics Engineers (IEEE): New York, NY, USA, 2012; pp. 10.2.1–10.2.4.
12. Ham, S.-J.; Mo, H.-S.; Min, K.-S. Low-Power $V_{DD}/3$ Write Scheme with Inversion Coding Circuit for Complementary Memristor Array. *IEEE Trans. Nanotechnol.* **2013**, *12*, 851–857. [CrossRef]
13. Lee, Y.K.; Jeon, J.W.; Park, E.-S.; Yoo, C.; Kim, W.; Ha, M.; Hwang, C.S. Matrix Mapping on Crossbar Memory Arrays with Resistive Interconnects and Its Use in In-Memory Compression of Biosignals. *Micromachines* **2019**, *10*, 306. [CrossRef]
14. Han, R.; Huang, P.; Zhao, Y.; Cui, X.; Liu, X.; Jin-Feng, K. Efficient evaluation model including interconnect resistance effect for large scale RRAM crossbar array matrix computing. *Sci. China Inf. Sci.* **2018**, *62*, 22401. [CrossRef]
15. Yakopcic, C.; Taha, T.M.; Subramanyam, G.; Pino, R.E. Memristor SPICE Modeling. In *Advances in Neuromorphic Memristor Science and Applications*; Springer Nature: London, UK, 2012; pp. 211–244.
16. Yakopcic, C.; Hasan, R.; Taha, T.; McLean, M.; Palmer, D. Memristor-based neuron circuit and method for applying learning algorithm in SPICE. *Electron. Lett.* **2014**, *50*, 492–494. [CrossRef]
17. Aguirre, F.L.; Pazos, S.M.; Palumbo, F.; Sune, J.; Miranda, E. Application of the Quasi-Static Memdiode Model in Cross-Point Arrays for Large Dataset Pattern Recognition. *IEEE Access* **2020**, *8*, 202174–202193. [CrossRef]
18. Miranda, E. Compact Model for the Major and Minor Hysteretic I–V Loops in Nonlinear Memristive Devices. *IEEE Trans. Nanotechnol.* **2015**, *14*, 787–789. [CrossRef]
19. Patterson, G.; Sune, J.; Miranda, E. Voltage-Driven Hysteresis Model for Resistive Switching: SPICE Modeling and Circuit Applications. *IEEE Trans. Comput. Des. Integr. Circuits Syst.* **2017**, *36*, 2044–2051. [CrossRef]
20. Yakopcic, C.; Taha, T.M.; Subramanyam, G.; Pino, R.E. Generalized Memristive Device SPICE Model and its Application in Circuit Design. *IEEE Trans. Comput. Des. Integr. Circuits Syst.* **2013**, *32*, 1201–1214. [CrossRef]
21. Kvatinsky, S.; Friedman, E.G.; Kolodny, A.; Weiser, U.C. TEAM: ThrEshold Adaptive Memristor Model. *IEEE Trans. Circuits Syst. I Regul. Pap.* **2013**, *60*, 211–221. [CrossRef]
22. Kvatinsky, S.; Ramadan, M.; Friedman, E.G.; Kolodny, A. VTEAM: A General Model for Voltage-Controlled Memristors. *IEEE Trans. Circuits Syst. II Express Briefs* **2015**, *62*, 786–790. [CrossRef]
23. Eshraghian, K.; Kavehei, O.; Cho, K.R.; Chappell, J.M.; Iqbal, A.; Al-Sarawi, S.F.; Abbott, D. Memristive device fundamentals and modeling: Applications to circuits and systems simulation. *Proc. IEEE* **2012**, *100*, 1991–2007. [CrossRef]
24. Biolek, D.; Biolek, Z.; Biolková, V.; Kolka, Z. Modeling of TiO_2 memristor: From analytic to numerical analyses. *Semicond. Sci. Technol.* **2014**, *29*, 125008. [CrossRef]
25. Biolek, Z.; Biolek, D.; Biolkova, V.; Kolka, Z. Reliable Modeling of Ideal Generic Memristors via State-Space Transformation. *Radioengineering* **2015**, *24*, 393–407. [CrossRef]

26. Kim, T.; Kim, H.; Kim, J.; Kim, J.-J. Input Voltage Mapping Optimized for Resistive Memory-Based Deep Neural Network Hardware. *IEEE Electron Device Lett.* **2017**, *38*, 1228–1231. [CrossRef]
27. Choi, S.; Lee, J.; Kim, S.; Lu, W.D. Retention failure analysis of metal-oxide based resistive memory. *Appl. Phys. Lett.* **2014**, *105*, 113510. [CrossRef]
28. Raghavan, N.; Frey, D.D.; Bosman, M.; Pey, K.L. Statistics of retention failure in the low resistance state for hafnium oxide RRAM using a Kinetic Monte Carlo approach. *Microelectron. Reliab.* **2015**, *55*, 1422–1426. [CrossRef]
29. Lin, Y.-D.; Chen, P.S.; Lee, H.-Y.; Chen, Y.-S.; Rahaman, S.Z.; Tsai, K.-H.; Hsu, C.-H.; Chen, W.-S.; Wang, P.-H.; King, Y.-C.; et al. Retention Model of TaO/HfO_X and TaO/AlO_X RRAM with Self-Rectifying Switch Characteristics. *Nanoscale Res. Lett.* **2017**, *12*, 407. [CrossRef] [PubMed]
30. Wong, H.S.P.; Lee, H.Y.; Yu, S.; Chen, Y.S.; Wu, Y.; Chen, P.S.; Lee, B.; Chen, F.T.; Tsai, M.J. Metal–oxide RRAM. *Proc. IEEE* **2012**, *100*, 1951–1970. [CrossRef]
31. Wu, W.; Wu, H.; Gao, B.; Yao, P.; Zhang, X.; Peng, X.; Yu, S.; Qian, H. A methodology to improve linearity of analog RRAM for neuromorphic computing. In Proceedings of the IEEE Symposium on VLSI Technology, Honolulu, HI, USA, 18–22 June 2018; pp. 103–104.
32. Kim, S.; Park, B.-G. Nonlinear and multilevel resistive switching memory in Ni/Si_3N_4/Al_2O_3/TiN structures. *Appl. Phys. Lett.* **2016**, *108*, 212103. [CrossRef]
33. Ciprut, A.; Friedman, E.G. Energy-Efficient Write Scheme for Nonvolatile Resistive Crossbar Arrays with Selectors. *IEEE Trans. Very Large Scale Integr. (VLSI) Syst.* **2018**, *26*, 711–719. [CrossRef]
34. Yao, P.; Wu, H.; Gao, B.; Tang, J.; Zhang, Q.; Zhang, W.; Yang, J.J.; Qian, H. Fully hardware-implemented memristor convolutional neural network. *Nat. Cell Biol.* **2020**, *577*, 641–646. [CrossRef]
35. Wang, C.; Feng, D.; Tong, W.; Liu, J.; Li, Z.; Chang, J.; Zhang, Y.; Wu, B.; Xu, J.; Zhao, W.; et al. Cross-point Resistive Memory. *ACM Trans. Des. Autom. Electron. Syst.* **2019**, *24*, 1–37. [CrossRef]
36. Chang, C.-C.; Chen, P.-C.; Chou, T.; Wang, I.-T.; Hudec, B.; Chang, C.-C.; Tsai, C.-M.; Chang, T.-S.; Hou, T.-H. Mitigating Asymmetric Nonlinear Weight Update Effects in Hardware Neural Network Based on Analog Resistive Synapse. *IEEE J. Emerg. Sel. Top. Circuits Syst.* **2018**, *8*, 116–124. [CrossRef]
37. Wang, W.; Song, W.; Yao, P.; Li, Y.; Van Nostrand, J.; Qiu, Q.; Ielmini, D.; Yang, J.J. Integration and Co-design of Memristive Devices and Algorithms for Artificial Intelligence. *iScience* **2020**, *23*, 101809. [CrossRef]
38. Milo, V.; Zambelli, C.; Olivo, P.; Perez, E.; Mahadevaiah, M.K.; Ossorio, O.G.; Wenger, C.; Ielmini, D. Multilevel HfO_2-based RRAM devices for low-power neuromorphic networks. *APL Mater.* **2019**, *7*, 081120. [CrossRef]
39. Tuli, S.; Rios, M.; Levisse, A.; Esl, D.A.; Tuli, S.; Rios, M.; Levisse, A. Rram-vac: A variability-aware controller for rram-based memory architectures. In Proceedings of the 25th Asia and South Pacific Design Automation Conference (ASP-DAC), Beijing, China, 13–16 January 2020; pp. 181–186.
40. LeCun, Y.; Cortes, C.; Burges, C.J.C. MNIST Handwritten Digit Database. Available online: http://yann.lecun.com/exdb/mnist/ (accessed on 28 January 2021).
41. Lee, A.R.; Bae, Y.C.; Im, H.S.; Hong, J.P. Complementary resistive switching mechanism in Ti-based triple TiO_X/TiN/TiO_X and TiOx/TiOxNy/TiOx matrix. *Appl. Surf. Sci.* **2013**, *274*, 85–88. [CrossRef]
42. Duan, W.J.; Song, H.; Li, B.; Wang, J.-B.; Zhong, X. Complementary resistive switching in single sandwich structure for crossbar memory arrays. *J. Appl. Phys.* **2016**, *120*, 084502. [CrossRef]
43. Yang, M.; Wang, H.; Ma, X.; Gao, H.; Hao, Y. Voltage-amplitude-controlled complementary and self-compliance bipolar resistive switching of slender filaments in Pt/HfO_2/HfOx/Pt memory devices. *J. Vac. Sci. Technol. B* **2017**, *35*, 032203. [CrossRef]
44. Chen, C.; Gao, S.; Tang, G.; Fu, H.; Wang, G.; Song, C.; Zeng, F.; Pan, F. Effect of Electrode Materials on AlN-Based Bipolar and Complementary Resistive Switching. *ACS Appl. Mater. Interfaces* **2013**, *5*, 1793–1799. [CrossRef]
45. Aguirre, F.; Rodriguez, A.; Pazos, S.; Sune, J.; Miranda, E.; Palumbo, F. Study on the Connection Between the Set Transient in RRAMs and the Progressive Breakdown of Thin Oxides. *IEEE Trans. Electron Devices* **2019**, *66*, 3349–3355. [CrossRef]
46. Frohlich, K.; Kundrata, I.; Blaho, M.; Precner, M.; Ťapajna, M.; Klimo, M.; Šuch, O.; Skvarek, O. Hafnium oxide and tantalum oxide based resistive switching structures for realization of minimum and maximum functions. *J. Appl. Phys.* **2018**, *124*, 152109. [CrossRef]
47. Lin, C.-Y.; Wu, C.-Y.; Wu, C.-Y.; Hu, C.; Tseng, T.-Y. Bistable Resistive Switching in Al_2O_3 Memory Thin Films. *J. Electrochem. Soc.* **2007**, *154*, G189–G192. [CrossRef]
48. Møller, M.F. A scaled conjugate gradient algorithm for fast supervised learning. *Neural Netw.* **1993**, *6*, 525–533. [CrossRef]
49. Prezioso, M.; Merrikh-Bayat, F.; Hoskins, B.D.; Adam, G.C.; Likharev, K.K.; Strukov, D.B. Training and operation of an integrated neuromorphic network based on metal-oxide memristors. *Nat. Cell Biol.* **2015**, *521*, 61–64. [CrossRef]
50. Hu, M.; Li, H.; Wu, Q.; Rose, G.S.; Chen, Y. Memristor Crossbar Based Hardware Realization of BSB Recall Function. In Proceedings of the 2012 International Joint Conference on Neural Networks (IJCNN), Brisbane, Australia, 10–15 June 2012; pp. 1–7.
51. Fouda, M.E.; Lee, S.; Lee, J.; Eltawi, A.M.; Kurdahi, F. Mask Technique for Fast and Efficient Training of Binary Resistive Crossbar Arrays. *IEEE Trans. Nanotechnol.* **2019**, *18*, 704–716. [CrossRef]

52. Hu, M.; Strachan, J.P.; Li, Z.; Grafals, E.M.; Davila, N.; Graves, C.; Lam, S.; Ge, N.; Yang, J.J.; Williams, R.S. Dot-Product Engine for Neuromorphic Computing. In Proceedings of the 53rd Annual Design Automation Conference, Austin, TX, USA, 5–9 June 2016; pp. 1–6.
53. Liang, J.; Yeh, S.; Wong, S.S.; Wong, H.-S.P. Effect of Wordline/Bitline Scaling on the Performance, Energy Consumption, and Reliability of Cross-Point Memory Array. *ACM J. Emerg. Technol. Comput. Syst.* **2013**, *9*, 1–14. [CrossRef]
54. Hagan, M.; Demuth, H.; Beale, M.; De Jesús, O. *Neural Network Design*, 2nd ed.; Hagan, M., Ed.; Oklahoma State University: Stillwater, OK, USA, 2014; ISBN 978-0971732117, 0971732116.
55. Truong, S.N. Compensating Circuit to Reduce the Impact of Wire Resistance in a Memristor Crossbar-Based Perceptron Neural Network. *Micromachines* **2019**, *10*, 671. [CrossRef] [PubMed]

Journal of
Low Power Electronics and Applications

MDPI

Article

A Morphable Physically Unclonable Function and True Random Number Generator Using a Commercial Magnetic Memory

Mohammad Nasim Imtiaz Khan [1,*], Chak Yuen Cheng [1,2], Sung Hao Lin [1], Abdullah Ash-Saki [1] and Swaroop Ghosh [1]

1 Department of Electrical Engineering, Pennsylvania State University, State College, PA 16801, USA; chakyuec@alumni.cmu.edu (C.Y.C.); frank19940124@gmail.com (S.H.L.); axs1251@psu.edu (A.A.-S.); szg212@psu.edu (S.G.)
2 Department of Electrical and Computer Engineering, Carnegie Mellon University, Pittsburgh, PA 15213, USA
* Correspondence: mohammad.nasim.imtiaz.khan@intel.com

Abstract: We use commercial magnetic memory to realize morphable security primitives, a Physically Unclonable Function (PUF) and a True Random Number Generator (TRNG). The PUF realized by manipulating the write time and the TRNG is realized by tweaking the number of write pulses. Our analysis indicates that more than 75% bits in the PUF are unusable without any correction due to their inability to exhibit any randomness. We exploit temporal randomness of working columns to fix the unusable columns and write latency to fix the unusable rows during the enrollment. The intra-HD, inter-HD, energy, bandwidth and area of the proposed PUF are found to be 0, 46.25%, 0.14 pJ/bit, 0.34 Gbit/s and 0.385 μm^2/bit (including peripherals) respectively. The proposed TRNG provides all possible outcomes with a standard deviation of 0.0062, correlation coefficient of 0.05 and an entropy of 0.95. The energy, bandwidth and area of the proposed TRNG is found to be 0.41 pJ/bit, 0.12 Gbit/s and 0.769 μm^2/bit (including peripherals). The performance of the proposed TRNG has also been tested with NIST test suite. The proposed designs are compared with other magnetic PUFs and TRNGs from other literature.

Keywords: MRAM; TRNG; PUF; morphable security primitive; hardware security primitive

Citation: Khan, M.N.I.; Cheng, C.Y.; Lin, S.H.; Ash-Saki, A.; Ghosh, S. A Morphable Physically Unclonable Function and True Random Number Generator Using a Commercial Magnetic Memory. *J. Low Power Electron. Appl.* **2021**, *11*, 5. https://doi.org/10.3390/jlpea11010005

Received: 23 December 2020
Accepted: 11 January 2021
Published: 14 January 2021

Publisher's Note: MDPI stays neutral with regard to jurisdictional clai-ms in published maps and institutio-nal affiliations.

1. Introduction

There has been a proliferation of Internet-of-Things (IoTs) edge devices, and cybersecurity aspects of such devices are becoming a concern. Cybersecurity techniques, securing only the upper layer of software stack, are not sufficient anymore as underlying hardware faces a plethora of security and trust issues such as cloning, reverse engineering, Trojan insertion, side channel attack [1], recycling/counterfeiting, and so on. Therefore, many techniques and countermeasures are explored to ensure security and trust of hardware systems at various levels. For example, security primitives like recycling sensor [2], Physically Unclonable Functions (PUF) [3,4], True Random Number Generator (TRNG) [5], tamper sensor [6], encryption engines [7], Trojan detection [8–10], etc., are proposed to secure hardware. The security solutions are mostly driven by CMOS -based technologies. However, the CMOS-based solutions can be limited by the small set of features that can be leveraged to develop security primitives such as process-variation (PV) and thermal noise. In this regard, emerging technologies can be promising. They offer new sources of randomness and noise that can be harnessed to design robust security primitives. Besides, the solutions can achieve low power, high density, and high speed.

Prior Work on PUF: PUF is one of the widely accepted hardware security primitives that finds application in authentication. A PUF exploits differences between two chips due to intrinsic variation during the manufacturing process [4] to generate chip-specific and unique signatures. Several conventional and emerging technologies such as CMOS [3,4], memristor [11] and spintronic technologies [12,13] are explored to design PUFs. The CMOS

PUFs include Static RAM (SRAM) based memory PUF, arbiter PUF and ring oscillator based PUFs [3]. Emerging technology based PUFs include memristor, spintronic memory, Resistive RAM (RRAM) [14,15], Domain Wall Memory (DWM), Magnetoresistive RAM (MRAM), etc. For example, DWM is used to design arbiter PUFs with exponential Challenge Response Pairs (CRP) which are resilient to machine learning attack [16]. Several PUFs based on Magnetoresistive RAM (MRAM) are also proposed [17–19]. In [17], the authors utilize unique energy-tilt of a Magnetic Tunnel Junction (MTJ) which stems from random geometric variations in the MRAM cells to generate PUF responses. The work in [18] identified the unreliable cells in a PUF to devise a zero bit-error-rate PUF. In Ref. [19], a strong PUF is proposed based on combining the resistances of a group of cells and generating their digital signature. The work exploits nano-scale analog disorders of MRAM, and this technique can be extended to other memory technologies.

Prior Work on TRNG: TRNGs exploit a source of randomness such as thermal noise, dynamic variations, etc. to generate random numbers. Ideally, the outputs of a TRNG must have high entropy and zero correlation. Several TRNGs are proposed using spintronic devices in prior work [20–23]. In [20,21], TRNG is implemented by manipulating the amplitude of the programming pulse. However, Ref. [20] requires controlling current in the order of μAs which is hard to achieve and Ref. [21] requires integration of analog circuit which is very sensitive to noise. A stochastic programming by current-driven STT using a Complementary Polarizer Spin Dice (CPSD) proposed in [22]. In [23], algorithms for PUF (based on read current) and TRNG (based on pulse width/amplitude manipulation) are proposed using MRAM. However, implementation details and results are not provided.

Proposed morphable PUF and TRNG: We propose a morphable security primitive using commercial magnetic RAM which can be used as both a PUF and a TRNG. To run it in the PUF mode, write time is controlled, and to run it in the TRNG mode, the number of write pulses is manipulated. Thus, it is named as *morphable*.

The magnetic tunnel junctions (MTJs) in the MRAM exhibit different write latencies owing to intrinsic and extrinsic PVs. For the same write time a bit may (or may not) flip in two different chips (extrinsic variation). This observation can be exploited to generate unique signatures from different chips which is useful for designing a PUF. We also notice that the same bit in a chip will *randomly* flip (intrinsic variation) if written multiple times with the same data. This is useful for designing a TRNG. The Figure 1 schematically shows the concept of re-purposing the MRAM in two different modes, i.e., PUF and TRNG. Thus, a 128 KB commercial MRAM chip can be converted to work solely as a 128 KB PUF or a 128 KB TRNG, or it can cohabitate a 64 KB PUF and a 64 KB TRNG.

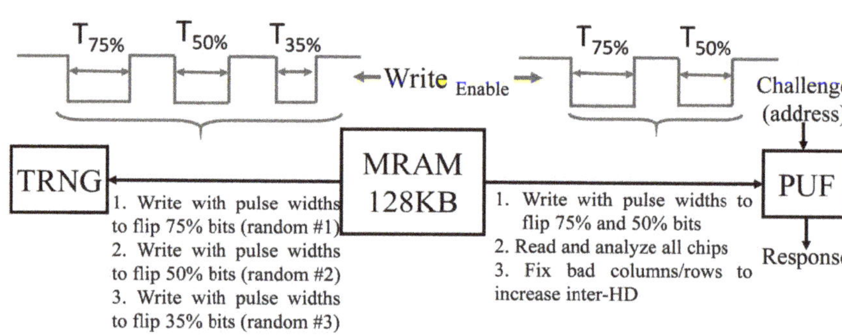

Figure 1. Morphable security primitive using Magnetoresistive RAM (MRAM).

Note that the data width of the MRAM chips we used for this work is 8-bit. Therefore, we call each address as a row and number the rows with the corresponding address. Each row produces 8 bit and we number them as column 0 to 7 from Most Significant Bit (MSB) to Least Significant Bit (LSB). We summarize our methodology to realize PUF and TRNG from the MRAM chip below.

Working principle of proposed PUF (wPUF): First, we flush the bits of the PUF (write with 0s). Then, we try writing 1 in all the bits. The write time of the pulse is set to 50% switching probability so that 50% of the bits flip. However, due to stochastic nature of the bitcell and process variation, each chip will be written with different data which can be used as the signature of the chip. However, our analysis shows that 4 columns of all rows of the chips are stuck to 0 (2 columns)/1 (2 columns) and do not show probabilistic switching as expected. Remaining 4 columns show the probabilistic switching and therefore, overall switching probability is around 50%. These severely limits PUF variation from chip to chip. We noted that the 4 columns are stuck because either they are very strong (stuck to 0, requires more write time) or weak (stuck to 1, requires less write time and always gets written to 1). We expect that the PUFs based on real memory chip implementation of any emerging NVMs might exhibit this type of behavior. Therefore, some of the bits of each address could be unusable for PUF. In this work, we propose techniques to fix these bad columns by exploiting the temporal randomness of good columns. Note that this is in contrast to MRAM and STT-MRAM PUFs presented in literature [17–19] that are specifically and carefully designed (bits, access transistor and peripherals) to amplify and capture the variability and to achieve high inter-HD and low intra-HD.

Working principle of proposed TRNG: Our analysis show that just biasing an address with 50% switching probability does not provide all possible outcomes. For example, the number of possible outcomes for a 4-bit TRNG is 16. However, we observed less number of outcomes due to strong/weak bits which limits the scope of the TRNG and makes it Pseudo Random Number Generator (PRNG). Therefore, we propose the following technique for TRNG: first, we write all 0 s in the cells of TRNG; then we write all bits to 1 s by selecting the write time to flip 75% of the bits (i.e., 75% switching probability) to extract the first random number. For generating the second and third random number from the same address, we propose to repeat the above steps with the write time to flip 50% of the bits (i.e., 50% switching probability) and with the write time to flip 25% of the bits (i.e., 25% switching probability) respectively. This way we get all 16 possible outcomes from the 4 good columns with tolerable standard deviation. For example, 75% switching probability will mainly generate 4'b0111, 4'b1011, 4'b1101 and 4'b1110, 50% will mainly generate 4'b0011, 4'b0101, 4'b1001, 4'b0110, 4'b1010 and 4'b1100 and 25% will mainly generate 4'b0001, 4'b0010, 4'b0100 and 4'b1000. The remaining ones (4'b0000/4'b1111) are also generated (with lower recurrence number) mainly with 25%/75% probability if all the four bits are either very strong or weak respectively. Note that we did not fix the 4 bad columns with the good ones in case of TRNG. This is to prevent machine learning attacks that can profile the TRNG outcomes with fewer iterations (since the fixing can make the bits correlated).

Morphable Security Primitive: To use the MRAM in the TRNG mode (i.e., to generate unique true random numbers), the MRAM needs to be re-written every time. On the other hand, to use it as a PUF, the MRAM needs to be written only once (during the enrollment phase). The data that get written to PUF addresses depends on the PV which cannot be replicated by a malicious entity. It should be noted that PUF can be morphed to TRNG if written multiple times as proposed for TRNG.

To the best of our knowledge, this is the first experimental demonstration of a PUF and TRNG using commercial MRAM chip. A work-in-progress version of this work has been published in [24] where the methodology is discussed and initial data were presented. However, this work explains the design and results in detail. In summary, we make the following contributions:

- We characterize the MRAM bit-to-bit write latency under voltage and temperature variations.
- We characterize the MRAM response under multiple write disturbs which can be useful for TRNG.
- We propose a write PUF (wPUF) by biasing the MRAM with a write latency with 50% switching probability. The proposed PUF exhibits excellent stability and uniqueness.
- We show that 75% of the bits could be unresponsive to a challenge and propose techniques to convert them into useful bits avoiding expensive row and columns masking.
- We propose a TRNG by exploiting random MRAM responses under multiple write disturbs.
- We benchmark the proposed PUF and TRNG with existing designs.

The rest of the paper is organized as follows: Section 2 provides the background of MRAM technology, details of the experimental setup and analysis of the MRAM responses to write latency and number of writes. Sections 3 and 4 describe the proposed PUF and TRNG. Section 5 presents discussion and Section 6 draws conclusions.

2. Background on MRAM and Its Variation

In this section, we present basics of the toggle MRAM and characterize its statistical and temporal behavior.

2.1. Basics of MRAM

MRAM bitcell (Figure 2) contains one MTJ and one NMOS access transistor. The MTJ lies between a pair of metal-lines named digit-line and bit-line to facilitate write operation. The metal-lines are parallel to the cell plane and placed orthogonal to each other. An induced magnetic field is created with appropriate polarity by passing current through the lines during write operation. The field exerts a torque on the free layer magnetic orientation, causing it to flip. During read operation, the access transistor is turned ON, and a fixed voltage is applied across the cell to sense the equivalent resistance.

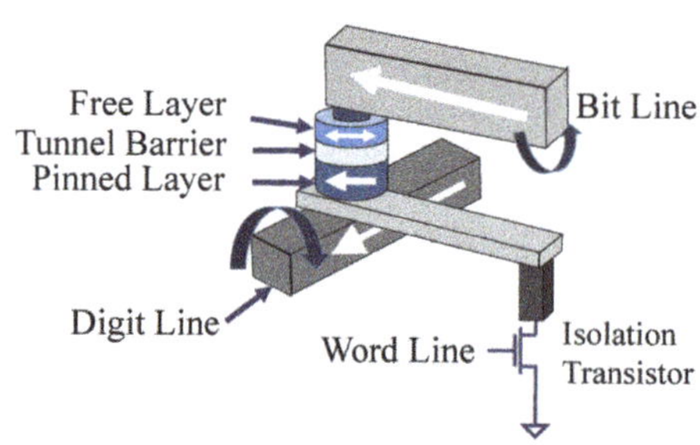

Figure 2. Toggle MRAM bitcell containing one Magnetic Tunnel Junction (MTJ) and one access transistor.

A commercial toggle MRAM chip (MR4A08B) with 2 million rows and 8 columns is used in this work to validate the proposed morphable security primitive. Table 1 captures the key features of the MRAM chip. It should be noted that we have used the first 128 KB (1 Mbit) out of 16 Mbit of the chip. Figures 3 and 4 presents the timing diagram of the write and read operations of the MRAM chip. From the data sheet, it is evident that the chip requires maximum of 15 ns of write enable (low) signal for successful write and data is valid after 25 ns of read enable signal. In this work, we modified the specified read and write times to realize the proposed PUF and TRNG.

Table 1. Characteristic of the MRAM Chip.

Parameter	Value
Capacity	16 Mbit
Read/Write Cycle	35 ns
Address/Data Bus Length	21/8
Retention Time	>20 years
AC stand by Current	9–14 mA
AC Active Current (Read/Write)	60–68 mA/152–180 mA

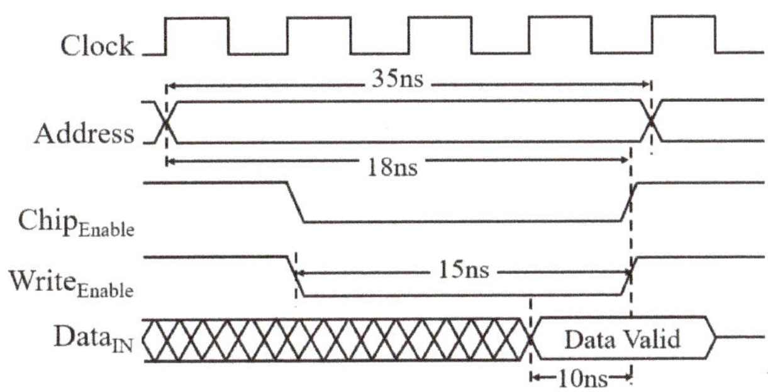

Figure 3. Write cycle of MRAM chip. Note that the chip requires maximum of 15 ns of write enable (low) signal for successful write.

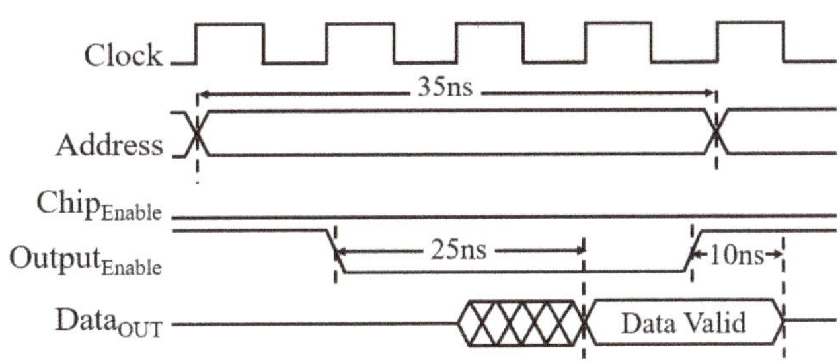

Figure 4. Read cycle of MRAM chip.

2.2. Experimental Setup

A four-layer PCB is designed to interface the MRAM chip with a logic analyzer and Digilent Basys3 [25]. The PCB is designed with minimum wire resistance, inductance and capacitance to stabilize the supply current for PUF power measurement. A 1 Ω thick film current sensing resistor is connected between the ground of the PCB and the MRAM chip to sense the MRAM current during read/write. A heat gun is used to change the ambient temperature. We have analyzed a total of 10 chips.

First, we interfaced the MRAM chip with logic analyzer and wrote all addresses at faster than 15 ns write time@V_{DD} = 3.3 V, 25 °C. We found that the data is written successfully for all chips at 12 ns (Figure 5). Therefore, the write time should be less than 12 ns at nominal operating condition to bias the cells with 50% write probability. We read the data at 30 ns in order to avoid read failures. Next, we interfaced the MRAM chip with the FPGA. The writing is done at much higher frequency to achieve partial write and reading is done at 25 MHz (40 ns). The read/write traces are captured by Keysight DSOS-804A Oscilloscope [26] with sampling frequency of 20 GSa/s and Bandwidth of 8 GHz. The experimental setup is shown in Figure 6. Data is sent to a PC using UART (baud rate 9600 bps) for MATLAB analysis.

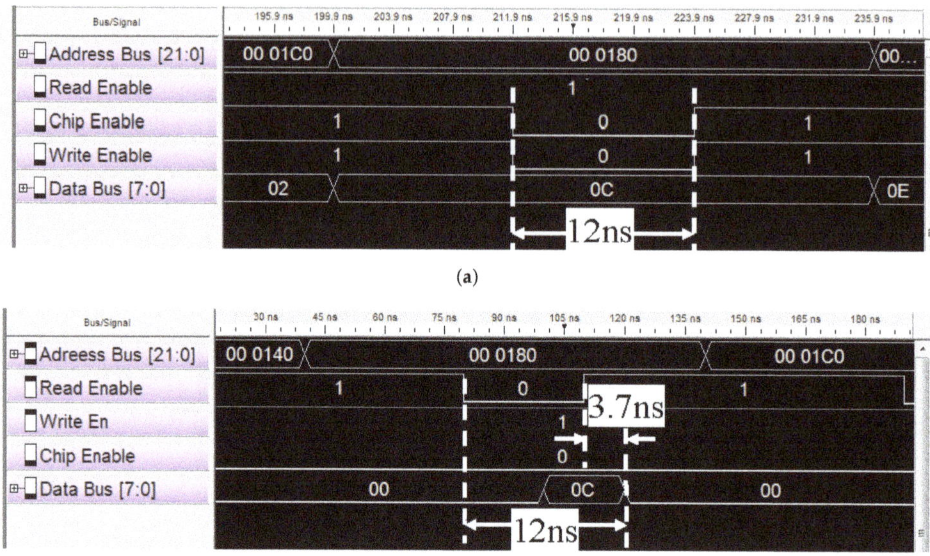

Figure 5. MRAM (**a**) write and (**b**) read waveforms. Note that with 12 ns of write enable (low) signal, data (0x0C) is written successfully (to address 0x00 0180).

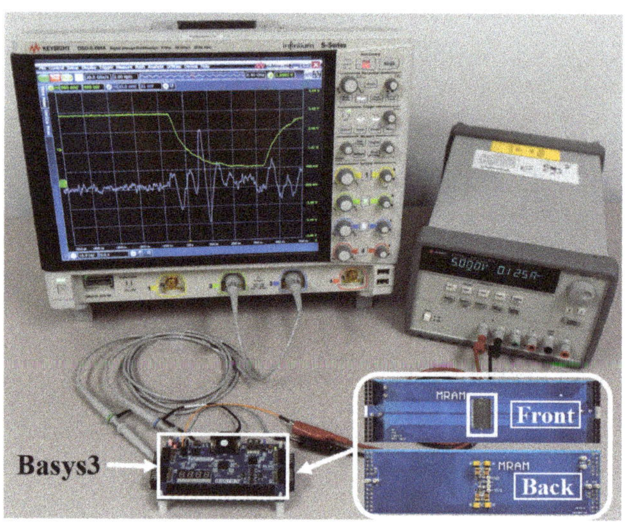

Figure 6. Experimental setup. The MRAM PCB (inset) is mounted on top of Basys3 for noise reduction.

2.3. Switching Variation of MRAM

We swept the write time of each of the 10 chips to identify the Cumulative Distribution Function (CDF) of write time for $0 \rightarrow 1$ switching. Figure 7 shows the CDFs of three chips that are found to possess highest/median/lowest write times. To realize PUF, the write latency of each chip should be chosen to write 50% of the bits to get uniform 0/1 distribution. From Figure 7 it is obvious that write time varies from chip to chip. This could be useful for TRNG application. From the experimental results and analysis, we note the following:

(a) The write latency can be arranged in increasing order as, $0 \rightarrow 0$, $1 \rightarrow 0$, $0 \rightarrow 1$, $1 \rightarrow 1$ (i.e., writing $0 \rightarrow 0$ takes shortest time and $1 \rightarrow 1$ takes longest time ($T_{(1\rightarrow1)}$)).

(b) For writing $1 \rightarrow 1$ with $t_{write} < 10$ ns and 10 ns$< t_{write} < T_{1\rightarrow1}$, the stored data becomes 0 and stochastic respectively. From this observation we conclude that the chip initializes the data to 0 before writing a 0 or 1. We believe that, this observation is rooted at the toggle MRAM implementation which toggles the bit (irrespective of the existing state) during write.

(c) Consecutive writes of 1 s with the same short write time reduces number of 1 s in the data (consequence of (b)). For example, if the addresses are flushed with all 0 s and then written with all 1 s with write time to flip 75% bits (which is $< T_{1\rightarrow1}$) and the writing is repeated with same time (without flushing in between), the number of 1 s in the data reduces to ≈49%, ≈34% and then remains fairly constant (Figure 8). This can be useful for TRNG (further explained in Section 4).

(d) Columns 0 and 7 are biased to data 0, columns 1 and 2 are biased to data 1 and column 3–6 mainly shows 50% switching probability (Figure 9) although we select the write time to flip 50% of the bits for the entire memory array. Therefore, columns 0, 1, 2 and 7 cannot be used for PUF and TRNG. If every cell needs to be biased at their own 50% switching probability, then write time for each cell in a row/column needs to be controlled individually which is practically impossible.

(e) Error-free read can be performed with 23.3 ns under 3.0 V–3.6 V and 25 C–60 C. The highest read latency is 23.3 ns@3.0 V, 60 C and lowest read latency is 22.47 ns@3.6 V, 25 C.

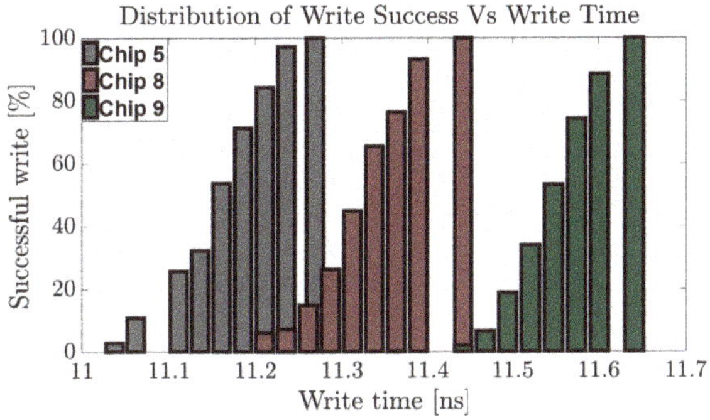

Figure 7. Cumulative Distribution Function (CDF) of write success of 3 chips (out of 10). These 3 chips exhibit maximum write time variation.

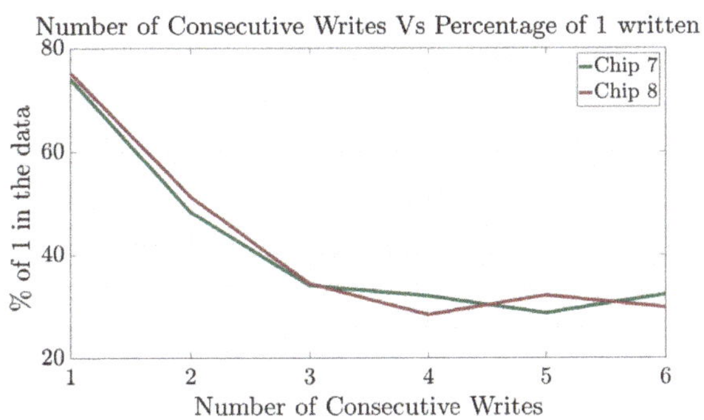

Figure 8. Number of consecutive writes vs. percentage of 1s in the data.

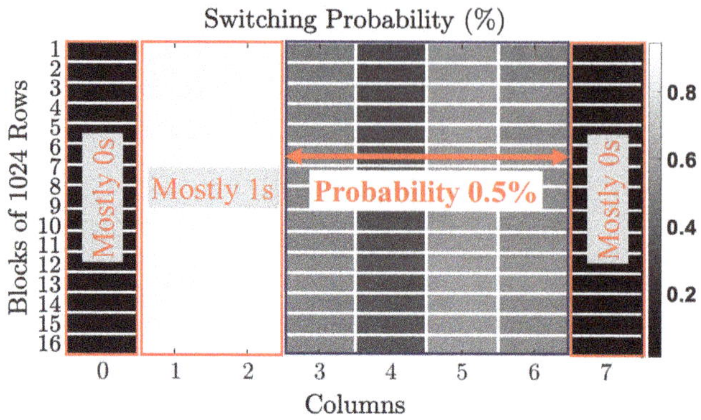

Figure 9. Switching probability distribution of 16 KB MRAM. First 16384 rows are grouped to 16 blocks (1024 rows/block).

3. PUF

Memory PUFs exploit random initialization of the memory bits due to PV. The address bits are used as challenge and the bits read from the memory array with the address is the response of the PUF. In this section, we present the proposed wPUF and analyze its performance using experimental results.

3.1. Proposed wPUF

The principle of the proposed wPUF is explained in algorithm (Figure 10). The wPUF will have two phases:

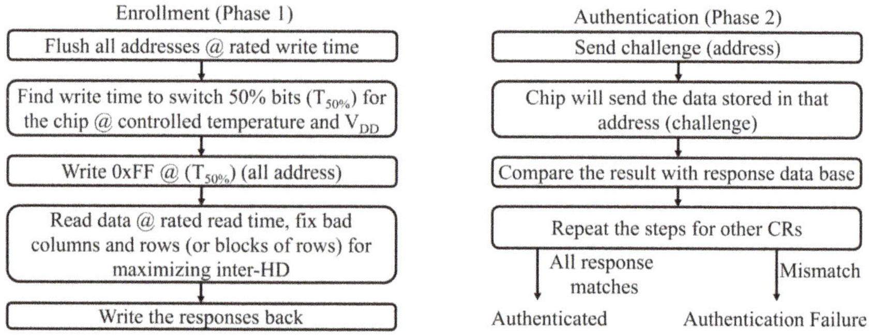

(**a**) Enrollment phase (**b**) Authentication phase

Figure 10. Two phases of wPUF, (**a**) Enrollment phase; (**b**) Authentication phase.

(a) Enrollment phase (Figure 10a): In this phase, unique signature from a chip is extracted using the method described below. First, the manufacturer prepares the memory in an initial state by writing all bits with 0 s (flushing of the MRAM). Next, the manufacturer try writing all 1 s with a write time $T_{50\%}$ to switch 50% bits to 1. Note that, $T_{50\%}$ is supply voltage (V_{DD}) and temperature dependent. As the manufacturer has access to a controlled-environment, managing target V_{DD} and temperature will not be an issue. Finally, the entire address space is read to extract unique signatures. The steps are repeated for several chips to create a pool of data for statistical analysis and subsequent correctional measures. Then, all chip responses will be compared to separate the unresponsive columns/rows (that show no variation) for all chips and which gives highest uniqueness for different chips. Based on this observation, the unresponsive columns and rows need further processing (discussed in Section 3.3).

(b) Authentication phase (Figure 10b): This phase involves sending challenges (memory addresses) to the PUF and getting back responses (data stored in those addresses). Sufficient number of CRPs should be matched so that a chip can be distinguished as unique and authenticate. If there are reasonable mismatch in the CRP pairs, authentication should fail.

It should be noted that the enrollment phase for a chip is just a one-time operation. Once enrollment phase is performed by the manufacturer, only authentication phase (involves read operation) needs to be performed in real time.

3.2. Performance Analysis

PUF performance is evaluated with mainly three parameters:

3.2.1. Uniqueness (Inter-Die HD)

Uniqueness (measured by inter-die HD) in the PUF response enables the identification of different chips uniquely. A 50% inter-die HD is desirable. In this work, the inter-die HD is calculated for 10 chips and the average is found to be only 22.5% which is very low. The reason for getting the low inter-HD is attributed to, (i) a smaller number of bits per challenge (4 bits) to compare the responses of different chips (only 16 combinations), and, (ii) inability to bias each bit at their 50% switching probability individually. The inter-die HD is improved in the next subsection.

3.2.2. Reliability (Intra-Die HD)

Reliability (measured by intra-die HD) is the measure of the dependency of PUF response to the intra chip voltage and temperature variations. A 0% intra-die HD is desirable. Intra-die HD is measured by XORing the responses of the PUF at various voltages and temperatures. In this work, we capture the responses of all chips for V_{DD} ranging 3.0 V to 3.6 V and temperature ranging from 25 °C to 60 °C with read time of 23.3 ns which gives intra-die HD = 0%. A perfect intra-die HD is achieved by relaxing the read time. System throughput can be increased by reducing the read time which incurs a non-zero intra-die HD for some operation condition. We propose to implement relaxed read time and compromise system throughput since achieving 0% intra-die HD is critical.

3.2.3. Uniformity

For uniformity in the PUF response, the probability of 1's and 0's in the response for possible challenges should be 50%. We evaluate the uniformity by the frequency metric in the NIST benchmark for all the possible 16 CRPs of 4 MRAM bits of the PUF and found it to be ≈48% with block frequency test. Entropy test on the responses show satisfactory *p*-value (>0.01) which ensures randomness.

3.3. Improving Inter-HD

Figure 11 shows the average inter-HD of all 10 chips for first 80 rows (8 rows per row block). From the result, we observe that inter-HD of Col 0, 1, 2 and 7 are poor as they are stuck to either 0 or 1. Furthermore, we also observe that certain address rows provide good inter-HD (green/blue) whereas other rows provide very poor inter-HD (red). We propose the following techniques to improve the inter-HD.

3.3.1. Improving Column Performance

We propose writing the same address twice with write time corresponding to 75% switching probability. The first/second response will have 75%/50% number of 1s. Since the two responses are purely random, we propose using first response of columns 3–6 as the response of bad columns (i.e., columns 0, 1, 2 and 7) and the second response of columns 3–6 as their own response. Therefore, we avoid masking of unusable columns and restore the PUF bandwidth to 8 bits per challenge. The response obtained for bad columns are enforced on them with relaxed write latency for subsequent authentication.

This technique exploits the fact that each bit of MRAM can produce more than one random bit of information.

3.3.2. Improving Row Performance

By improving column performance, inter-die HD of stuck columns improves by 1.7X-4X (individually). However, total inter-HD remains poor (≈28%) due to poor performance of some rows that exhibit 0% inter-HD. Further, we observe that the rows with poor inter-HD are mainly stuck to either all 0s or all 1s. The rows stuck to 0s (perhaps due to higher thermal stability) are fixed by re-writing them with higher write time to trigger statistical flipping. The row stuck to all 1s (perhaps due to low thermal stability) are fixed by re-writing them with lower write time.

After improving row/column performances, we obtain inter-HD of 46.25% which is close to the ideal value. The proposed improvement technique incurs one-time energy (≈2.7X) and computational overhead during the enrollment phase. Furthermore, it should be noted that, alternative improvement technique of masking the rows/blocks of rows incurs significant area overhead. For the case of Figure 11, 75% bits are lost if masking is implemented. The energy, bandwidth and area of the wPUF are 0.14 pJ/bit, 0.34 Gbit/s, 0.385 μm^2/bit (including all peripheral circuits). Table 2 benchmarks proposed wPUF with existing MRAM/STTRAM PUFs. We can observe that the proposed PUF is comparable to prior works. It should be noted that (i) the proposed PUF is based on a commercial chip which gives practical process variation; and, (ii) the data bus is 8 bit long, therefore the bandwidth is comparatively low.

Row Block	Column 1	2	3	4	5	6	7	8
1	0.05	0.125	0.125	0.275	0.25	0.25	0.25	0
2	0.05	0.125	0.15	0.35	0.3	0.45	0.475	0
3	0.05	0.125	0.125	0.35	0.275	0.375	0.35	0
4	0.125	0.075	0.075	0.05	0.25	0.05	0.1	0.1
5	0.05	0.375	0.375	0.225	0.15	0.35	0.25	0
6	0.05	0.15	0.15	0.425	0.375	0.475	0.45	0
7	0.05	0.15	0.15	0.475	0.35	0.475	0.325	0
8	0.125	0.125	0.125	0.175	0.225	0.05	0.175	0.075
9	0	0.175	0.175	0.2	0.125	0.35	0.125	0
10	0.05	0.05	0.05	0.425	0.275	0.475	0.4	0

Figure 11. Average inter-HD of 10 chips for first 80 rows (8 rows/block).

Table 2. Performance comparison of different MRAM/STTRAM PUFs with the proposed wPUF.

PUFs	Inter-Die HD (%)	Inter-Die HD (%)	Entropy	Area (MTJ) (μm^2)	Bandwidth (Gbit/s)	Energy/bit (pJ)	Experimental
[15]	-	50.1	0.985	0.046	6.4	-	No
[17]	0.02	47	0.99	6.74 (64 bit)	12.8	-	Yes
[27]	7.76	60.6	-	0.065	6.4	2.42	No
[28]	-	49.89	0.95	0.005	6.4	0.001	No
wPUF (This Work)	0	22.5 (before) 46.26 (after)	0.95	0.385 [1]	0.34	0.14 [1]	Yes

[1] includes decoder, sense amp, and other peripheral circuitry and considers 100 mil × 100 mil of die size for TSOP-II IC package.

4. TRNG

TRNG generates true random number based on some random inherent noise. For memory-based TRNG, the memory bitcell is biased at the 50% switching probability point at which the bit stabilizes to either 1 or 0 depending on the noise. Therefore, the written value is highly unpredictable, varies from chip to chip and strongly depends on the operating conditions and noise. In this section, we present the proposed TRNG and analyze its performance using experimental results.

4.1. Proposed TRNG

Figure 12 presents the algorithm to realize TRNG from the MRAM chip. First, the memory address is flushed, i.e., 0 s are written to all bits. Next, the write time is set to 75% switching probability and 1 s are written to the addresses. The 75% switching probability means the ratio of 1 s to 0 s will be 75%. The data written is considered as the first random number. Then, the same address is written again with all 1 s with the same write time. We notice that the ratio of 1 s to 0 s reduces to ≈50% (observation (b), Section 2) on the second write. This is the second random number. Similarly, when we write for the 3rd time the 1 s to 0 s ratio drops to ≈35% which gives out the 3rd random number. By repeating the same process for different write times and different addresses, we can generate more random numbers.

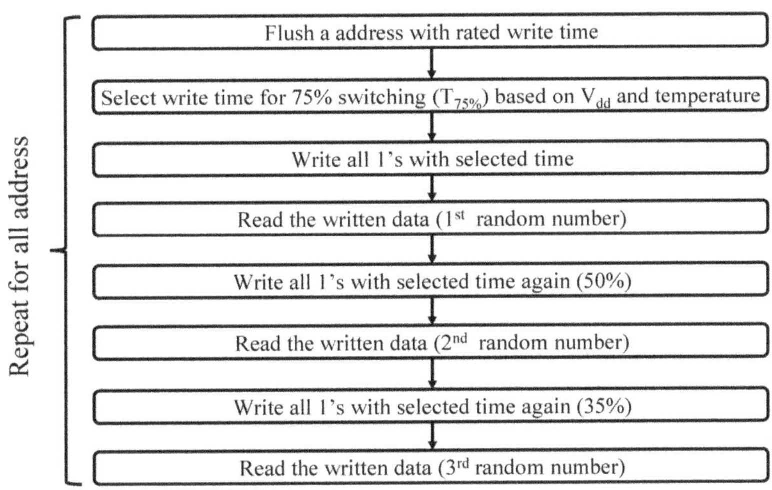

Figure 12. Algorithm of the proposed True Random Number Generator (TRNG).

Since it is impossible to bias individual MRAM bits to their 50% switching probability using a common write latency, we bias the MRAM using three different switching probabilities and extract all possible outcomes from n-bits. For example, biasing 4 bits at 50% switching probability means that on an average, 2 bits out of 4 will be set to '1'. This will produce 6 outcomes out of 16 possibilities (i.e., 1100, 0110, 0011, 1010, 1001 and 0101). Biasing at 75% switching probabilities mean 3 bits will be set to 1s producing 4 new outcomes (1110, 0111, 1011 and 1101). Biasing at 35% switching probabilities mean 1 bit will be set to 1s producing 4 new outcomes (0001, 0010, 0100 and 1000). Note that the remaining two outcomes (0000/1111) are also produced by TRNG but with lower frequency.

Write operation of any spintronic memory suffers due to temperature and V_{DD} variation. Therefore, TRNG can be biased to some preferential values and behaves as Pseudo Random Number Generator (PRNG). Tracking of V_{DD} and temperature are needed for magnetic TRNG to change the biasing conditions accordingly and achieve desired switching probability.

We assume that the processor can track V_{DD} and temperature to select appropriate write time. Figure 13a shows the proposed circuit and Figure 13b shows the timing waveform to implement variable write time. For this example, we have assumed that the processor runs at 1 GHz (time period 1 ns) and write time can vary from 9 ns to 13 ns depending on operating condition. Therefore, from this circuit we can select write time with a step size of 0.5 ns in that range (by asserting WR_ENABLE and any of the other 8 pulses). If finer step size is required, more pulses can be generated with less duty cycle. However, in that case, the MUX will have more inputs (4:1 MUX (16 pulses in total) for 0.25 ns granularity).

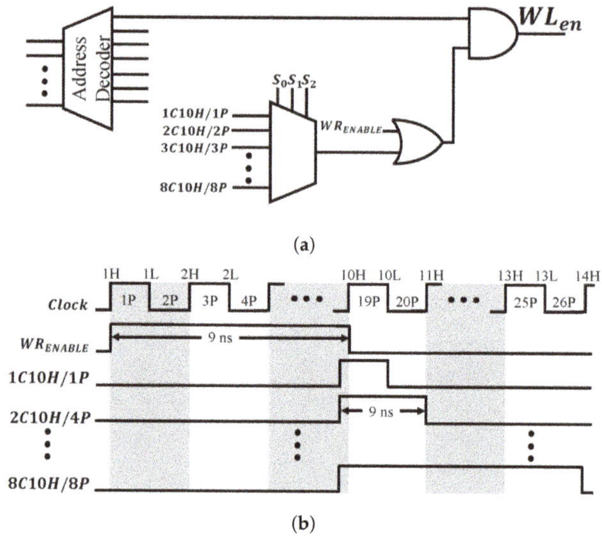

Figure 13. (**a**) Circuit and (**b**) timing waveform to implement specific write time selection for achieving desired switching probability.

4.2. Performance Analysis

(a) Entropy: Entropy of TRNG defines the randomness of the generated data. Entropy can be calculated with the following equation [29]:

$$E = -\sum_{i=2}^{2} p_i log_2(p_i) \tag{1}$$

where p1 and p2 are probability of 1 s and 0 s respectively in a n-bit long data stream. For an ideal TRNG, the entropy is 1. We have calculated entropy of all chips column-wise and the average entropy is shown in Figure 14a. It is evident that, columns 0, 1, 2, and 7 show poor randomness (explained in previous section) and therefore needs to be masked. After masking, the proposed TRNG offers an entropy of 0.95.

(b) Repeatability: Repeatability is another important metric to evaluate TRNG performance. Ideally, a TRNG should only repeat when all other possible cases are already covered. For example, for a 4-bit TRNG, the generated value should only repeat when all the 16 possible values are generated. However, practically this is very difficult to achieve. In the proposed TRNG algorithm, we are able to get all possible outcomes with small standard deviation (0.0062) and the correlation coefficient is calculated as 0.05. Figure 14b shows the outcomes of 10,000 cases for the proposed TRNG.

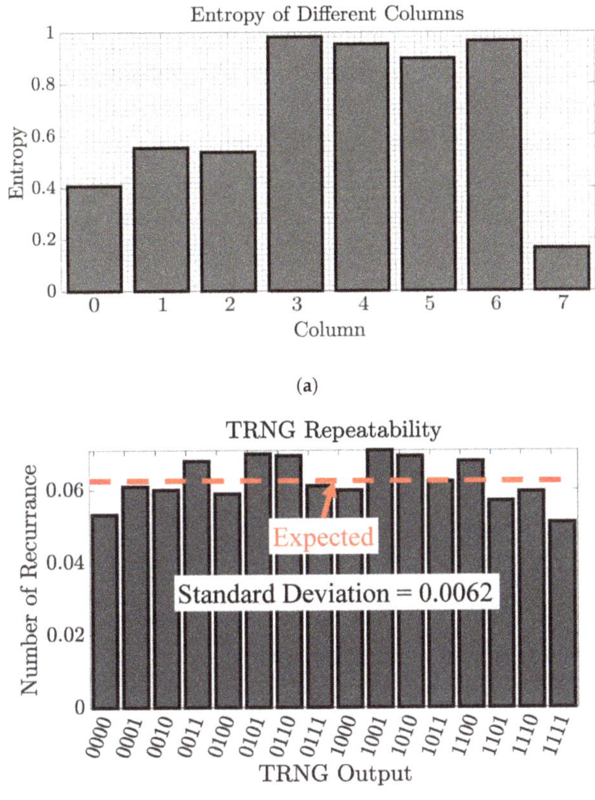

Figure 14. (**a**) Average entropy of all chips measured column-wise; (**b**) repeatability of TRNG outcomes out of 10,000 responses.

The energy, bandwidth and area of the proposed TRNG are found to be 0.41 pJ/bit, 0.12 Gbit/s and 0.769 μm^2/bit (including all peripheral circuits). We have also tested our TRNG outputs with NIST STS randomness test (summarized in Table 3). It is evident from the results that the proposed TRNG achieves excellent quality. Table 4 benchmarks the proposed TRNG with existing MRAM/STTRAM TRNG. We can observe that the proposed PUF is comparable to prior works and its energy (bandwidth) is significantly lower (higher) than others.

Table 3. Summary of NIST Suite Statistical Result.

NIST Statistical Test	*p*-Value	Proportion	Result
Frequency	0.349865	199/200	Pass
Block Frequency	0.257217	199/200	Pass
Cumulative Sums	0.393322	199/200	Pass
Discrete Fourier Transform	0.476393	199/200	Pass
Approximate Entropy	0.844361	200/200	Pass

Table 4. Performance comparison of different MRAM/STTRAM TRNG with proposed TRNG.

TRNG	Correlation	Entropy	Area (MTJ) (μm^2)	Bandwidth (Gbit/s)	Energy/bit (pJ)	Experimental
[20]	0.003	-	0.014	0.0005	14.97	Yes
[21]	-	-	0.0085 [1]	0.0833 [1]	0.3386 [1]	Yes
This Work	0.05	0.95	0.769 [2]	0.12	0.41 [2]	Yes

[1] Considering 3 ns of read time; [2] Includes all peripheral circuitry and considers only 4 bits of each row. Thus, area is 2× of wPUF.

5. Discussions

V_{DD} and temperature tracking: Any biasing technique to achieve a particular switching probability (pulse width/duration) is susceptible to V_{DD}/temperature. Therefore, V_{DD}/temperature tracking is required to select appropriate biasing condition which can be designed based on statistical data.

Considerations to other magnetic memory architecture: Consecutive writes of 1 s with less than $T_{1\to 1}$ write time gives 75%, 50% and 35% of number of 1 s in the data (observation (c)) for the toggle MRAM chip. However, three different write times ($T_{75\%}/T_{50\%}/T_{35\%}$) can be implemented for other memory (that does not show this behavior).

Novelty of this work: Prior works consider bitwise normal distribution of switching probability. However, there are several practical challenges with the real memory implementation. First, it offers a narrow distribution with some columns/rows stuck at 0/1. Besides, the real memory chip does not provide granular access to each individual bits and for that biasing of the bits cannot be bitwise tailored. Even custom biasing for each row is impractical. To the best of our knowledge, we make the first attempt to systematically understand and address these practical challenges in this paper.

Note that we have selected MRAM since it is very promising due to its low static and read power consumption. However, the proposed post-processing techniques to improve the inter-HD of PUF and entropy of TRNG are applicable to other memory technologies.

TRNG robustness to Machine Learning Attack: Random Number Generator (RNG) can be vulnerable to machine learning attack [30]. However, an RNG can be robust against such attack if the non-linearity is very high (i.e., no repetitive patterns in outcomes of RNG). Since the response of the proposed TRNG is non-linearly dependent on numerous parameters, e.g., write pulse width, write voltage and temperature due to non-linear magnetization dynamics of MRAM free layer, the proposed MRAM TRNG is expected to be robust against machine learning attack.

6. Conclusions

We have investigated TRNG and PUF implementation using magnetic memory and manipulation of write time and number of writes. We have analyzed the practical implications of designing TRNG/PUF using commercial MRAM and addressed these issues to achieve high quality.

Author Contributions: Conceptualization, M.N.I.K. and S.G.; methodology, M.N.I.K.; software, A.A.-S.; validation, M.N.I.K., C.Y.C., S.H.L. and A.A.-S.; formal analysis, M.N.I.K.; investigation, M.N.I.K.; resources, S.G.; data curation, C.Y.C. and S.H.L.; writing—M.N.I.K., A.A.-S. and S.G.; original draft preparation, M.N.I.K. and A.A.-S.; writing—review and editing, M.N.I.K., A.A.-S. and S.G.; visualization, M.N.I.K.; supervision, M.N.I.K. and S.G.; project administration, M.N.I.K. and S.G.; funding acquisition, S.G. All authors have read and agreed to the published version of the manuscript.

Funding: This work is supported by Semiconductor Research Corporation (2847.001 and 3011.001) and National Science Foundation (CNS-1722557, CCF-1718474, DGE-1723687, OIA-2040667 and DGE-1821766).

Conflicts of Interest: The authors declare no conflict of interest.

References

1. Khan, M.N.I.; Bhasin, S.; Yuan, A.; Chattopadhyay, A.; Ghosh, S. Side-Channel Attack on STTRAM Based Cache for Cryptographic Application. In Proceedings of the 2017 IEEE International Conference on Computer Design (ICCD), Boston, MA, USA, 5–8 November 2017; pp. 33–40.
2. Lin, C.W.; Ghosh, S. Novel self-calibrating recycling sensor using Schmitt-Trigger and voltage boosting for fine-grained detection. In Proceedings of the Sixteenth International Symposium on Quality Electronic Design, Santa Clara, CA, USA, 2–4 March 2015; pp. 465–469.
3. Herder, C.; Yu, M.; Koushanfar, F.; Devadas, S. Physical Unclonable Functions and Applications: A Tutorial. *Proc. IEEE* **2014**, *102*, 1126–1141. [CrossRef]
4. Suh, G.E.; Devadas, S. Physical Unclonable Functions for Device Authentication and Secret Key Generation. In Proceedings of the 2007 44th ACM/IEEE Design Automation Conference, San Diego, CA, USA, 4–8 June 2007; pp. 9–14.
5. Tsoi, K.H.; Leung, K.H.; Leong, P.H.W. Compact FPGA-based true and pseudo random number generators. In Proceedings of the 11th Annual IEEE Symposium on Field-Programmable Custom Computing Machines, FCCM 2003, Napa, CA, USA, 9–11 April 2003; pp. 51–61.
6. Magnetic Tamper Detection Using Low-PowerHall Effect Sensors. Available online: http://www.ti.com/lit/ug/tidub69/tidub69.pdf (accessed on 12 January 2021).
7. Miura, N.; Fujimoto, D.; Tanaka, D.; Hayashi, Y.-I.; Homma, N.; Aoki, T.; Nagata, M. A local EM-analysis attack resistant cryptographic engine with fully-digital oscillator-based tamper-access sensor. In Proceedings of the 2014 Symposium on VLSI Circuits Digest of Technical Papers, Honolulu, HI, USA, 10–13 June 2014; pp. 1–2.
8. Ghosh, S.; Basak, A.; Bhunia, S. How Secure Are Printed Circuit Boards Against Trojan Attacks? *IEEE Design Test* **2015**, *32*, 7–16. [CrossRef]
9. Khan, M.N.I.; Nagarajan, K.; Ghosh, S. Hardware Trojans in Emerging Non-Volatile Memories. In Proceedings of the 2019 Design, Automation & Test in Europe Conference & Exhibition (DATE), Florence, Italy, 25–29 March 2019; pp. 396–401.
10. Khan, M.N.I.; De, A.; Ghosh, S. Cache-Out: Leaking Cache Memory Using Hardware Trojan. *IEEE Trans. Very Large Scale Integr. (VLSI) Syst.* **2020**, *28*, 1461–1470. [CrossRef]
11. Mazady, A.; Rahman, M.T.; Forte, D.; Anwar, M. Memristor PUF—A Security Primitive: Theory and Experiment. *IEEE J. Emerg. Sel. Top. Circuits Syst.* **2015**, *5*, 222–229. [CrossRef]
12. Iyengar, A.; Ghosh, S.; Ramclam, K.; Jang, J.-W.; Lin, C.-W. Spintronic PUFs for Security, Trust, and Authentication. *J. Emerg. Technol. Comput. Syst.* **2016**, *13*, 1–5. [CrossRef]
13. Ghosh, S.; Govindaraj, R. Spintronics for associative computation and hardware security. In Proceedings of the 2015 IEEE 58th International Midwest Symposium on Circuits and Systems (MWSCAS), Fort Collins, CO, USA, 2–5 August 2015; pp. 1–4.
14. Chen, A. Utilizing the Variability of Resistive Random Access Memory to Implement Reconfigurable Physical Unclonable Functions. *IEEE Electron Device Lett.* **2015**, *36*, 138–140. [CrossRef]
15. Zhang, L.; Fong, X.; Chang, C.; Kong, Z.H.; Roy, K. Highly reliable memory-based Physical Unclonable Function using Spin-Transfer Torque MRAM. In Proceedings of the 2014 IEEE International Symposium on Circuits and Systems (ISCAS), Melbourne, VIC, Australia, 1–5 June 2014; pp. 2169–2172.
16. Chen, A.; Hu, X.S.; Jin, Y.; Niemier, M.; Yin, X. Using emerging technologies for hardware security beyond PUFs. In Proceedings of the 2016 Design, Automation & Test in Europe Conference & Exhibition (DATE), Dresden, Germany, 14–18 March 2016; pp. 1544–1549.
17. Das, J.; Scott, K.; Rajaram, S.; Burgett, D.; Bhanja, S. MRAM PUF: A Novel Geometry Based Magnetic PUF With Integrated CMOS. *IEEE Trans. Nanotechnol.* **2015**, *14*, 436–443. [CrossRef]
18. Vatajelu, E.I.; Natale, G.D.; Prinetto, P. Zero bit-error-rate weak PUF based on Spin-Transfer-Torque MRAM memories. In Proceedings of the 2017 IEEE 2nd International Verification and Security Workshop (IVSW), Thessaloniki, Greece, 3–5 July 2017; pp. 128–133.
19. Khaleghi, S.; Vinella, P.; Banerjee, S.; Rao, W. An STT-MRAM based strong PUF. In Proceedings of the 2016 IEEE/ACM International Symposium on Nanoscale Architectures (NANOARCH), Beijing, China, 18–20 July 2016; pp. 129–134.
20. Seki, A.F.T.; Kubota, K.Y.H.; Imamura, H.; Yuasa, S.; Ando, K. Spin dice: A scalable truly random number generator based on spintronics. *Appl. Phys. Express* **2014**, *7*, 083001.
21. Oosawa, S.; Konishi, T.; Onizawa, N.; Hanyu, T. Design of an STT-MTJ based true random number generator using digitally controlled probability-locked loop. In Proceedings of the 2015 IEEE 13th International New Circuits and Systems Conference (NEWCAS), Grenoble, France, 7–10 June 2015; pp. 1–4.
22. Fong, X.; Chen, M.; Roy, K. Generating true random numbers using on-chip complementary polarizer spin-transfer torque magnetic tunnel junctions. In Proceedings of the 72nd Device Research Conference, Santa Barbara, CA, USA, 22–25 June 2014; pp. 103–104.
23. Vatajelu, E.I.; Natale, G.D.; Prinetto, P. Security primitives (PUF and TRNG) with STT-MRAM. In Proceedings of the 2016 IEEE 34th VLSI Test Symposium (VTS), Las Vegas, NV, USA, 25–27 April 2016; pp. 1–4.
24. Khan, M.N.I.; Cheng, C.Y.; Lin, S.H.; Ash-Saki, A.; Ghosh, S. A Morphable Physically Unclonable Function and True Random Number Generator using a Commercial Magnetic Memory. In Proceedings of the 2020 21st International Symposium on Quality Electronic Design (ISQED), Santa Clara, CA, USA, 25–26 March 2020; p. 197.

25. Basys3^{TM} FPGA Board Reference Manual. Available online: reference.digilentinc.com/_media/basys3:basys3_rm.pdf (accessed on 12 January 2021).
26. The Standard for Superior Measurements. Available online: https://www.keysight.com/us/en/assets/7018-04261/data-sheets/5991-3904.pdf (accessed on 12 January 2021).
27. Zhang, X.; Sun, G.; Zhang, Y.; Chen, Y.; Li, H.; Wen, W.; Di, J. A novel PUF based on cell error rate distribution of STT-RAM. In Proceedings of the 2016 21st Asia and South Pacific Design Automation Conference (ASP-DAC), Macau, China, 25–28 January 2016; pp. 342–347.
28. Zhang, L.; Fong, X.; Chang, C.; Kong, Z.H.; Roy, K. Optimizating Emerging Nonvolatile Memories for Dual-Mode Applications: Data Storage and Key Generator. *IEEE Trans. Comput. Aided Des. Integr. Circuits Syst.* **2015**, *34*, 1176–1187. [CrossRef]
29. Shannon, C.E. A mathematical theory of communication. *Bell Syst. Tech. J.* **1948**, *27*, 379–423. [CrossRef]
30. Kim, J.; Nili, H.; Truong, N.D.; Ahmed, T.; Yang, J.; Jeong, D.S.; Sriram, S.; Ranasinghe, D.C.; Ippolito, S.; Chun, H.; et al. Nano-Intrinsic True Random Number Generation: A Device to Data Study. *IEEE Trans. Circuits Syst. Regul. Pap.* **2019**, *66*, 2615–2626.

Journal of
Low Power Electronics and Applications

MDPI

Article

Continuous-Time Programming of Floating-Gate Transistors for Nonvolatile Analog Memory Arrays †

Brandon Rumberg, Spencer Clites, Haifa Abulaiha, Alexander DiLello and David Graham *

Lane Department of Computer Science and Electrical Engineering, West Virginia University, Morgantown, WV 26506, USA; brandon@aspinity.com (B.R.); sclites@mix.wvu.edu (S.C.); hmabulaiha@mix.wvu.edu (H.A.); adilello@mix.wvu.edu (A.D.)

* Correspondence: david.graham@mail.wvu.edu; Tel.: +1-304-293-9692

† This paper is an extended version of our paper published in Rumberg, B.; Graham, D. A floating-gate memory cell for continuous-time programming. In Proceedings of the IEEE Midwest Symposium on Circuits and Systems, Boise, ID, USA, 5–8 August 2012; pp. 214–217.

Abstract: Floating-gate (FG) transistors are a primary means of providing nonvolatile digital memory in standard CMOS processes, but they are also key enablers for large-scale programmable analog systems, as well. Such programmable analog systems are often designed for battery-powered and resource-constrained applications, which require the memory cells to program quickly and with low infrastructural overhead. To meet these needs, we present a four-transistor analog floating-gate memory cell that offers both voltage and current outputs and has linear programming characteristics. Furthermore, we present a simple programming circuit that forces the memory cell to converge to targets with 13.0 bit resolution. Finally, we demonstrate how to use the FG memory cell and the programmer circuit in array configurations. We show how to program an array in either a serial or parallel fashion and demonstrate the effectiveness of the array programming with an application of a bandpass filter array.

Keywords: floating-gate transistor; nonvolatile memory; continuous-time programming; floating-gate memory array; FPAA; reconfigurable

Citation: Rumberg, B.; Clites, S.; Abulaiha, H.; Dilello, A.; Graham, D. Continuous-Time Programming of Floating-Gate Transistors for Nonvolatile Analog Memory Arrays. *J. Low Power Electron. Appl.* **2021**, *11*, 4. https://doi.org/10.3390/jlpea11010004

Received: 13 November 2020
Accepted: 11 January 2021
Published: 13 January 2021

1. Introduction

In an effort to reduce the power consumption of battery-powered devices, analog signal processing is being reinvestigated to supplement and/or replace digital systems for making early decisions regarding incoming sensor information. However, analog systems are extremely sensitive to biasing conditions and, thus, need accurate control over their parameters to achieve the desired performance. Particularly in systems consisting of arrays of analog elements, such as field-programmable analog arrays [1–3] and other programmable analog arrays [4–6], a large number of analog parameters must be precisely established to achieve the desired performance.

Floating-gate (FG) transistors can serve as key enabling devices for such low-power analog systems. An FG transistor is a MOSFET that has no resistive connection to its gate; instead, a "control gate" couples capacitively onto the transistor's "floating gate." As a result, the charge on the FG is held fixed under nominal conditions but can be modified via Fowler-Nordheim tunneling and hot-electron injection, which both require elevated voltages. Because of its nonvolatile memory characteristics, FG transistors are ubiquitous in digital systems in the form of EEPROM and Flash memory. However, the ability to finely tune the amount of programmed charge on the FG allows these devices to be used as nonvolatile *analog* memory elements, as well. Consequently, FG transistors have found applications as variable threshold-voltage devices, programmable voltage/current sources, analog trimming for device matching, and within adaptive/learning circuits [7].

Modifying the stored charge on an FG transistor, which is often referred to as "programming" the FG, requires large voltages (typically 2–3 times greater than the rated V_{dd} of the process) to be applied to the FG transistor terminals in a controlled manner. Figure 1 depicts the two primary methods for programming an FG transistor—pulsed and continuous methods. Pulsed methods operate by iteratively applying short, high-voltage pulses to modify the charge and then measuring the FG after each pulse, repeating until a desired target is reached. Continuous methods, on the other hand, apply a constant high voltage and leverage feedback to ensure that the FG charge converges to the desired value. Pulse-based methods have dominated analog applications (e.g., [8]) because of the simplicity of design and high accuracy that has been demonstrated. However, continuous programming promises to be faster and require less peripheral circuitry than pulse-based programming, which are critical features in resource-constrained systems that need to save power and area.

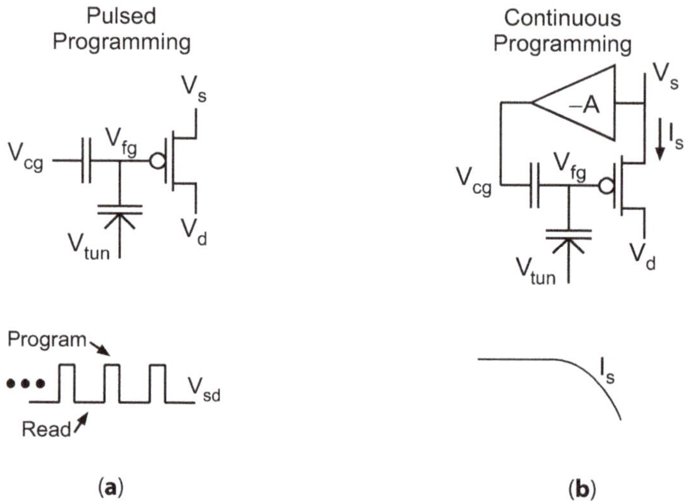

Figure 1. Pulsed programming and continuous programming. (**a**) In pulsed programming, the source-to-drain potential is alternately pulsed high for injection, and then placed at a nominal value to measure the floating gate's performance. (**b**) In continuous programming, injection occurs constantly, and a terminal (in this case the source current) is adjusted to decrease—and eventually shut off—injection as the target is approached.

In this paper, we describe a compact FG cell for continuous programming, which when combined with our simple programmer circuit, converges to target voltages with 13.0 bit resolution. This FG cell/programmer combination is primarily designed to work with battery-powered applications. This work is an extension of our early results in [9] in which the circuit was constructed using custom discrete elements and achieved far lower accuracy. Here, we present a fully integrated solution with far better programming accuracy and resolution. We also extend the results of a single memory cell in [9] to array applications. We describe how to build and program FG arrays in a serial (i.e., one at a time) fashion as well as present a method for parallelizing the programming of FG cells in an FG memory array to improve overall programming speed (first described in Masters Thesis [10]). This paper serves to provide a description on how to build easy-to-use programmable arrays of analog non-volatile memory for low-power applications.

Our basic memory cell uses an FG transistor in a source-follower configuration and linearizes injection via negative feedback to the control gate, as shown in Figure 1b. Such linear source-feedback injection has been used previously in [11], but we accomplish the same characteristics with the smaller current conveyor circuit. In addition to being smaller, this current conveyor memory cell also offers more flexible control over the injection rate since the FG source voltage, V_s, can be modified using either a voltage or a current input.

We describe the development of this system in the remainder of this paper. Section 2 provides an overview of FG programming. Section 3 describes various methods of continuous-time FG programming. Section 4 discusses our current-conveyor-based memory cell. Section 5 then describes the programmer circuit that is used to achieve specific target values. Section 6 discusses the two major methodologies to program an FG array—serial and parallel programming. Section 7 demonstrates the use of an FG array to precisely establish the corner frequencies of a programmable bandpass filter bank. Finally, Section 8 provides concluding remarks.

2. Overview of Floating-Gate Programming

Two phenomena are typically used to program FG transistors: hot-electron injection and Fowler-Nordheim tunneling. Injection occurs when a large source-to-drain potential (typically $V_{sd} > V_{dd}$) is applied to a PMOS FG transistor, thus causing high-energy carriers to impact-ionize at the drain. A fraction of the resulting ionized electrons disperse toward the surface with enough energy to overcome the oxide barrier and inject onto the FG. In the subthreshold region, which is our target operational region for low-power applications and high injection efficiency, the injection current from V_{fg} to V_d can be approximated as

$$I_{inj} \approx \beta I_s{}^{\alpha} e^{V_{sd}/V_{inj}} \tag{1}$$

where β, α, and V_{inj} are device-dependent fits [12]. Tunneling, on the other hand, requires high voltages (typically $V_{ox} > 2V_{dd}$). To avoid write disturbs during tunneling, unselected array elements must either be disconnected from the tunneling voltage using high-voltage switches or the FGs of the unselected elements must be raised to a sufficient voltage that tunneling does not occur. Due to this difficulty in isolating tunneling within an array, tunneling is typically used only for global erasure in analog memory arrays, while injection is used to write to individual elements. Consequently, we focus mainly on injection in this paper.

Due to their ability to provide dense, low-power, analog biases, FGs are elemental in large-scale programmable analog systems—such as filter banks [4], classifiers [5], and field-programmable analog arrays [1,13]. In these systems, circuit parameters (e.g., corner frequencies) are controlled by the charge on the FGs; as a result, system performance depends strongly on the programming accuracy. Prior pulse-based programming techniques have achieved high accuracy [8,11]. One advantage that pulse-based techniques have in terms of accuracy is that the FG is measured in a state that is similar to run-mode: with no high programming voltages applied to the cell, and with the same current levels that will be used in run-mode. Unfortunately, pulsing is inherently slow due to the time spent reading, during which the high programming voltages are stepped down and the FG is allowed to settle before the measurement is taken; if measuring low currents, which is necessary in low-power applications, then the read time further increases due to the long integration time that is necessary for accurate measurement. Methods to increase the programming speed with pulse-based programming rely on precise knowledge of each FG's characteristics, so that each pulse can move more aggressively toward the target [8], but this adds to the complexity. Additionally, pulsing techniques require high-precision data conversion and pulse timing, and possibly large-range current measurement [14], all of which complicate the inclusion of analog FG memory in simple, resource-constrained systems. Thus, there is a need for fast, compact, low-overhead, and accurate programming: we posit that continuous-time programming is more appropriate for resource-constrained systems.

3. Continuous-Time Floating Gate Programming

Continuous-time FG programming is accomplished by using feedback to stop programming when the memory cell reaches its target value. A variety of continuous programming circuits have been presented, ranging from a single-transistor circuit [12] that self-converges due to the negative feedback of injection current from the FG to the drain, to more complex circuits with improved speed and accuracy. Figure 2 shows the two primary types of FG cells for self-converging, continuous programming that use the inherent feedback in MOSFETs undergoing hot-electron injection to converge to a final value. For both the circuits, as electrons are injected onto the FG, the FG voltage, V_{fg}, decreases. As a result, the source-to-drain potential also decreases, and, according to Equation (1), the injection current decreases and will eventually shut off. The circuit in Figure 2a, which was presented in [12], provides repeatable results and can be programmed to different targets either by using different values of I_1 for a constant V_{cg}, or by using different values of V_{cg} for a constant I_1. However, the convergence time depends on the initial conditions; if the initial charge on the FG is too high, then the device cannot initially produce I_1. As a result, injection starts very slowly, and the total time to converge can be long (even though the final stages of convergence occur quickly). While the circuit of Figure 2b will typically start injecting quickly, it will slowly converge to its final value, often on the order of minutes to converge. Whereas the circuit of Figure 2a starts slow and finishes quickly, the circuit of Figure 2b starts quickly and finishes slowly. Both circuits suffer from potential long convergence times because they only rely on internal feedback to converge.

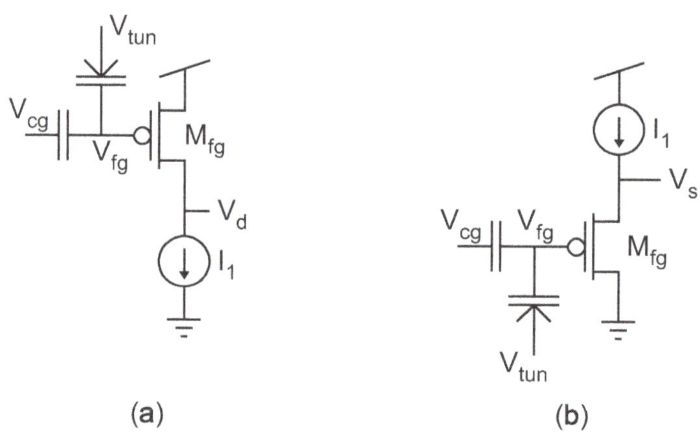

Figure 2. Programming cells without explicit feedback to keep all of the terminals of M_{fg} constant, thereby resulting in situations that have slow injection. (**a**) Programming cell with constant current applied to the drain. (**b**) Programming cell with constant current applied to the source.

External feedback, on the other hand, can be used to linearize the injection and tunneling processes to ensure that the FG charge is programmed to the desired value in a reasonable and predictable amount of time. External feedback can be used to keep the drain, source, and floating-gate potentials at constant values during the programming process so that injection and/or tunneling occur at a constant rate. As a result, the external feedback can prevent the FG cell from entering a region in which its conditions are not amenable to programming, which would lengthen the programming duration. However, linearizing the programming process means that the FG cell no longer self converges. Instead, systems that linearize the programming rates require additional external circuitry to stop the injection and/or tunneling at the appropriate time.

Several previous circuits were presented that use feedback to linearize the programming process, including memory cells that use a comparator to terminate programming when a target has been reached [15,16], memory cells that use differential amplifiers to linearize programming [17], and a system that uses both hot-electron and hot-hole injection to bidirectionally converge on a target [18]. In each of these systems, the programming rate is held constant until the target value is reached, and then the programming is abruptly stopped. Such programming faces a severe tradeoff between programming speed and accuracy [16]. In contrast, the programming circuitry that we present in this work adjusts the FG transistor's channel current in order to reduce the programming rate as the target value is approached; this adaptation of the current in the FG transistor allows our programmer and FG memory cell to achieve a better tradeoff between programming speed and accuracy.

Figure 3 depicts the two primary concepts behind using negative feedback to V_{cg} to keep all the terminals of M_{fg} constant through the programming process, and thus keep injection and/or tunneling rates constant. However, as previously mentioned, these memory cells no longer converge on their own, but require additional programming circuitry. In both of these circuits, V_{fg} is constant and V_{cg} ramps linearly up during injection, or down during tunneling, to compensate for the change in charge on the FG—see Figure 4. V_{cg} thus provides our measure of the charge on the FG. We found the high gain around the loop of the circuit in Figure 3a to cause stability problems, and so we will not consider it any further. The source follower circuit in Figure 3b is the same configuration that has been used in pulse-based source-feedback injection to achieve 13-bit precision with program times on the order of 50 s/200 mV [11]. This circuit has good stability and offers good control over injection and tunneling through the manipulation of both V_{sT} (which sets V_s) and I_1. Our memory cell has the same basic characteristics as this circuit, but is smaller, which is important for large array applications.

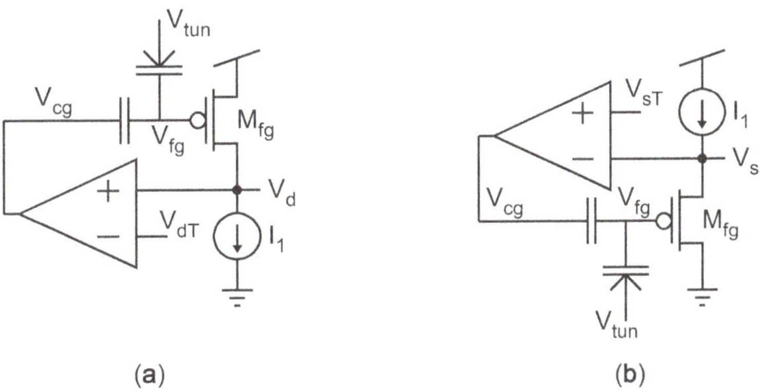

Figure 3. Programming cells that employ negative feedback to the gate to hold the terminals and current of M_{fg} constant, thus resulting in linear injection and tunneling. (**a**) Programming cell with constant current applied to the drain. (**b**) Programming cell with constant current applied to the source.

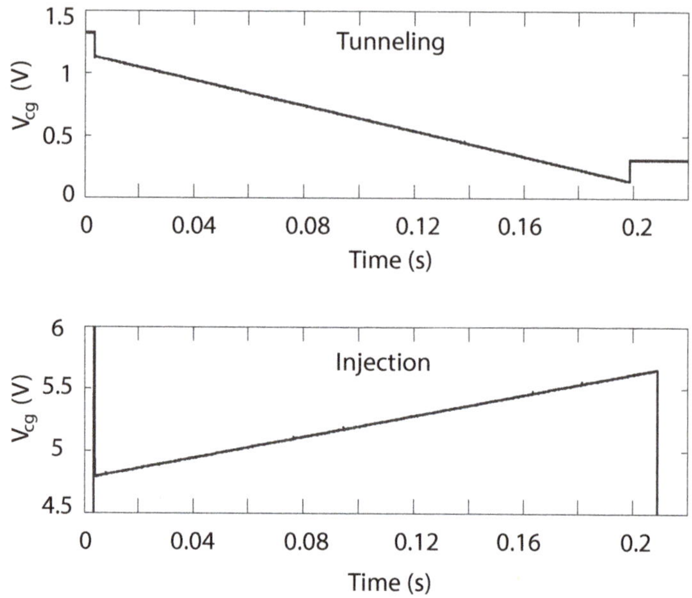

Figure 4. Demonstration of the linear programming characteristics of the circuit in Figure 3b.

4. Current-Conveyor-Based Memory Cell

To achieve the good characteristics of the circuit of Figure 3b but reduce the size, we developed the circuit in Figure 5. For simplicity, current sources are shown for I_1 and I_2, but in the actual implementation, each current source is implemented by a single transistor. In this memory cell, the inverting amplifier M_1–I_2 replaces the op-amp in Figure 3b. The resulting circuit structure is the current-controlled current conveyor, the details of which can be found in [19]. In this circuit, the negative feedback adjusts V_{cg} in order to force both V_{fg} and V_s to fixed voltages. The equilibrium point for V_s is controlled by both the voltage V_X and the current I_2. The equilibrium point of V_{fg} depends on both V_s and I_1. Thus, we maintain independent control of the source current and drain-to-source potential (the two main injection parameters) with this four-transistor circuit.

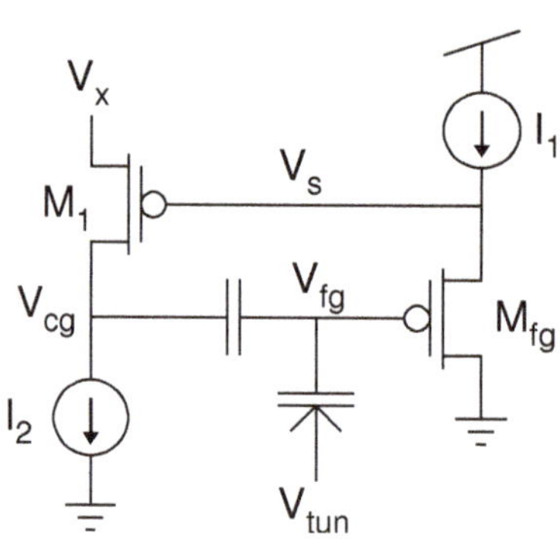

Figure 5. Our floating-gate memory cell, which is based on the current-controlled current conveyor circuit.

This memory cell offers three control terminals for modifying injection: two currents (I_1 and I_2) and one voltage (V_X). Using the subthreshold injection approximation in

Equation (1), we can solve for the injection current as a function of the control terminals in subthreshold operation as

$$I_{inj} \approx \beta I_1{}^{\alpha} \left(\frac{I_2}{I_0}\right)^{-\frac{U_T}{\kappa V_{inj}}} e^{\frac{V_X-(1-\kappa)V_{dd}}{\kappa V_{inj}}} \quad (2)$$

where I_0 is the pre-exponential current scaler for M_1, κ is related to the subthreshold slope for M_1, and $U_T = kT/q$ is the thermal voltage. Figure 6 shows measured injection rates as a function of each of these control terminals. The injection rate was measured by determining the slope of V_{cg} during injection experiments that were similar to the injection pane in Figure 4; this slope is equal to the injection current normalized by the control-gate capacitance. When not being swept, V_X, I_1, and I_2 were held fixed at 5 V, 860 nA, and 2 nA, respectively. Additionally, since the feedback holds V_{fg} constant, this cell has linear tunneling characteristics. Figure 7 shows the dependence of the tunneling current on V_X while all other terminals were held fixed.

The experiments shown in Figures 6 and 7 demonstrate the ability to adjust the cell's programming rate over a large range using either voltage or current inputs. Additionally, the weak dependence on I_2—approximately an inverse fifth root dependence—makes I_2 appropriate for fine rate adjustment. Furthermore, the cell works well in the subthreshold region, where power consumption is low and Equation (2) holds true.

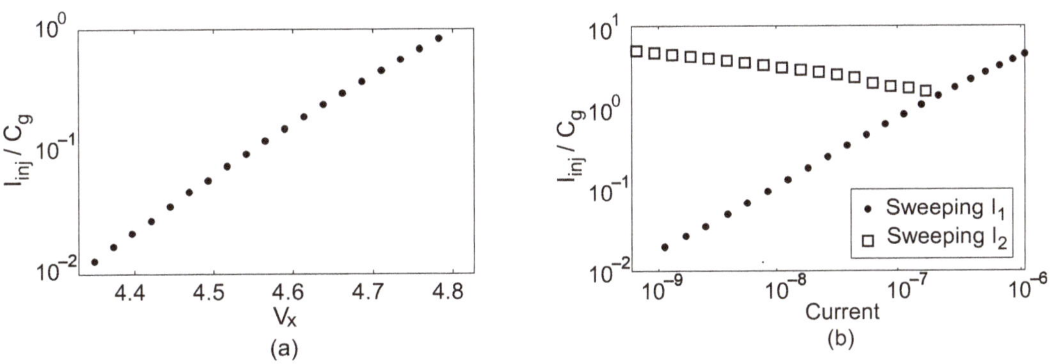

Figure 6. Measured dependence of the injection current on the three control terminals of the circuit: (**a**) V_X, (**b**) I_1, and I_2.

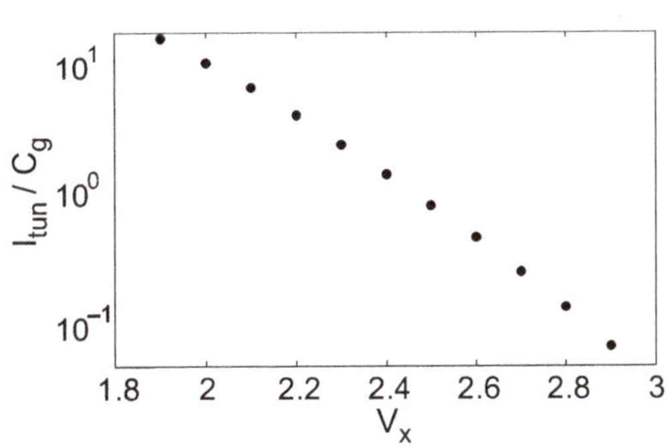

Figure 7. Measured dependence of tunneling current on terminal V_X.

5. Programmer Circuit

The combination of control terminals makes the memory cell very flexible in terms of creating a "programming circuit" to inject the memory cell to a desired value. Figure 8 illustrates one possible programming circuit that uses I_1 as the control terminal, and we will use this programmer circuit throughout the rest of this work. This programmer circuit consists of an operational transconductance amplifier (OTA) and a current mirror. In program mode, the programmer circuit is connected to the FG memory cell in the configuration shown in Figure 8. The OTA converts the difference between V_{cg} and a target value, V_{targ}, into a current. This current is rectified by the current mirror M_2–M_3 and is forced into the source terminal of the FG transistor. Accordingly, this current is able to precisely control the programming of the FG memory cell.

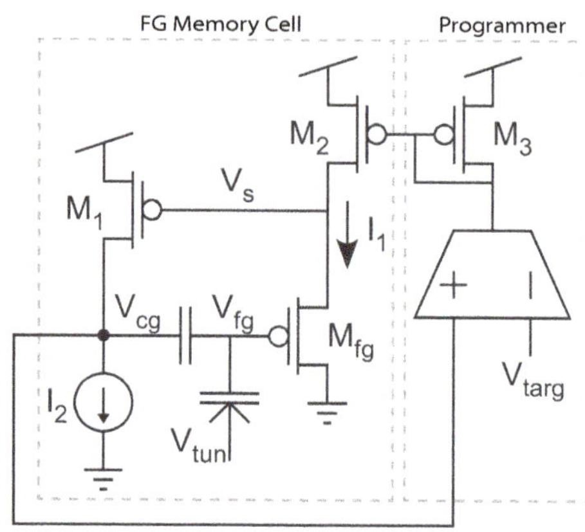

Figure 8. Our memory cell programming circuit.

Figure 9 illustrates the programming procedure with measured data from an integrated circuit fabricated in a standard 0.5 µm CMOS process. Prior to "writing" a value to the FG memory cell, the memory cell is erased by tunneling the FG transistor—a large voltage is applied to V_{tun} until the control-gate voltage, V_{cg}, of the FG memory cell drops to a voltage near ground. With V_{tun} reduced to its run-time voltage and with V_{dd} held at the nominal supply voltage, the value of V_{cg} maintains a low (near-ground) voltage. This is the "erased" state of the FG memory cell.

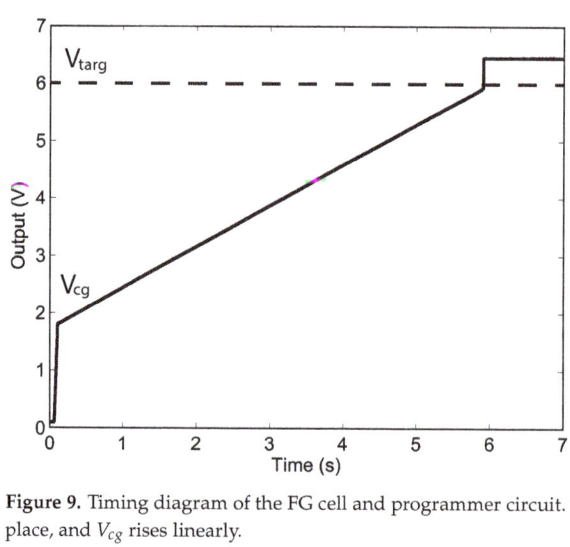

Figure 9. Timing diagram of the FG cell and programmer circuit. While $V_{targ} > V_{cg}$, injection takes place, and V_{cg} rises linearly.

To initiate injection, the supply voltage is ramped up to an elevated value, $V_{dd,fg}$, which pulls V_{cg} up to a non-zero value. The exact value of $V_{dd,fg}$ necessary for injection is process-dependent, and [20] provides details on how this voltage changes with technology nodes. While V_{cg} is well below the target value, the OTA output current is saturated, and the injection rate is constant because the current through the FG transistor is constant, while also maintaining constant voltages on the source, drain, and floating gate. As V_{cg} approaches V_{targ}, the OTA enters its linear input range, and the current in the FG transistor becomes proportional to the difference between V_{cg} and V_{targ}. Consequently, as the target is approached, the injection rate is reduced, and eventually stopped, by the reducing I_1. When I_1 shuts off, the current conveyor structure stops operating, and V_{cg} is pulled high. At this point, injection no longer occurs. The FG memory cell and programmer have their supply voltage lowered to the nominal V_{dd}, and programming of the FG memory cell has been completed.

After injecting the FG memory cell, it is placed in read mode by disconnecting it from the programmer OTA and current mirror—i.e., it is configured as Figure 5. The cell's voltage output is read from V_{cg} while constant currents are applied to I_1 and 1_2.

Alternatively, the FG transistor can be disconnected from the rest of the memory cell for current readout mode. In current readout mode, V_{cg} is connected to a fixed potential, the source of M_{fg} is connected to V_{dd}, and the drain is connected directly to the circuit that it biases. In short, M_{fg} is configured as a current source, with the exact value of current dependent on the charge programmed on the FG.

The combination of this programmer circuit and the FG memory cell is able to provide a linear mapping between the target voltages (V_{targ}) and the output voltages (V_{cg}) of the memory cell after being programmed. Figure 10 shows the measured relationship between the target voltages and the corresponding output voltages after the FG memory cell has been disconnected from the programmer and the supply voltage has returned to the nominal V_{dd}. These measurements were taken from an integrated circuit fabricated in a standard 0.35 μm CMOS process. As can be seen, linearly spaced target voltages result in linearly spaced output voltages. The bottom pane shows the deviation from a straight line with a slope of 1.0025 (the linear fit to the data). Over a target range of 0.9 V–2.1 V (1.2 V total), the worst-case deviation from the straight line was only 0.49 mV.

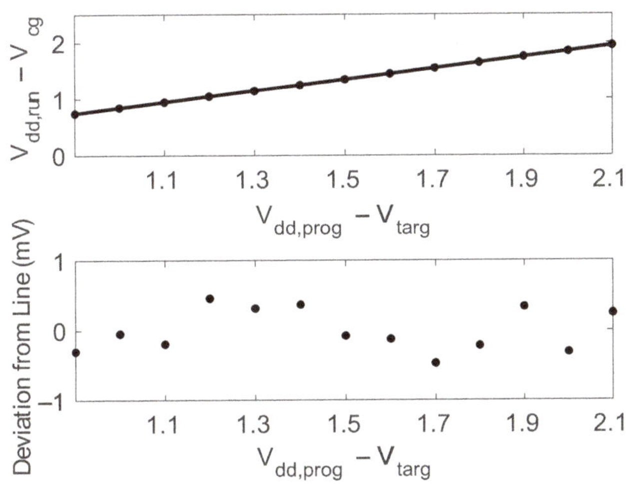

Figure 10. Measured programming accuracy.

The programmer/memory-cell combination is capable of programming to a larger range of voltages, but the relationship begins to deviate slightly from a straight line with a larger V_{targ} range. Figure 11 shows the V_{targ} to V_{cg} relationship for a voltage range of 2.2 V. The V_{cg} values deviate as much as 7 mV from the ideal straight line. However, a simple calibration step can be used to correct for these deviations from the straight line.

The curvature of the output V_{cg} values has an approximately third-order relationship. Therefore, using a third-order polynomial to calibrate the V_{targ} to V_{cg} relationship results in a worst-case deviation of 0.9 mV from a straight line, as is shown in Figure 11.

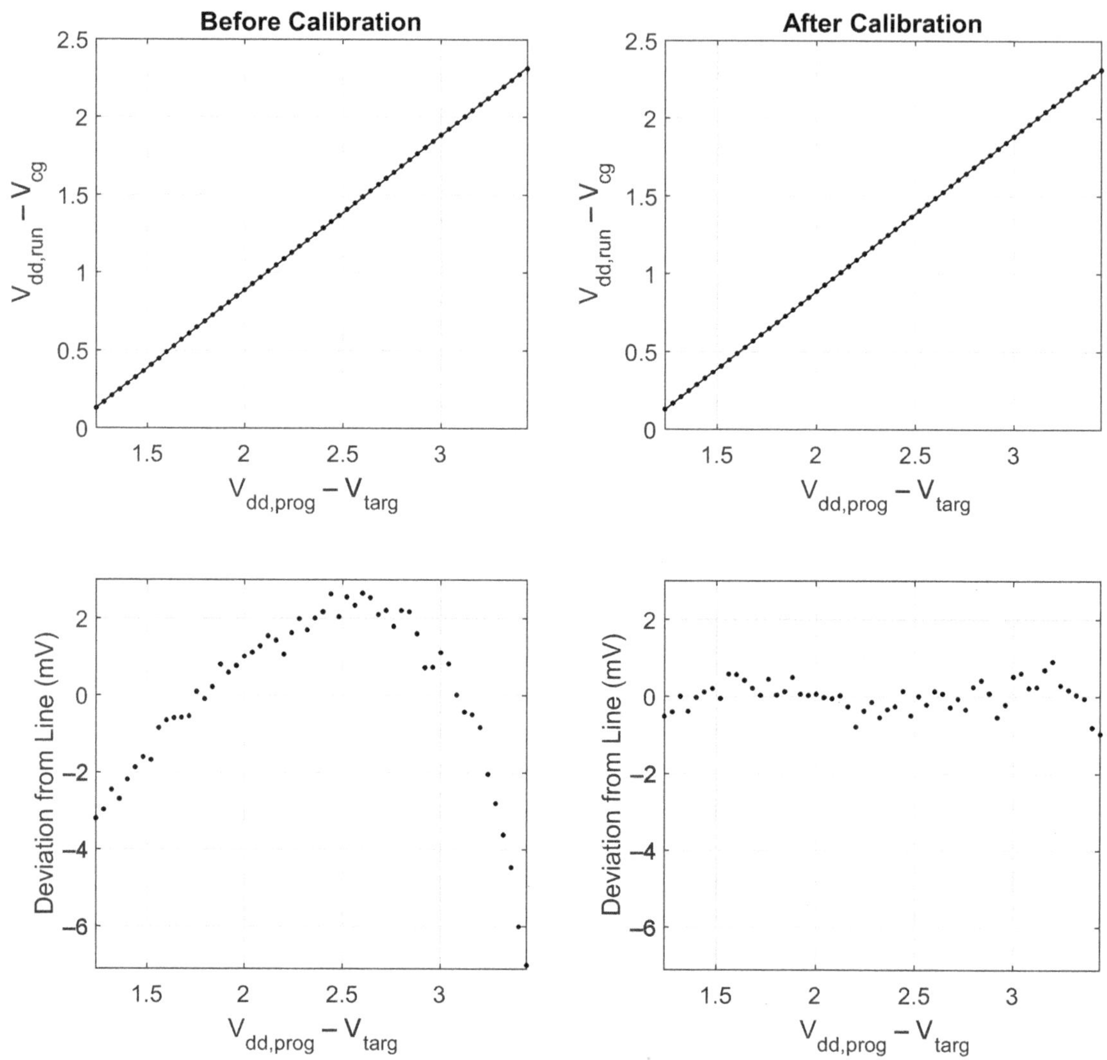

Figure 11. Measured programming accuracy before and after calibration.

To verify the repeatability and precision of the programming process, the memory cell was programmed using the programmer circuit for linearly spaced values of V_{targ}. The FG memory cell was programmed 100 times to each target value ranging from 1.24 V–3.56 V (2.32 V total range), with a full erasure after each write/measurement. Each programming cycle was 100 ms in duration and used $V_{dd,fg} = 6$ V. Figure 12 shows the standard deviation of the 100 measurements of V_{cg} for each V_{targ}, which had a worst-case value of 280 μV. Using the worst-case standard deviation of the repeatability measurements as the minimum detectable change that can be distinguished over the 2.32 V range, then this FG cell/programmer combination is capable of 13.0 bits of resolution when programming.

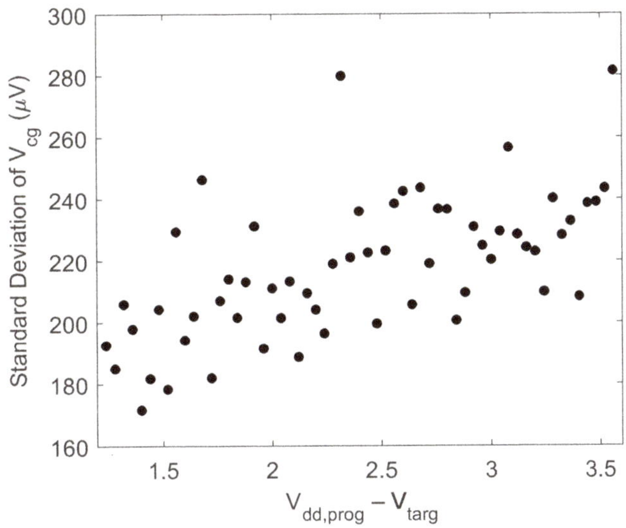

Figure 12. Measured programming precision.

Previous work in similar technology nodes has shown that charge retention after programming is very good (e.g., [21–23]), with results indicating that 10-year lifespans can be achieved from FG memory for analog applications with little charge loss. For example, ref. [23] has shown drift of less than 0.5 µV over 10 years at room temperature, which is sufficient for maintaining the 13.0 bit resolution in our system. One item that should be pointed out is that high fields through the oxide due to tunneling and injection can cause some damage to the oxide and also result in charge trapped in the oxide. However, in contrast to digital systems in which nonvolatile memory undergoes frequent write/erase cycles, nonvolatile analog memory write/erase cycling is often quite sporadic in comparison—with nonvolatile analog memory having a spectrum of needs in terms of the frequency of write/erase cycles. At one end of the spectrum of nonvolatile analog memory applications, the FG device only needs to be programmed once to account for process variations, biasing conditions, and/or mismatch compensation. In these cases, the nonvolatile memory does not need to undergo any extra write/erase cycles, so damage to the oxide will be inconsequential. Even at the other end of the spectrum where applications require more write/erase cycles, the frequency of write/erase cycling is still likely dramatically less than in digital systems, so it is likely that these systems will not undergo significant stresses either. However, if they do suffer some effects of oxide degradation and/or charge trapping, then the calibration curve from Figure 11 can be retaken periodically to reassess the target values needed to achieve the desired accuracy from the system. Analysis of long-term retention and effects of charge trapping in nonvolatile analog memory is ongoing research work and will continue to be studied in further detail. Because of the varied needs of programming nonvolatile analog memory in terms of how fast and how often they need to be programmed, we provide two methods for programming arrays of FG transistors in the next Section—(1) serial programming for area-constrained systems that do not need frequent re-writes and (2) parallel programming for systems that either need frequent re-writes or cannot handle long outage times during write/erase cycling.

Since the main application of the FG memory cell introduced here is to be used in low-power analog systems, the power dissipation of the memory cell in run mode should be small. In voltage readout mode, the cell in Figure 5 was biased with $I_1 = 20$ nA and $I_2 = 2$ nA, yielding a low power consumption of 66 nW/cell. If the FG transistor is configured for current readout, then the current is part of the circuit that it is biasing and does not contribute any additional power consumption beyond that of the circuit.

6. FG Array Programming

Since a benefit of FG transistors is that they allow for dense memory arrays, we employed our FG memory in several analog memory arrays. In an array configuration, FG memory cells are arranged in M rows by N columns, depending on the size of the application. Multiplexing circuitry at each of the FG memory-cell terminals is used to select/deselect cells and to apply the required voltages for the read/program processes.

The two main methodologies for programming an array of FGs are serial and parallel programming. As their names imply, serial programming involves programming one floating gate at a time, while parallel programming involves programming multiple floating gates simultaneously. Serial programming is suitable for applications where the chip area is constrained, since only one programmer circuit is required for the entire chip. On the other hand, parallel programming is preferable for analog applications that require faster write times for a large number of FGs; however, parallel programming requires larger area overhead for additional programming circuitry. We will discuss these two programming methodologies in the following subsections.

6.1. Serial Programming of FG Arrays

In serial programming, FGs are programmed one transistor at a time. Therefore, only one programmer circuit is needed per chip, which helps to keep the infrastructural circuitry compact. During injection, one specific FG cell is selected and connected directly to the programmer circuit of Figure 8. When the supply voltage is elevated to a voltage large enough to induce injection, which we will denote as $V_{dd,fg}$, injection starts, and only the selected FG memory cell is injected. All unselected FG memory cells are configured to prevent injection by pulling the unselected cells' V_{cg} to $V_{dd,fg}$ and by setting the drain-to-source voltage, V_{sd}, of the FG transistor to be low ($\sim$0 V) by connecting the source and drain to ground. Once the selected FG memory cell is injected to the desired target, it is disconnected from the programmer and a new FG memory cell is selected and connected to the programmer circuit. The process is repeated for each of the FG memory cells needing to be injected.

Since only one programmer circuit is used in serial programming, $M \times N$ programming cycles are required to program an $M \times N$ array. For large arrays, the $M \times N$ programming cycles could cause a significant unwanted down-time in which the system is not operational while it is being programmed, which could also be impractical for large arrays that require frequent reprogramming. However, not all applications require constant up-time or frequent rewrites. For example, in [1], the serial programming method was used to program the FG transistors on a reconfigurable analog processor. A die photograph of this programmable analog system is shown in Figure 13.

A signal-flow block diagram of the serial programming method is shown in Figure 14. A serial peripheral interface (SPI) is used to select the particular FG memory cell to be programmed, connect the programmer circuit to the appropriate cell, enable read/write mode, and control a voltage scaling digital-to-analog converter (DAC) to apply the desired target voltage. The programming process for this scheme operates as follows:

1. Globally erase all floating gates using tunneling
2. Raise the supply voltage to its elevated injection level $V_{dd,fg}$
3. Set the DAC output voltage (V_{targ}) and select a specific row/column combination
4. Connect the programmer circuit to the corresponding FG memory cell
5. Programming starts immediately—hold for approximiately 100 ms to ensure that injection completes—injection will automatically shut off when the FG memory cell reaches the desired target
6. Repeat steps 3–5 for all FG memory cells in the array that need to be injected, one at a time
7. Lower the supply voltage to V_{dd} when all FG cells have been programmed

Figure 13. A die photograph of a large-scale programmable analog system chip employing serial programming.

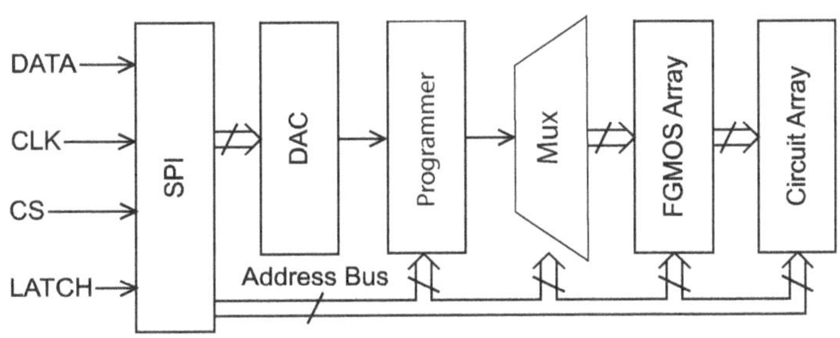

Figure 14. Signal flow diagram of the presented serial programming architecture.

6.2. Parallel Programming of FG Arrays

To accelerate the process of programming a large number of FG cells, a parallel programming technique can be used. In parallel programming, more than one cell is programmed at a time. Consequently, multiple programmer circuits are required per chip to accomplish parallel programming and will result in faster programming, since one programming cycle is able to program multiple FGs in the array. However, in order to program X floating gates in parallel, there must be X programmer circuits available—one for each FG that will be programmed simultaneously. In this subsection, we present an FG array with a parallel programmer scheme. In this particular system for an $M \times N$ array, N programmer circuits are used—one for each column.

A block diagram of the whole parallel programming system is shown in Figure 15. Like with the serial programming system, a digital interface for programming via SPI is used to set all parameters needed for programming. Consequently, only four digital input signals are required to program the full array of FG memory cells, which minimizes the number of pins required to interface with the chip, reduces programming overhead, and removes some of the programming details from the end user.

A voltage-scaling DAC is again used to generate analog target voltages. However, differently from the serial programming scheme, the output of the DAC is sampled by an array of sample-and-hold (S/H) circuits which provide the actual target voltages that are applied to the array of FG memory cell circuits to perform programming. This array of S/H circuits permits one single DAC to be used instead of N DACs for each FG memory cell being programmed in parallel. The DAC is stepped through N different values, and each of the N outputs are sampled—one output by each S/H circuit. These N voltages are held at the outputs of the S/H circuits (i.e., V_{targ} values of each of the N programmer circuits) until all N FG memory cells are programmed.

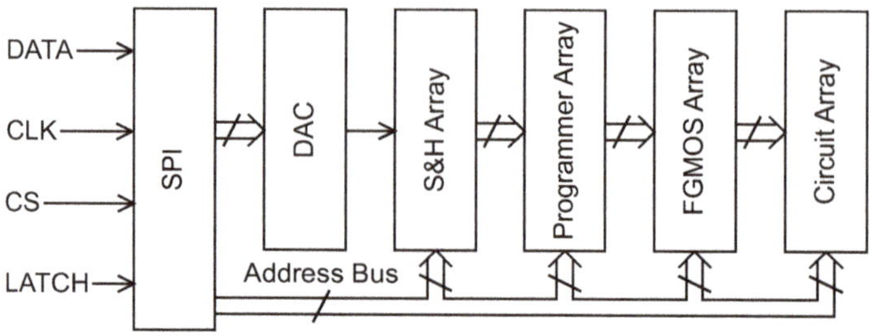

Figure 15. Signal flow diagram of the presented parallel programming architecture.

The requirements of the S/H circuits used in this application are that they must have long hold times to reliably maintain a constant target voltage while all N FG memory cells are programmed in parallel. Additionally, the S/H should have low pedestal error so as to not introduce offset between the DAC and the individual V_{targ} values.

To achieve the long hold times and low pedestal error, we employed a S/H topology based on [24]. This S/H employs Miller feedback in its hold-mode configuration to increase the effective hold capacitance, C_{hold}, without requiring larger drawn capacitors. This configuration reduces the droop rate of the S/H. A simplified version of the S/H schematic is shown in Figure 16a. The two switches, S_1 and S_2, are comprised of transmission gates which include half-sized dummy transmission-gate switches on each node except for V_{in}, since this charge injection error gets absorbed by the input source and does not affect V_{out}. Also, note that switch S_1 is clocked using Φ_{1d}, a delayed version of Φ_1. Consequently, S_2 opens slightly before S_1 when transitioning to hold mode, further reducing charge injection [25].

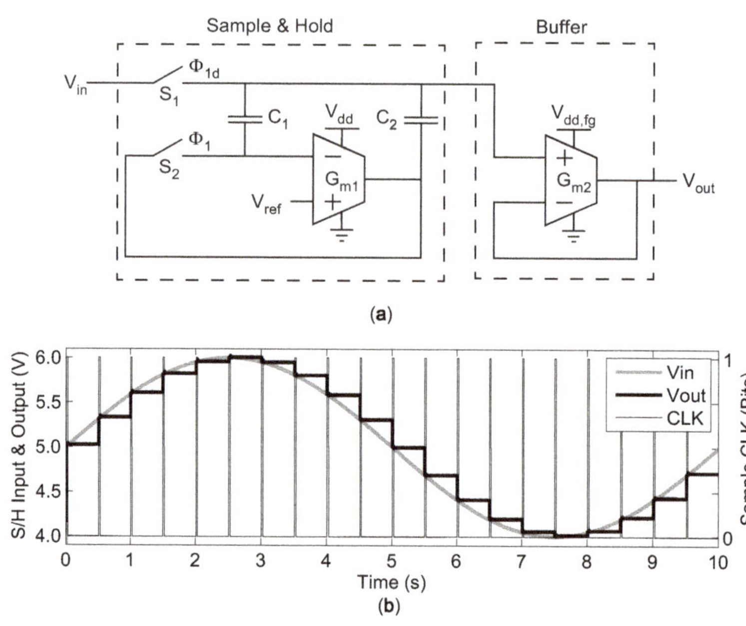

Figure 16. (**a**) Schematic of the sample-and-hold circuit with Miller hold capacitance. (**b**) Transient response to a sinusoidal waveform.

In sample mode, the S/H OTA, G_{m1}, is connected as a unity-gain buffer, forcing its inverting input to equal V_{ref} as C_1 and C_2 are charged to V_{in}. In hold mode, S_1 and S_2 are opened, and Miller feedback through C_1 and C_2 forces the capacitance on the hold node to be $C_{hold} \approx C_2(1 + A)$, where A is the open-loop gain of the S/H OTA, G_{m1}. Figure 16b demonstrates the S/H's operation by showing a transient plot of the S/H sampling a sine wave. The time scale is large, illustrating the long duration of hold times achievable by this circuit (hold time of 500 ms in this example).

The addition of the S/H array necessitates extra considerations to initiate injection in each device. Before programming, the output of the DAC is forced to ~0 V. Each S/H is sequentially selected, and ~0 V is sampled onto each of them. Next, the supply voltage of the buffer OTA, G_{m2}, is raised to $V_{dd,fg}$ (note that G_{m1} does not need to be raised up to the higher supply voltage), and each FG in row M is connected to its corresponding programmer. Since a low V_{targ} was set for each programmer, all of the FG memory cells connected to the programmers have their V_{cg} latched to a high voltage, and no FG memory cell undergoes injection. The DAC then sequentially applies the desired V_{targ} for each FG memory cell. However, injection still does not happen because V_{cg} is still latched high. To initiate injection, an extra switch is connected to the output of the programmer OTAs of Figure 8. A short-duration "start" pulse briefly shorts the output of the OTAs to ground, causing current to flow in the current mirrors biasing the FG transistors. This current resets the V_{cg} values of the memory cells to a low value. At the conclusion of the start pulse, the FG memory cells inject to the targets. Once the last programmer has started injecting its FG memory cell, the supply voltage is left at a high value of $V_{dd,fg}$ for a set period of time to ensure that all FGs have reached their targets.

Figure 17 shows the die photograph of an example system employing parallel programming of an array of FG memory cells. This integrated circuit was fabricated on a standard 0.5 µm CMOS process, and system-level results from this integrated circuit will be shown in the following Section. Figure 18 shows two FG memory cells on this chip being programmed in parallel using our programmer. Note that while 8 memory cells are programmed in parallel on this chip, the voltage outputs of only two cells were observable at any time—one was observable via selection by the the SPI, and another was hard-wired to output pins for debugging purposes. In this Figure at 100 ms, the first S/H is clocked, sampling the DAC output to set V_{targ1}. Shortly after, the start pulse is applied to initiate injection. Then, the next column is selected, and the process is repeated. A very long programming time-scale is shown in this example to illustrate the long hold times available by the S/H circuitry. However, typical programming hold times are 100 ms.

Figure 17. Die photograph of a programmable bandpass array chip using parallel programming.

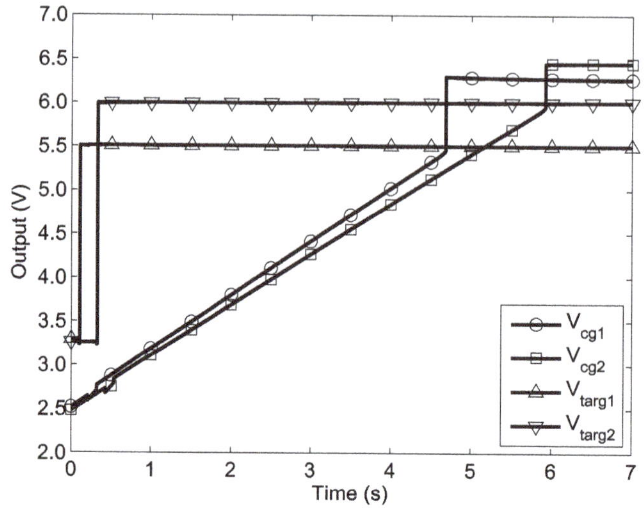

Figure 18. Transient response of node V_{cg} on two FG memory cells being programmed in parallel using our parallel programmer.

A summary of the parallel programming procedure is as follows:

1. Globally erase all floating gates using tunneling
2. Sample ~0 V on each of the S/H circuits
3. Raise the supply voltage to its elevated injection level, $V_{dd,fg}$
4. Set the DAC output voltage and select a specific row/column combination
5. Sample the target voltage from the DAC to set V_{targ}
6. Initiate injection with the start pulse
7. Repeat steps 4–6 for each subsequent FG memory cell in the column
8. Repeat steps 4–7 for each row in the array
9. Lower the supply voltage to V_{dd} when all FG cells have been programmed

6.3. Serial vs. Parallel Programming

As a comparison between the two methods for programming arrays, Figure 19 demonstrates the programming time for serial and parallel programming. As the Figure illustrates, to program an array of FGs serially, one FG is programmed at a time, which makes the programming process linearly proportional to the overall size of the array. Using the serial method for programming results in overall programming time of:

$$t_{tot,serial} = M \times N \times (t_s + t_i) \tag{3}$$

where M is the number of rows, N is the number of columns, t_i is the injection time, and t_s is pre-injection time (time to select the FG cell, connect the programmer, and apply the start pulse if needed) which is very short compared to t_i. On the other hand, our parallel programming method dramatically reduces the overall programming time by staggering FG programming through time, as shown in Figure 19. To program an $N \times M$ array using our parallel programming method, the overall programming time is:

$$t_{tot,parallel} = M \times (N \times t_s + t_i) \tag{4}$$

Generally, comparing the two methods, our parallel programming compromises between minimizing die area (by using a programmer circuit per each column) and programming time, which makes it the most appropriate programming method for FG-dense analog applications. Serial programming is most appropriate for systems that are size-constrained and/or can handle longer down times between write cycles.

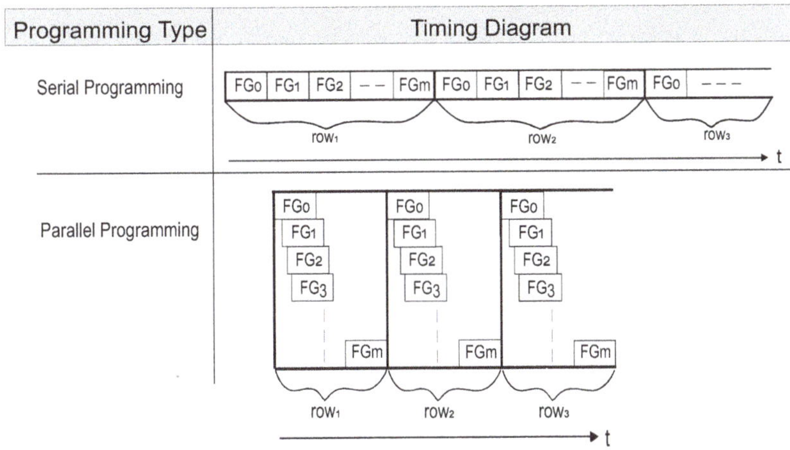

Figure 19. Serial vs. parallel programming.

7. System Application

FG transistors have a wide range of analog applications that require a large number of FGs to be integrated on a single die. These applications range from simple filter banks to more complicated field-programmable analog arrays (FPAAs). FGs are used to provide biasing voltages/currents for those analog applications. A proof-of-concept system was fabricated in a standard 0.5 µm CMOS process, and it consists of a programmable filter array employing our parallel programmer. This chip contains 8 sample-and-holds, 8 programmer circuits, a 2×8 array of floating-gate transistors (16 total), and 8 bandpass filters. This chip also contains the SPI, DAC, and miscellaneous peripheral circuitry. Each bandpass filter requires two FGs for biasing—one for the low corner frequency and one for the high corner frequency. The FGs are distributed in an array of 2 rows and 8 columns. In this configuration, the chip allows for one row of FGs to be programmed in parallel. Thus, two programming sequences, one for each row, are required to program the full chip. A die photograph of the chip is shown in Figure 17.

To demonstrate the programmer's ability to directly tune circuit parameters, we use the capacitively-coupled current conveyor (C^4) presented in [4] and shown in Figure 20a. The C^4 is a transconductance-capacitance (G_m-C) filter whose corner frequencies are proportional to the transconductances of two OTAs—$G_{m,L}$ and $G_{m,H}$ in Figure 20a. Since these transconductances are directly proportional to the bias currents of each OTA, the corner frequencies can be directly tuned using the FG memory cell as a current reference to bias them. Figure 20b shows how the FG transistors are configured as current sources and are connected to the OTAs. Figure 21a shows the effect on frequency response holding $G_{m,H}$ constant and programming different values of $G_{m,L}$; Figure 21b shows the effect of increasing $G_{m,H}$. As can be seen, the two corner frequencies can be tuned orthogonally, so that the current in one OTA does not impact the corner frequency set by the other OTA. In this example system, the biases providing the low corner frequencies for each of the bandpass filters were contained on one row (and programmed simultaneously), and all the biases for the high corner frequency were on the other row.

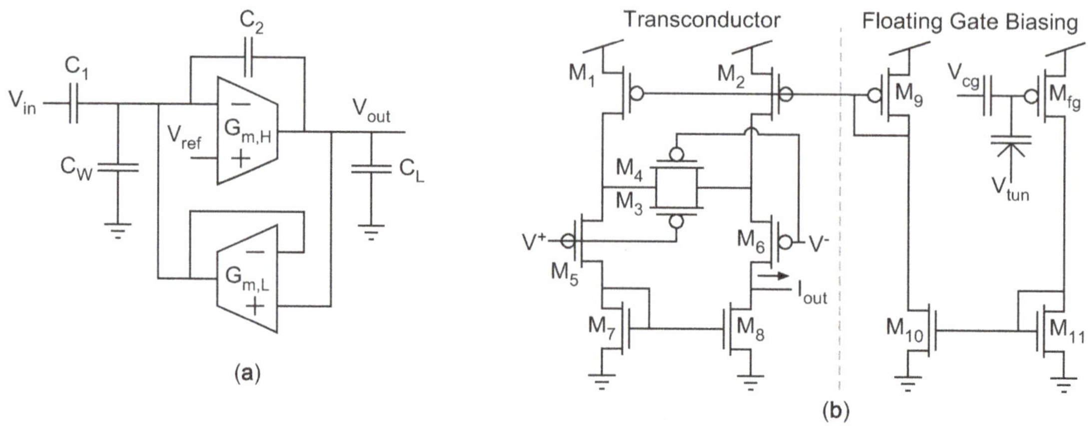

Figure 20. (**a**) Schematic of the OTA-based C^4. (**b**) Schematic of the OTA used for both $G_{m,L}$ and $G_{m,H}$.

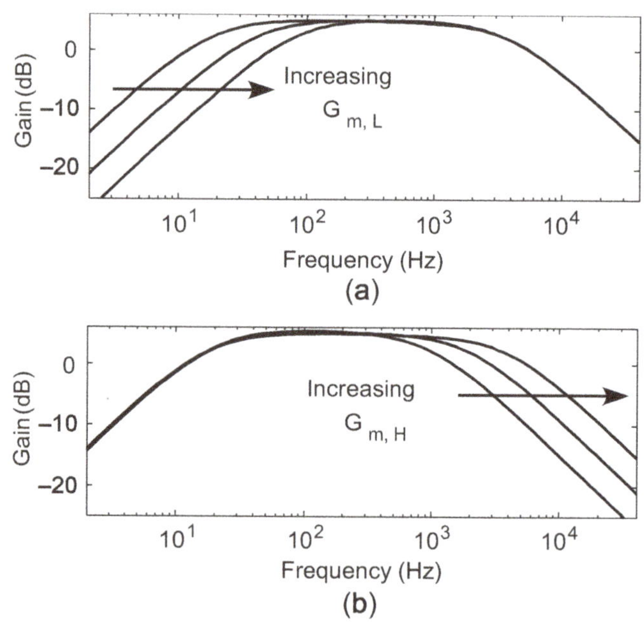

Figure 21. (**a**) Independent tuning of the low corner frequency. (**b**) Independent tuning of the high corner frequency.

Figure 22 demonstrates the capability of accurately programming the parameters of a filter array using the parallel programming structure presented in this paper. Three filter spacings are demonstrated: full-octave spacing, half-octave spacing, and third-octave spacing. The value of the quality factor, Q, for each of these configurations was chosen according to fractional-octave spacing rules, such that the filters cross at their −3 dB points. Therefore, $Q \sim 1.4$ for octave spacing, $Q \sim 2.9$ for half-octave spacing, and $Q \sim 4.3$ for third-octave spacing. Figure 22 (top) shows the results of programming the C^4s to octave spacing starting at $f_c = 88$ Hz, Figure 22 (middle) shows half-octave spacing beginning at $f_c = 300$ Hz, and Figure 22 (bottom) shows third-octave spacing beginning at $f_c = 445$ Hz.

As can be seen from Figure 22, the programming structure presented in this paper is able to precisely tune the circuit parameters to achieve exponentially spaced center frequencies. Additionally, the use of FG transistors provides a mechanism to tune the circuit's operation for multiple conditions (e.g., different center frequency spacings, bandwidths, etc.).

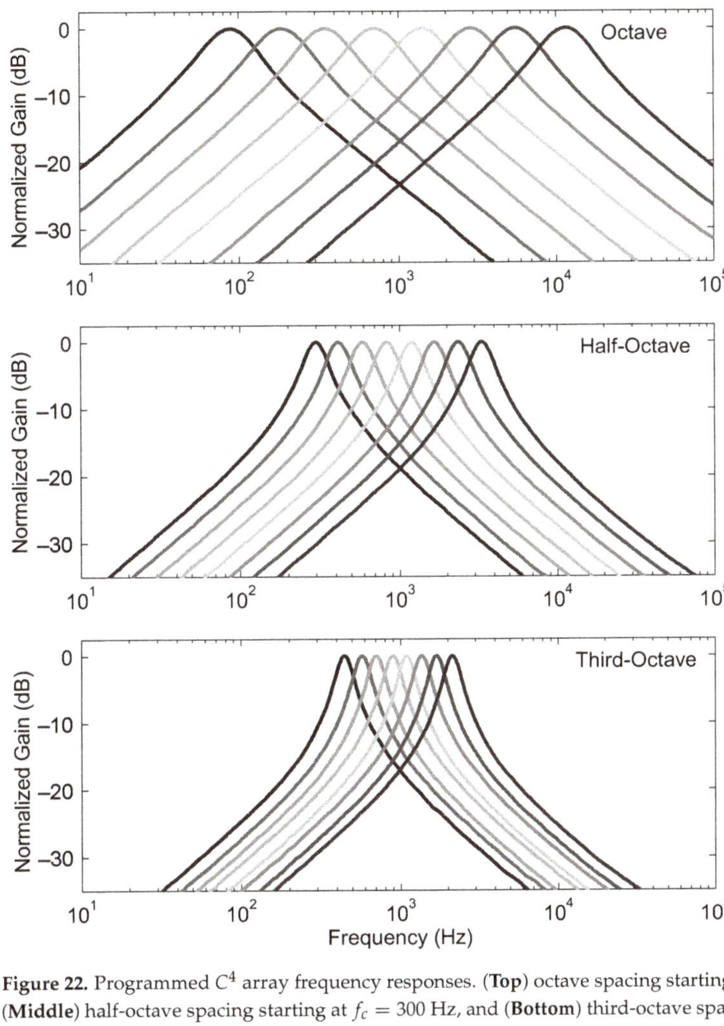

Figure 22. Programmed C^4 array frequency responses. (**Top**) octave spacing starting at $f_c = 88$ Hz, (**Middle**) half-octave spacing starting at $f_c = 300$ Hz, and (**Bottom**) third-octave spacing starting at $f_c = 445$ Hz.

8. Conclusions

We presented a compact analog FG memory cell that uses a continuous-time programming technique. Two different integrated circuits with memory cells and the programmer have been fabricated to characterize the design—one in standard 0.35 µm CMOS with serial array programming and the other in standard 0.5 µm CMOS with parallel programming. The FG memory cell and the programmer circuit were characterized and tested for repeatable programming. We demonstrated that the FG memory cell could be programmed to have a linear relationship with a target voltage over a range of 2.32 V with a resolution of 13.0 bits, all while being programmed in under 100 ms.

The FG memory cell was used to build a memory array that can be used in analog applications. A parallel programming technique was presented that significantly reduces the time required to inject all FG cells in the array. Finally, as a proof of concept, the FG memory array was used as programmable current sources to program a C^4 bandpass filter array.

Arrays of this FG memory cell are ideally suited to low-power applications that require analog processing of information—particularly applications that pre-process sensor information to make early classification and detection of events, such as [26]. Either a serial or parallel programming paradigm, as described in this paper, could be used, and the determination between the two should be made based on area constraints, allowable duration of down-time while reprogramming, and frequency of write/erase cycles. Future work in the area of programmable non-volatile analog memory will include (1) circuits to linearize the V_{targ} to V_{cg} transfer function better to reduce the need for a calibration phase, (2) circuitry to permit negative voltages to be used in the injection process to eliminate the need for a voltage ramp-up phase for injection (early work has been presented in [27]), (3) infrastructural circuits to support generating the write/erase cycles, (4) demonstration of performance in newer technology nodes, etc.

Author Contributions: Conceptualization, B.R. and D.G.; Data curation, D.G.; Funding acquisition, D.G.; Investigation, B.R., S.C., H.A. and A.D.; Methodology, B.R., S.C., H.A., A.D. and D.G.; Supervision, D.G.; Writing—original draft, B.R., S.C. and D.G.; Writing—review and editing, H.A. and D.G. All authors have read and agreed to the published version of the manuscript.

Funding: This material is based on work supported by the National Science Foundation under Award No. 1148815 and by the United States Army Research Laboratory under Contract W911NF-10-2-0109.

Conflicts of Interest: The authors declare no conflict of interest.

References

1. Rumberg, B.; Graham, D.; Clites, S.; Kelly, B.; Navidi, M.; Dilello, A.; Kulathumani, V. RAMP: Accelerating wireless sensor hardware design with a reconfigurable analog/mixed-signal platform. In Proceedings of the 14th International Conference on Information Processing in Sensor Networks, Seattle, WA, USA, 13–17 April 2015; pp. 47–58.
2. Hasler, J. Large-Scale Field-Programmable Analog Arrays. *Proc. IEEE* **2020**, *108*, 1283–1302. [CrossRef]
3. Becker, J.; Henrici, F.; Trendelenburg, S.; Ortmanns, M.; Manoli, Y. A Field-Programmable Analog Array of 55 Digitally Tunable OTAs in a Hexagonal Lattice. *IEEE J. Solid-State Circuits* **2008**, *43*, 2759–2768. [CrossRef]
4. Rumberg, B.; Graham, D. A Low-Power and High-Precision Programmable Analog Filter Bank. *IEEE Trans. Circuits Syst. II Exp. Briefs* **2012**, *59*, 234–238. [CrossRef]
5. Lu, J.; Young, S.; Arel, I.; Holleman, J. A 1 TOPS/W Analog Deep Machine-Learning Engine with Floating-Gate Storage in 0.13 µm CMOS. *IEEE J. Solid-State Circuits* **2015**, *50*, 270–281. [CrossRef]
6. Shah, S.; Töreyin, H.; Güngör, C.B.; Hasler, J. A Real-Time Vital-Sign Monitoring in the Physical Domain on a Mixed-Signal Reconfigurable Platform. *IEEE Trans. Biomed. Circuits Syst.* **2019**, *13*, 1690–1699. [CrossRef] [PubMed]
7. Hasler, P.; Minch, B.; Diorio, C. Floating-gate devices: They are not just for digital memories anymore. In Proceedings of the 1999 IEEE International Symposium on Circuits and Systems (ISCAS), Orlando, FL, USA, 30 May–2 June 1999; pp. 388–391.
8. Bandyopadhyay, A.; Serrano, G.; Hasler, P. Adaptive Algorithm Using Hot-Electron Injection for Programming Analog Computational Memory Elements Within 0.2% of Accuracy Over 3.5 Decades. *IEEE J. Solid-State Circuits* **2006**, *41*, 2107–2114. [CrossRef]
9. Rumberg, B.; Graham, D.W. A floating-gate memory cell for continuous-time programming. In Proceedings of the 2012 IEEE 55th International Midwest Symposium on Circuits and Systems (MWSCAS), Boise, ID, USA, 5–8 August 2012; pp. 214–217.
10. Clites, S. A Parallel Programmer for Non-Volatile Analog Memory Arrays. Master's Thesis, West Virginia University, Morgantown, WV, USA, 2015.
11. Huang, C.; Sarkar, P.; Chakrabartty, S. Rail-to-Rail, Linear Hot-Electron Injection Programming of Floating-Gate Voltage Bias Generators at 13-Bit Resolution. *IEEE J. Solid-State Circuits* **2011**, *46*, 2685–2692. [CrossRef]
12. Diorio, C. A p-Channel MOS Synapse Transistor with Self-Convergent Memory Writes. *IEEE Trans. Electron Dev.* **2000**, *47*, 464–472. [CrossRef]
13. Wunderlich, R.; Adil, F.; Hasler, P. Floating Gate-Based Field Programmable Mixed-Signal Array. *IEEE Trans. Very Large Scale Integr. (VLSI) Syst.* **2013**, *21*, 1496–1505. [CrossRef]
14. Basu, A.; Hasler, P.E. A Fully Integrated Architecture for Fast and Accurate Programming of Floating Gates Over Six Decades of Current. *IEEE Trans. Very Large Scale Integr. (VLSI) Syst.* **2011**, *19*, 953–962. [CrossRef]

15. Diorio, C.; Mahajan, S.; Hasler, P.; Minch, B.; Mead, C. A high-resolution nonvolatile analog memory cell. In Proceedings of the Proceedings of ISCAS'95-International Symposium on Circuits and Systems, Seattle, WA, USA, 30 April–3 May 1995; Volume 3, pp. 2233–2236.
16. Román, H.; Serrano, G. A system architecture for automated charge modifications of analog memories. In Proceedings of the 2010 53rd IEEE International Midwest Symposium on Circuits and Systems, Seattle, WA, USA, 1–4 August 2010; pp. 1069–1072.
17. Kim, K.H.; Lee, K.; Jung, T.S.; Suh, K.D. An 8-Bit-Resolution, 360-μs Write Time Nonvolatile Analog Memory Based on Differentially Balanced Constant-Tunneling-Current Scheme (DBCS). *IEEE J. Solid-State Circuits* **1998**, *33*, 1758–1762.
18. Wu, Y.D.; Cheng, K.C.; Lu, C.C.; Chen, H. Embedded Analog Nonvolatile Memory With Bidirectional and Linear Programmability. *IEEE Trans. Circuits Syst. II* **2012**, *59*, 88–92. [CrossRef]
19. Andreou, A.; Boahen, K.; Pouliquen, P.; Pavasović, A.; Jenkins, R.; Strohbehn, K. Current-Mode Subthreshold MOS Circuits for Analog VLSI Neural Systems. *IEEE Trans. Neural Netw.* **1991**, *2*, 205–213. [CrossRef]
20. Navidi, M.; Graham, D. A regulated charge pump for injecting floating-gate transistors. In Proceedings of the 2017 IEEE International Symposium on Circuits and Systems (ISCAS), Baltimore, MD, USA, 28–31 May 2017; pp. 2270–2273.
21. Ma, Y.; Gilliland, T.; Wang, B.; Paulsen, R.; Pesavento, A.; Wang, C.-H.; Nguyen, H.; Humes, T.; Diorio, C. Reliability of pFET EEPROM with 70-Angstrom tunnel oxide manufactured in generic logic CMOS processes. *IEEE Trans. Device Mater. Reliab.* **2004**, *4*, 353–358. [CrossRef]
22. St. John, I.; Fox, R. Leakage effects in metal-connected floating-gate circuits. *IEEE Trans. Circuits Syst. II* **2006**, *53*, 577–579. [CrossRef]
23. Srinivasan, V.; Serrano, G.; Gray, J.; Hasler, P. A Precision CMOS Amplifier Using Floating-Gate Transistors for Offset Cancellation *IEEE J. Solid-State Circuits* **2007**, *42*, 280–291. [CrossRef]
24. Lim, P.; Wooley, B. A High-Speed Sample-and-Hold Technique Using a Miller Hold Capacitance. *IEEE J. Solid-State Circuits* **1991**, *26*, 643–651. [CrossRef]
25. Carusone, T.; Johns, D.; Martin, K. *Analog Integrated Circuit Design*, 2nd ed.; John Wiley & Sons, Inc.: Hoboken, NJ, USA, 2012.
26. Bhattacharyya, S.; Andryzcik, S.; Graham, D. An Acoustic Vehicle Detector and Classifier Using a Reconfigurable Analog/Mixed-Signal Platform. *J. Low Power Electron. Appl.* **2020**, *10*, 6. [CrossRef]
27. Navidi, M.; Graham, D.; Rumberg, B. Below-Ground Injection of Floating-Gate Transistors for Programmable Analog Circuits. In Proceedings of the 2017 IEEE International Symposium on Circuits and Systems (ISCAS), Baltimore, MD, USA, 28–31 May 2017; pp. 1–4.

Journal of
Low Power Electronics and Applications

MDPI

Article

Logic-in-Memory Computation: Is It Worth It? A Binary Neural Network Case Study

Andrea Coluccio *, Marco Vacca and Giovanna Turvani

Department of electronics and telecommunication (DET), Politecnico di Torino, Corso Castelfidardo 39, 10129 Torino, Italy; marco.vacca@polito.it (M.V.); giovanna.turvani@polito.it (G.T.)

* Correspondence: andrea.coluccio@polito.it

Received: 23 December 2019; Accepted: 4 February 2020; Published: 22 February 2020

Abstract: Recently, the Logic-in-Memory (LiM) concept has been widely studied in the literature. This paradigm represents one of the most efficient ways to solve the limitations of a Von Neumann's architecture: by placing simple logic circuits inside or near a memory element, it is possible to obtain a local computation without the need to fetch data from the main memory. Although this concept introduces a lot of advantages from a theoretical point of view, its implementation could introduce an increasing complexity overhead of the memory itself, leading to a more sophisticated design flow. As a case study, Binary Neural Networks (BNNs) have been chosen. BNNs binarize both weights and inputs, transforming multiply-and-accumulate into a simpler bitwise logical operation while maintaining high accuracy, making them well-suited for a LiM implementation. In this paper, we present two circuits implementing a BNN model in CMOS technology. The first one, called Out-Of-Memory (OOM) architecture, is implemented following a standard Von Neumann structure. The same architecture was redesigned to adapt the critical part of the algorithm for a modified memory, which is also capable of executing logic calculations. By comparing both OOM and LiM architectures we aim to evaluate if Logic-in-Memory paradigm is worth it. The results highlight that LiM architectures have a clear advantage over Von Neumann architectures, allowing a reduction in energy consumption while increasing the overall speed of the circuit.

Keywords: Logic-in-Memory (LiM); Von Neumann's bottleneck; memory-wall

1. Introduction

Nowadays Logic-in-Memory (LiM) architectures are widely studied in order to solve the memory-wall problem, which is a bottleneck due to the communication between processing units and memories. A LiM implementation consists of very small computational units placed near a memory element. This enables a distributed computation instead of a classical Von-Neumann one. The big advantage of this design procedure is the reduction of the Von Neumann bottlenecks (such as fetching latency and the wasted power due to the communication between CPU-Memory), which enables also a very fast and energy-efficient structure. From a theoretical perspective, they bring many advantages, but are they worth it?

To answer to this important question, we have to consider implementing a LiM architecture by modifying the original structure of the memory, creating a structure that merges computation and memory Figure 1. As a natural consequence, the overall complexity of the customized design flow increases. To explore the features of a LiM implementation more in depth, two architectures have been designed, an Out-Of-Memory (OOM) that follows a classical Von Neumann approach and the derived LiM novel alternative. The performance obtained in both cases is subsequently compared.

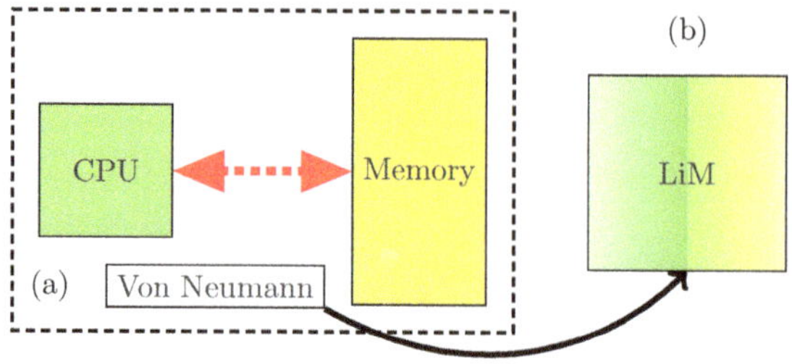

Figure 1. Von Neumann's classical architecture (**a**) composed of CPU and memory. Logic-in-Memory (LiM) novel architecture (**b**) that merges computation and memory.

As a case study, a memory-intensive application, a Neural Network (NN), was chosen, since it is a good candidate to demonstrate the benefits of a LiM architecture. NNs are used to perform very complex tasks such as speech and image recognition in a very efficient and accurate way. Convolutional Neural Network (CNN) and Multi-Layer Perceptron (MLP) models are employed and both can achieve very high accuracy. In literature, many CNNs have been proposed: they can be distinguished by their task, complexity and achieved accuracy. Considering image classification applications, the most common CNNs are LeNet-5[1] and AlexNet [2]. LeNet-5 is a very small network which is able to achieve a TOP-1 error rate of 0.35% on the Modified National Institute of Standards and Technology dataset (MNIST dataset). AlexNet is a more complex structure largely used for recognizing RGB images, which achieves a TOP-5 error rate of 16.4% on the ImageNet dataset. Also GoogLeNet [3], VGG-Net [4] and ResNet-152 [5] can be used on the same dataset and achieve 6.67%, 7.3% and 3.6% TOP-5 error rates respectively. In general, these models require a lot of computational resources implying very high energy consumption, thus making them inoperable in low energy contexts like embedded applications.

In this work, a binarized NN is chosen. Binary Neural Network (BNN) approximations have been proposed in several works like BinaryConnect (BC) [6], Binary-Weight Network (BWN) and XNOR-Net [7], in order to reduce the computational complexity by changing the weight-inputs precision, by means of a binarization process. Weights, and eventually inputs, are approximated with only two values (-1,+1), that can be represented on a single bit, '0' indicates -1 and '1' indicates +1. The chosen approximation for this work is the XNOR-Net [7]. The XNOR-Net reaches high accuracy rates compared to the original floating-point model and is particularly well suited for a LiM solution, since the binary multiplication can be performed by a simple XNOR gate. While a specific Neural Network model was chosen, the architectures were developed with reconfigurability in mind, meaning that most NNs can be implemented by the hardware. Our goal is to demonstrate the effectiveness of a LiM design, so our contributions in this work can be summarized as:

1. Realization of a reconfigurable OOM architecture implementing the XNOR-Net model and proposal of a possible design approach for a LiM alternative.
2. Identification of strong and weak points of a LiM solution with many NN models of different sizes and complexities.
3. Detailed performance evaluations with 45 nm @ 1.1 V CMOS technology. The estimations are performed with a synthesis and **.vcd**-based post place and route simulations for two models: a CNN and an MLP network respectively. The tools involved in this step are Synopsys Design Compiler for the synthesis, Mentor Modelsim for the simulation and Cadence Innovus for the place and route phase.
4. Generalized performance estimations for both architectures by means of parametric sweeps obtained from several synthesis processes with 45 nm @ 1.1 V technology. The aim is to compare the implementations with several different parameters and to identify the main differences. Discussion of the obtained results are provided and a qualitative comparison between the implementations are reported in Section 6.3.1.

5. A state-of-the-art comparison between our LiM and the Content addressable memory based implementation proposed in [8]. In our designs, memories are implemented as register files and each memory cell is a flip flop, since we didn't have the possibility to implement a custom memory. Consequently, the performance values obtained represent an overestimation of a real case. To determine how a real memory model impacts the results obtained, parameters from [8] are taken into account. Reference [8] implements a XNOR-Net LiM design with 65 nm CMOS technology, so in our synthesis procedure we used CMOS 65 nm technology @ 1.0 V to have a fair comparison.
6. Conclusions and discussions for future work.

The rest of the paper is organized as follows: Section 2 gives a brief explanation on what a LiM architecture is and recalls a useful classification from [9]. Section 3 discusses briefly NN background, giving an overview on what its main components are; binary approximations are compared and explained in more details. Section 4 reports the detailed design flow adopted for both OOM and LiM architectures and Section 5 makes an initial qualitative comparison between them. In Section 6, performance evaluations are reported, firstly taking two NN models as a case study and then by performing parametric sweeps. Lastly, Section 7 presents conclusions and future work.

2. LiM Background

A Quick Overview

LiM concept is widely discussed in the literature and a lot of different approaches have been adopted. In [9] an interesting classification of the various types of LiM paradigms is presented. Four possible typologies can be found.

1. Computation near Memory [9] where part of the computing blocks are moved in the memory proximity proposing solutions such as WIDE-IO2, which is a 3D stacked DRAM memory [10] with a logical layer placed at the bottom of the stack. Data are moved from the DRAM layers to the logical one employing Through-Silicon Vias (TSVs) and the result is then written back to one of the available DRAM layers. 3D stacked DRAM combined with TSVs allow to shorten the paths' lengths that data have to travel to reach the computational core, reducing the Von Neumann bottlenecks and improving efficiency.
2. Computation in Memory [9] paradigm is used in solutions with resistive arrays, based on technologies such as Magnetic Tunnel Junction (MTJ) devices [11]. MTJ is a component that can have two discrete resistance values, according to the direction of the magnetizations of its ferromagnets: if they are parallel, the MTJ is in low resistance state (R_P) meaning more current flowing through it, otherwise, they are in antiparallel configuration (R_{AP}) with highest resistance. These resistance states can be mapped in a logic fashion as logic '0' if they are antiparallel, logic '1' otherwise. By arranging multiple MTJs in a matrix configuration, both memory and logic operations can be performed analogically. Several works use MTJ devices. In [12], Generative Adversarial Network (GAN) implementation has been proposed. This Neural Network consists of a discriminator (D), that works as a detective in the training process, and a generator (G) as a deceiver in a semi-supervised fashion. In these networks, training is a critical issue so hardware accelerators are demanded. Reference [12] improves the so-called adversarial training process by using an array made of MTJs which simplifies the calculation of multiply-accumulate operations with ternary weights ($W \in \{-1, 0, 1\}$), transforming them into bulk In-Memory additions and subtractions. This work achieves remarkable results in term of efficiency and processing speed with respect to GPUs and ASICs. In [13], authors have developed a MTJ-based convolution accelerator in which the memory array is capable of performing bulk AND operations. They have included a small external logic which is in charge of computing the accumulations. Based on a similar working principle, Resistive Random-Access-Memory (RRAM) [14] are devices in which the logic data is encoded in two or multiple resistive states. Differently from MTJs, resistance is

determined by the conductivity of a conduction path that can be broken (high resistance state) or reformed (low resistance state). Sometimes it is used in a 1 transistor 1 RRAM (1T1R) configuration, to avoid unwanted or sneak current paths. In [15], authors have presented a memristor-based implementation of a BNN able to achieve both high accuracy on MNIST and IRIS dataset and low power consumption. In some others, improvements in memristor architectures have been proposed that enable multiple bits per cell. Reference [16] has exploited the frequency dependence of GeSeSn-W memristor devices to obtain multiple conductance values representing different weights. In [17], the memory array has been modified, including up to 4 memristors arranged in parallel in the same cell, in order to have multiple resistance values and so higher precision weights. Based on a similar approach to [12], a GAN training accelerator has been discussed in [18] which is able to efficiently perform approximated add/sub operations in a memristor array, achieving both speed-up and high energy efficiency.

3. Computation with Memory [9] concept consists of memory arrays that intrinsically perform calculations. Possible examples can be Content Addressable Memories (CAM) and Look-up tables.
4. Logic-in-Memory [9] is the concept that we are analyzing in this work, in which small computational units are placed inside or near a memory cell, to perform distributed computation.

As can be deducted, LiM is a widely studied and heterogeneous topic, and it is becoming increasingly important over the years. A lot of works presented in literature implement an application-specific LiM solution. The discussed emerging technologies are very promising, especially in Neural Networks applications, because of their high efficiency to compute multiply-accumulate operations [16]. In our work, we concentrated on CMOS technology because, while it is not the best available, RRAM and MTJ devices are still under development. As future task, we will focus our attention on them once these circuits are optimized.

3. Neural Networks: An Introduction

3.1. Neuron's Model

A NN is a computational model that is able to perform very complex tasks. It is composed of "neurons", which are the basic building blocks. By organizing them in an interconnected network, the NN can take decisions and learn when these decisions are wrong [19].

In Figure 2 a neuron structure example is depicted. As it is possible to see, it is made of two main parts which are *net*, which is in charge of weighted sum computation, and $f(net)$, which is an activation function applied to the neuron's output. In general, *net* expression can be written as:

$$net = \sum_{i=0}^{N} X_i \times W_i + Bias \tag{1}$$

where X_i is the input value, W_i is the corresponding weight and *Bias* is an additive term. Neurons' weights and biases can be adjusted to achieve the desired output with a procedure called training.

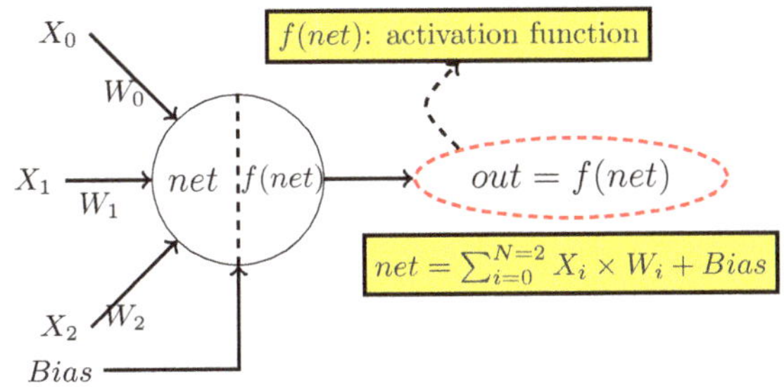

Figure 2. Schematic of a neuron, representing its structure. Three inputs example [19].

In Figure 2, it is indicated another part which is the activation function $f(net)$. Usually, this is a nonlinear function. The most important activation functions are Rectified Linear Unit (ReLU), hyperbolic tangent (tanh) and sigmoid function, which are discussed in great details in [20].

3.2. Neural Network's Structure

Usually, NNs are made up by layers, which are composed of a set of arranged neurons. The most common structures are Convolutional Neural Network and Multi-Layer Perceptron.

In Figure 3 it is reported the LeNet5 CNN as example. The network is composed of 2 convolutional, 2 pooling and 3 fully connected (FC) layers. Each of them is discussed in detail:

- Convolutional layers perform the convolution operation of the input feature map (IFMAP) with a set of weights called kernel. An example of a convolution computation is depicted in Figure 4. The parameter taken into account are the kernel's weights, the input feature map and the stride. After the first convolution is finished, the kernel window is moved by a step equal to stride, and a new convolution can start. In this example, the convolution computation match perfectly the neuron's equation reported in Equation (1), in fact after a convolutional layer is usually used an activation function to normalize the results. In the LeNet 5 CNN [1] example in Figure 3, all the convolutional layers have the same 5×5 kernel sizes. The first one produces six output feature maps (OFMAPs), meaning that the same IFMAP has been convolved with six different kernels. The second convolutional layer instead produces 16 OFMAPs, starting from 6 IFMAPs: for each input, there are 16 kernels that produce 16 outputs, so 16 from the first IFMAP, 16 for the second IFMAP and so on. This implies a total number of OFMAPs equals to

$$\#OFMAPs = 6 \times 16 \tag{2}$$

  To obtain 16 OFMAPs indicated by LeNet 5 scheme, the obtained OFMAPs of each layer are added together.

  These considerations bring to the following formula for a convolutional layer, derived from [21]:

$$y_o(j,i) = Bias_o + \sum_{c_{in}=0}^{\#C_{in}-1} \sum_{k=0}^{W_y-1} \sum_{p=0}^{W_x-1} k_{o,c_{in}}(k,p) \times X_{o,c_{in}}(j \times stride + k, i \times stride + p) \tag{3}$$

  where i, j are the indexes for the OFMAP corresponding pixel, c_{in} is the input channel index, $\#C_{in}$ the total number of input channels, W_x, W_y are the kernel's matrix size indicating number of rows and columns respectively, o subscript refers to the OFMAP considered and p, k are the kernel's indexes.
- Pooling layers have a similar behavior to convolutional layers. In the literature, different kind of poolings are used such as average or max pooling [22]. They perform the maximum (or the average) of the selected input pixels and returns only one value, performing the so-called

subsampling operation. Pooling, and more specifically max pooling, is widely used to reduce the size and the complexity of the CNN. In Figure 3, the kernel size is 2×2 for all the cases.

- FC layers are MLP subnetworks included in the CNN to perform the classification operation. They are made of layers of fully interconnected neurons, as shown in Figure 3.

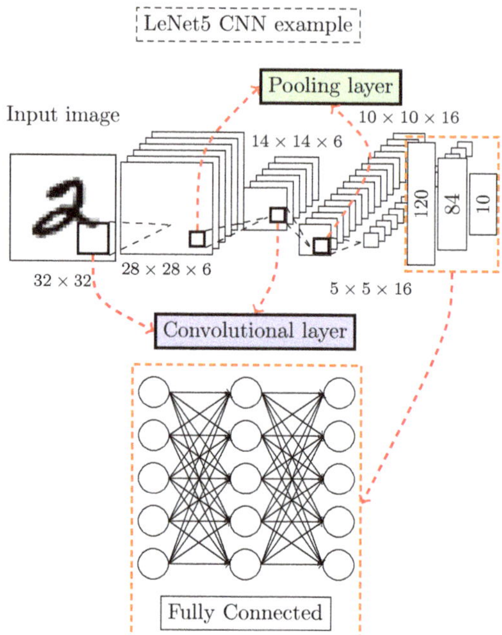

Figure 3. Structure of LeNet 5 Convolutional Neural Network (CNN) [1], composed of 2 convolutional, 2 pooling and 3 fully connected layers and their sizes are indicated in the model.

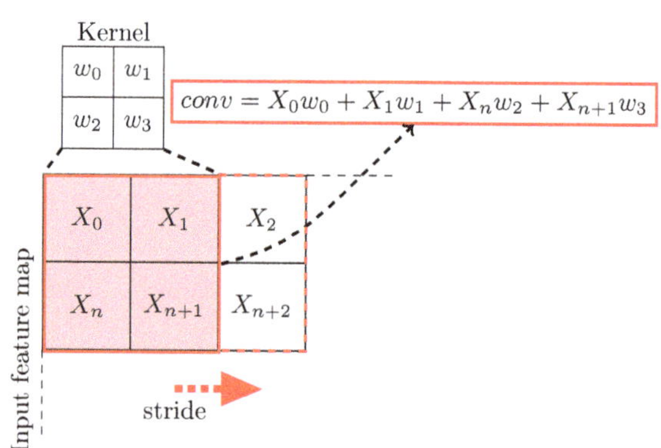

Figure 4. Convolution computation example with a 2×2 kernel.

There are also normalization layers (not reported in Figure 4). One of the most used is the Batch Normalization (BatchNorm) [23] that is very useful in BNNs to recover a portion of the accuracy lost from the binarization [24]. BatchNorm equation is reported from [23]:

$$\tilde{X} = \frac{X - \mu}{\sqrt{\sigma^2 + \epsilon}} \times \gamma + \beta \tag{4}$$

where μ, σ are the batch mean and variance, while γ, β are correction values. These four variables are trainable, meaning that during training procedure they are modified in order to increase the accuracy. ϵ is usually added to the variance to avoid 0 division if the variance is 0. ϵ is a very small number, so the following approximation for non-zero variance can be made:

$$\tilde{X} \approx \frac{X-\mu}{\sigma} \times \gamma + \beta = X \times \frac{\gamma}{\sigma} + \left(-\frac{\mu \times \gamma}{\sigma} + \beta\right) = X \times A + B \tag{5}$$

3.3. Binary Approximation

Since NN are very complex models, they can be very power hungry and implementing them on low energy budget systems, like in embedded application, can be challenging [25]. For this reason, a BNN approximation is chosen, trying to reach a good trade-off between complexity and accuracy. In [7] is presented an interesting comparison between some BNN approximations, introducing also XNOR-Net. The values are recalled in Figure 5. In the plot, TOP5 is intended as the accuracy classification rate to hit one out of five most probable classes. In the plot, TOP5 is intended as the accuracy rate to hit 1 out of 5 most probable classes. The BNNs accuracy are compared with the original floating-point implementation (FP) of AlexNet neural network [2].

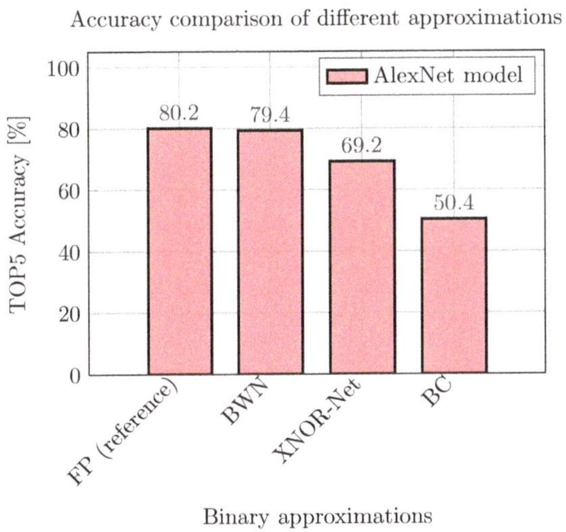

Figure 5. TOP5 accuracy comparison between different binary approximations [7].

In the considered approximation, all the weights are in binary format, meaning that $w \approx w_b \in \{-1, 1\}$ where w_b is the binary weight value. The binarization techniques are now briefly summarized from [7].

- BWN [7] binarizes only weights of the NN, keeping at full precision the activations and the inputs. By binarizing only weights, the convolution operation can be performed only with adds and subtractions, avoiding multiplication as reported in Equation (6) [7].

$$Conv_{out,BWN} = X * w + Bias \approx \alpha(X * w_b) + Bias \tag{6}$$

An extra factor α is multiplied to the convolution result, in order to compensate precision losses [7]:

$$\alpha = \frac{\sum_{i=0}^{N} \|w_i\|}{N} \tag{7}$$

where w_i is the considered full precision weight and N is the number of weights. BWN is a very good alternative useful to reduce CNN's complexity. However it requires full precision inputs and activations.

- XNOR-Net [7] binarizes both weights and inputs. The convolution result is obtained by performing the binary convolution and multiplying by a correction factor α (the same in Equation (7)) and a matrix $\mathbf{K}$. $\mathbf{K}$ is defined in Equation (8).

$$\mathbf{K} = \overbrace{\frac{\sum_{c_{in}=0}^{\#C_{in}-1} |X(:,:,c_{in})|}{\#C_{in}}}^{\text{First term}} * \overbrace{\begin{bmatrix} \frac{1}{W_x \times W_y} & \frac{1}{W_x \times W_y} & \cdots \\ \frac{1}{W_x \times W_y} & \frac{1}{W_x \times W_y} & \cdots \\ \vdots & \vdots & \ddots \end{bmatrix}}^{\text{Second term}} \tag{8}$$

In Equation (8), the first term indicates the absolute punctual sum of the multiple IFMAPs divided by the number of input channels, thus the number of IFMAPs. The second term is a regular matrix of $W_x \times W_y$ size, which contains $\frac{1}{W_x W_y}$ in all positions. Finally, the XNOR-Net convolution can be rewritten as [7]:

$$Conv_{out,XNOR-Net} \approx (X_b \circledast w_b) \cdot \mathbf{K} \times \alpha \tag{9}$$

where X_b is the binarized input, $\circledast$ is the binary convolution, $\cdot$ is punctual multiplication and $\times$ is a simple product. In [7] the binary convolution is performed considering the XNOR pop-counting of binary inputs/weights. XNOR truth table matches to the multiplication if -1 is mapped to logic '0' and +1 is logic '1'. Pop-counting computes the difference between the number of 1s and the number of 0s of the input sample.

- BC [6] binarizes both inputs and weights, without applying any correction factor to the final convolutional equation. This implies less recognition accuracy as shown in Figure 5. Taking into account all the considerations on the binarization techniques, we chose XNOR-Net [7] as reference model since it represents a very good trade-off between accuracy and complexity.

3.4. NN Implementations Based on LiM Concept

The LiM approach is often applied in NNs' implementations. Some of them are considering the binary approximations, choosing an implementation based on emerging technologies. Some works [12,13,26,27] are based on MTJ technology while [15–18,28,29] have used RRAM. In each of these works the resistive element is used to perform simple logical operations based on current sensing technique. In [26,27,30,31] several Binary Convolutional Neural Networks (BCNNs) implementations are discussed: they achieve very good results in terms of energy and power, thanks to the intrinsic low power nature of the MTJ and RRAM devices. Reference [28] proposes a BNN design based on SRAM array. The logic parts perform the computations and are disposed below the memory array. The memory parts enable to store the required parameters for the NN computation (like weights and biases) and the logic parts compute the results for the next layer, forming an alternation between memory-logic. This architecture achieves very good performance in terms of energy and speed, thanks to its pipelined-like structure. In [29], the NN has been mapped in a Wide-IO2 DRAM, using TSVs as high speed communication link obtaining remarkable results in terms of execution time.

4. OOM and LiM Architectures

Here we discuss the adopted processing flow more in detail. The goal of a LiM architecture is to move part of the computation inside a memory array, which already contains the needed logical elements to complete the calculations. Using as a case study the XNOR-Net, we can derive that the main part of the algorithm is the calculation of the XNOR products combined with pop-counting to determine the result of the binary convolution. The adopted design flow is the following:

- Design of a classical architecture, called OOM, capable of implementing the XNOR-Net model;
- Derive a LiM alternative, defining what are the building blocks inside a memory cell;
- Qualitative comparison of the architectures, describing the advantages and disadvantages of both of them;
- Performance estimation and comparison by means of synthesis and place and route procedures of two NN models;
- Performance estimation of different NN models.

4.1. OOM Architecture Design

4.1.1. Single Input/Multiple Output Channels Design

By looking at Equation (9) and Figure 6, the core part of the OOM architecture is composed of a set of XNOR gates and a pop-counter. In Figure 6, a tiny example of a 2×2 convolution is reported: since the dimension of the kernel is 4, the total number of XNOR gates required are 4, because each of them performs a multiplication. In general, their number must be at least equal to $W_x \times W_y$. NNs' kernel sizes depend on the model chosen, for example in AlexNet the maximum kernel size is 11×11 [2] or in LeNet5 is 5×5 [1]. The flexibility of the hardware circuit depends strictly on how big the kernel size is considered, so a worst-case analysis must be taken into account. To best of our knowledge, kernel sizes higher than 11×11 are very seldom, since accuracy usually decreases with bigger filter sizes. Binarized inputs/weights are fed directly to XNOR inputs from a memory implemented as a register file (RF) in the design, named Binary Input RF in Figure 7. In Binary Input RF, each row contains all the input elements required for a convolutional window computation, implying a bitwidth size equals to $W_x \times W_y$ bits. Regarding the number of rows, they have to be at least equal to the total number of convolutional windows D_{out}^2 required, which is also the dimension of the OFMAP. D_{out} can be computed considering kernel, IFMAP sizes (D_{in}) and the stride.

$$D_{out} = \frac{D_{in} - W_x}{stride} + 1 \tag{10}$$

Also in this case, the number of memory rows D_{out}^2 has to consider the maximum OFMAP size of the NN model considered. When a different OFMAP has to be computed, the weight set is simply switched by using a multiplexer. The total number of multiplexer is equal to the maximum number of OFMAPs, called number of output channels (#C_{out}) of the NN. In Equation (10), it is indicated only W_x, since usually the kernels are regular matrices with $W_x = W_y$.

Regarding the pop-counting computation, handling many parallel inputs requires too many hardware resources. For this reason, the outputs of the XNOR gates are multiplexed and only one of them is processed per clock cycle. A pop-counter can be simply implemented with an adder, a NOT gate and a register as shown in Figure 8. Together with the pop-counting circuit, the main computational part has been called XNOR-Pop Unit, as shown in Figure 7.

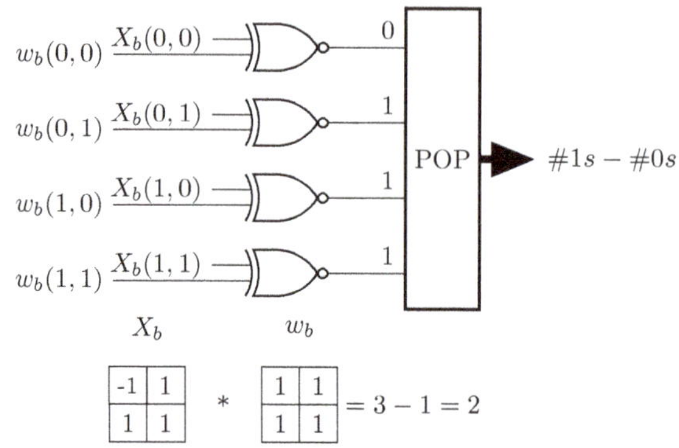

Figure 6. Example of Binary convolution based on XNOR-Pop procedure.

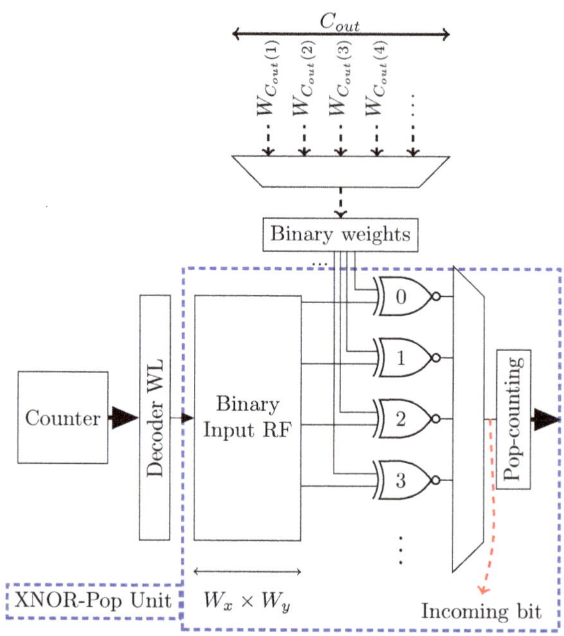

Figure 7. Out-Of-Memory (OOM) main computational part. Each Binary Input RF's row holds the binary inputs required for a convolutional window computation, while weights are provided by an external memory. The outputs of the XNOR gates have been multiplexed to reduce the computational overhead of the pop-counting part.

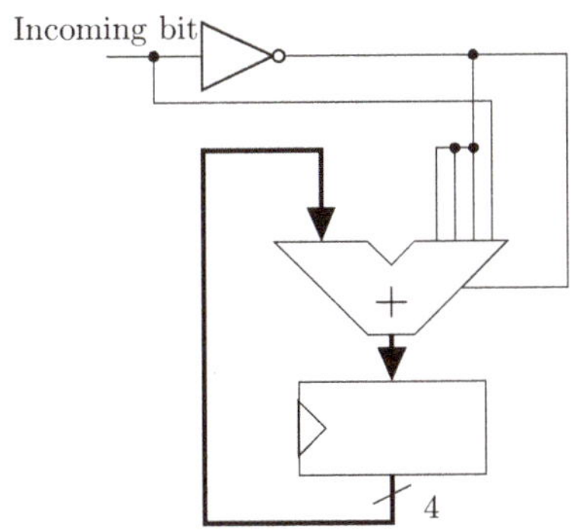

Figure 8. Four bits example of a pop-counter circuit.

4.1.2. Multiple Input Channels Design

Many CNNs have multiple IFMAPs in input. Each convolutional window must be computed separately and, in the end, summed together to get the resulting OFMAP. This can be obtained by increasing the level of parallelism of the architecture, having multiple XNOR-Pop Units working at the same time. As can be seen in Figure 9, a C_{in} number of XNOR-Pop Units are required and a final accumulation circuit computes the sum of the single channels. XNOR-Pop Units are multiplexed to reduce the hardware complexity for bigger networks.

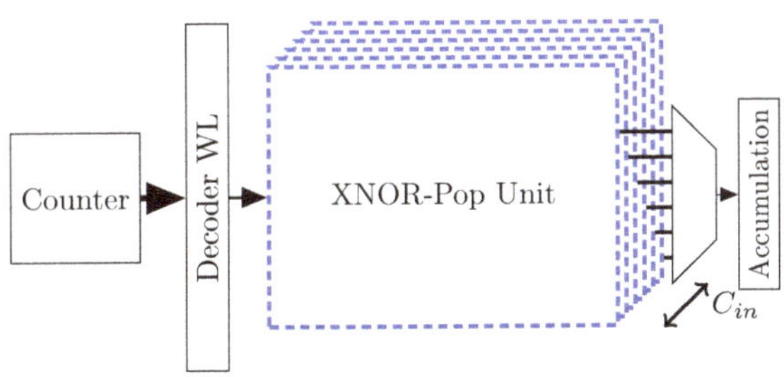

Figure 9. Multiple input channels OOM design.

4.1.3. FC Layer Integration

Until this point, the convolution algorithm has been mapped on the hardware architecture described so far. To implement the fully connected layer the same circuit can be reused by simply inverting weights and inputs sources. To better understand this concept, the example reported in Figure 10 is considere. The weights values for each input neuron are w_0^0, w_1^0, w_2^0 for X_0, w_0^1, w_1^1, w_2^1 for X_1 and w_0^2, w_1^2, w_2^2 for X_2. The output O_0 can be computed considering Equation (11).

$$O_0 = \text{pop-count}(\overline{X_0 \oplus w_0^0}, \overline{X_1 \oplus w_0^1}, \overline{X_2 \oplus w_0^2}) \tag{11}$$

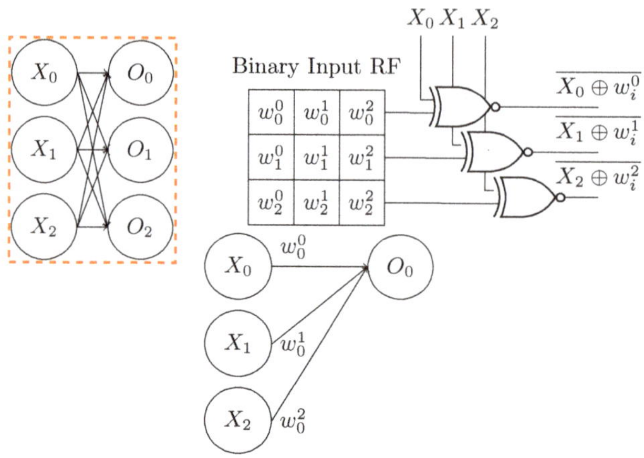

Figure 10. Example of a 3-3 FC network mapping.

As depicted in Figure 10, the Binary Input Register File (RF) contain the binary weights instead of the inputs, in fact by addressing each line the multiplication of the weights with inputs is performed, and then pop-counted.

The size of the Binary Input RF is also bounded to the FC network's characteristics, so the relations of width-height of the Binary Input RF are the following:.

$$\begin{cases} \text{Memory size}_x = max(W_x \times W_y, \text{\#input neurons}_{FC}) \\ \text{Memory size}_y = max(D_{out}, \text{\#output neurons}_{FC}) \end{cases} \tag{12}$$

Although this is the straight forward way to map an FC algorithm on the architecture, this can be very complex with a high number of input neurons. Considering LeNet5 [1] depicted in Figure 3, the first FC layer has 120 output neurons, that can be acceptable, but for more sophisticated algorithms like AlexNet [2], which has 4096 input neurons, makes this kind of scheduling very inefficient. A generic output neuron's equation O_i is given by

$$O_i = \sum_{j=0}^{4095} X_j \times w_i^j + Bias = X_0 \times w_i^0 + X_1 \times w_i^1 + ... + X_{4095} \times w_i^{4095} + Bias \tag{13}$$

this sum can be computed by performing fewer number of adds per each clock cycle. The partial result is stored and added in each clock cycle. The algorithm steps become:

$$\begin{aligned} &\text{Store temp}(0) = 0 \\ &\text{Store temp}(1) = \sum_{j=0}^{n} X_j \times w_i^j + \text{Store temp}(0) \\ &\text{Store temp}(2) = \sum_{j=n+1}^{2n} X_j \times w_i^j + \text{Store temp}(1) \\ &\quad ... \end{aligned} \tag{14}$$

where n is the total number of considered terms for each summation and Store temp holds temporary additions partial results.

Figure 11 shows an example of serialization of 2 input neurons per cycle, meaning that only a subset of weights (highlighted by the dashed lines in the figure) are stored inside the Binary Input RF.

The partial result is computed and then temporarily stored in each algorithmic step. In Equation (14), n is 2 and consequently the Memory size values can be rewritten as

$$\begin{cases} \text{Memory size}_x = max(W_x \times W_y, n) \\ \text{Memory size}_y = max(D_{out}, \#\text{output neurons}_{FC}) \end{cases} \tag{15}$$

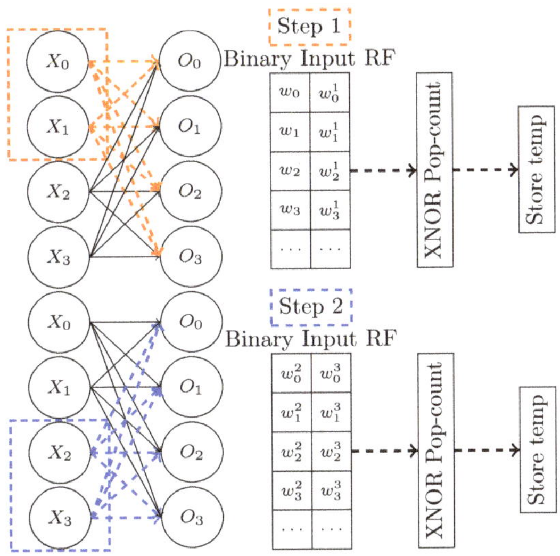

Figure 11. Example of serialization of the FC computation.

4.1.4. OOM Convolution-FC Unit

A scheme of the complete OOM architecture is now provided in Figure 12 and each element's functionality is reported in details.

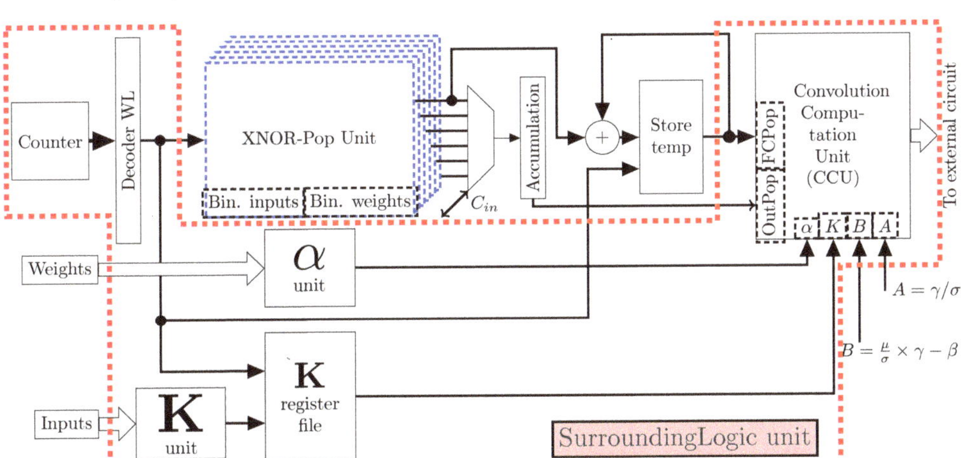

Figure 12. Complete OOM architecture. The thicker red dashed line frames the units which are the main components of the SurroundingLogic unit. Inputs are provided to the SurroundingLogic unit from the external world. Outputs are processed and saved outside in the testbench.

- XNOR-Pop Unit is the block described before, which takes Binarized weights/inputs and computes the resulting binary convolution;
- Store temp is a register file that holds the partial FC values, resulting from the scheduling described in Section 4.1.3. As the XNOR-Pop Unit's Binary Input RF, which is depicted in Figure 7, it is addressed by the same counter, since only one neuron is processed per time. Only the output coming from the first XNOR-Pop Unit is taken, because FC requires only one input channel to be executed;
- **K** and α units are in charge of computing **K** and α values, as required by XNOR-Net convolution approximation expressed in Equation (9). Since **K** are matricial values, a register file has been inserted in the design to hold them;
- Convolution Computation Unit (CCU) performs the final calculation to provide the convolutional result, which is the formula reported in Equation (9). Moreover, it applies BatchNorm, if required by the algorithm: its coefficients are computed offline and provided by the external testbench.

4.2. LiM Architecture

4.2.1. XNOR-Pop LiM Unit

The design driving concept of a LiM architecture is to increase as much as possible the level of parallelism. Starting from the OOM standard implementation, we designed two LiM arrays that perform XNOR bitwise and pop-counting operations. Since we didn't have the possibility to implement a custom memory, we used as memory element a flip flop and a static CMOS based logical part.

These choices imply a higher power and area estimation in the synthesis phase, that will be discussed more in details in Section 6. Regarding the XNOR part, the idea is to put a XNOR gate inside each memory cell and to perform the binary product between the content of the cell and an external binary input. An example is depicted in Figure 13, in which is shown how a simple 2×2 convolution is mapped inside a LiM array. In order to perform the bitwise multiplication between the binary input and the corresponding weight, as we can see from the example in Figure 13 the highlighted portion of IFMAP has to be convolved with the kernel in the following way:

$$\text{Incoming bit}_0 = \text{pop-count}(\overline{X_0 \oplus w_0}, \overline{X_1 \oplus w_1}, \overline{X_4 \oplus w_3}, \overline{X_5 \oplus w_3}) \tag{16}$$

Since one of the XNOR inputs is hardwired to an external connection, it is sufficient to store inside the memory array the input required to perform the convolution. The same for the following row line: the convolution is performed with the same kernel, so each memory row corresponds to a convolutional window.

Regarding pop-counting procedure, in order to reduce the complexity of the memory cell, we can simplify the pop-count equation in the following way:

$$\text{pop-count} = \#1s - \#0s = 2 \times \#1s - \text{length(word)} \tag{17}$$

where length(word) is intended as the size of the array entering in the pop-counter, which is 4 in Figure 13. A ones counter is simply made of half adders (HA) connected as depicted in Figure 14, so in the pop-counting part there will be a HA for each memory cell. Figure 15 provides an overview of the entire LiM implementation. It is possible to distinguish between LiM XNOR part and the LiM ones counter whose detailed architectures are depicted in Figures 13 and 14 respectively. Together, with the multiplexer depicted in Figure 15, they form the LiM XNOR-Pop unit.

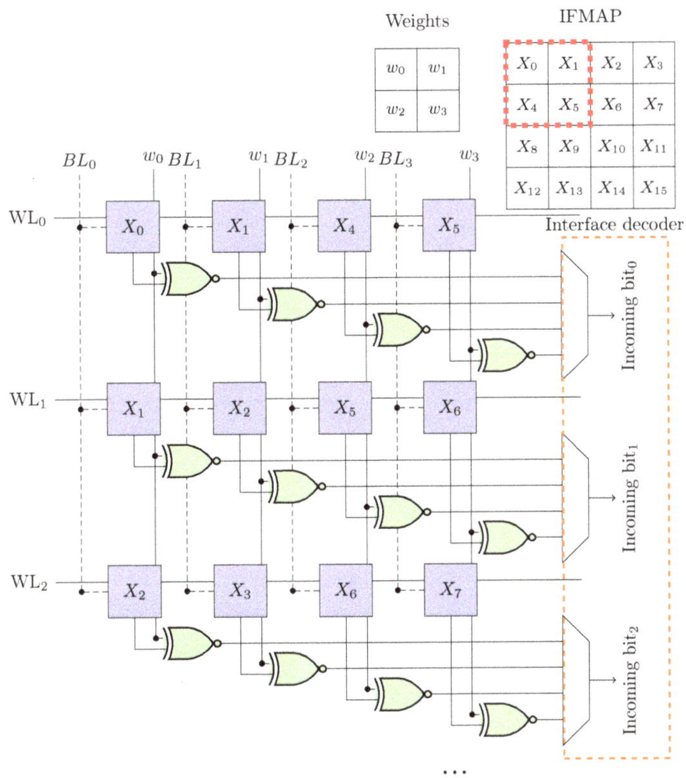

Figure 13. XNOR part of the XNOR-Pop Unit LiM implementation: example of 2×2 kernel and 4×4 IFMAP sizes with stride 1.

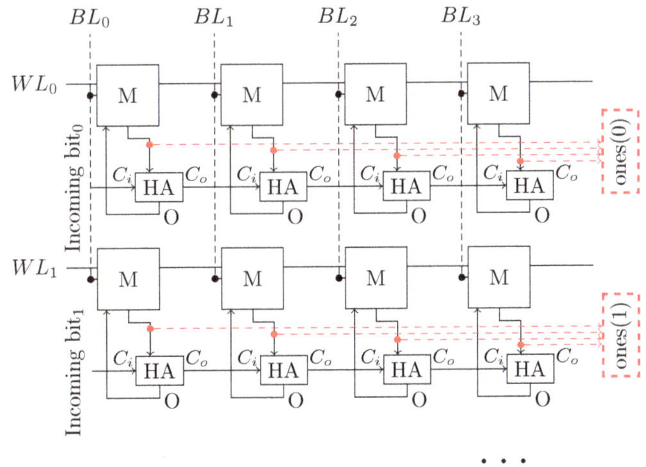

Figure 14. Example of a 4 bits ones counter LiM implementation.

4.2.2. LiM convolution-FC Unit

From the previous considerations, the entire LiM architecture can be designed as shown in Figure 15.

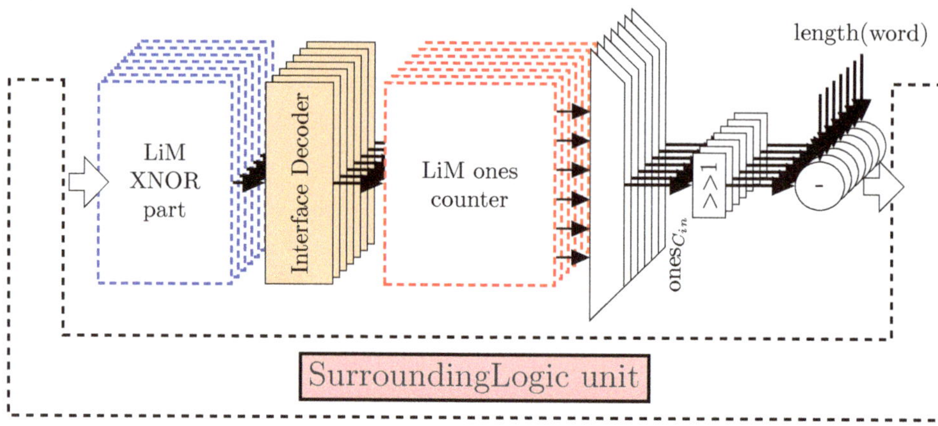

Figure 15. LiM entire architecture. The main blocks of the LiM implementation are the LiM XNOR part, interface decoder, LiM one-counter, and shifters-subtractors for the pop-counting computation, that are replicated for C_{in} number of times. The surrounding logic is the same as the OOM case reported in Figure 12.

The Surrounding logic unit remains the same, since the interface has been kept between OOM-LiM XNOR-Pop units. The other units are replicated $\#C_{in}$ times, depending on the total number of input channels required by the algorithm. The LiM alternative can achieve a higher level of parallelism, because XNOR-ones counter parts can perform the operations in parallel. In Figure 15, there are also "$<< 1$" blocks: they perform the shift by 1 position, corresponding to multiplication by 2.

4.3. Top-Level Entity

The top-level entity contains both the Convolution-FC, Pooling circuits. Pooling is simply made of a multiplexed comparator that takes the maximum out of $W_x \times W_y$ number of inputs. The top-level entity contains also an Interface, which is in charge of dispatching the inputs coming from the testbench and to provide the results of Pooling/Convolution-FC to the outside. The top-level entity can be schematized in Figure 16.

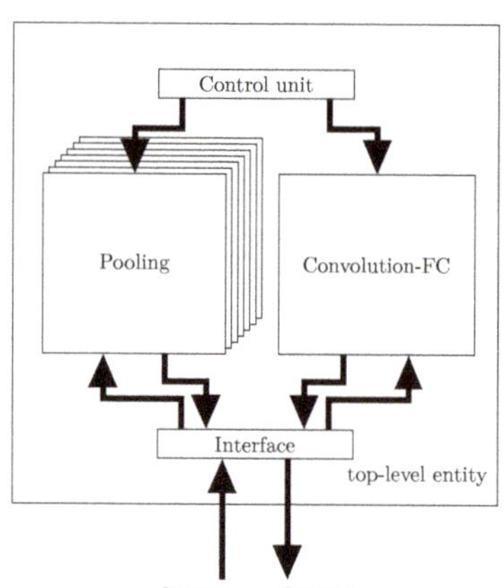

Figure 16. Top-level entity of both LiM and OOM architectures.

5. Qualitative Comparison OOM-LiM Architectures

In order to make a qualitative comparison, algorithm execution time was considered as a benchmark parameter. We can distinguish between convolution, fully connected and max pooling execution times, since they are completely different. The computation is based on a CNN, since it generally contains all of those layers. The CNN's parameters are not specified, since we are doing a parametric estimation.

5.1. OOM Execution Time

5.1.1. Pooling Layer

In our analysis we start from Pooling layer. As said in Section 4.3, Pooling is made of a simple multiplexed comparator. The input scanning ends when all them have been considered, so after an entire pooling window content is evaluated. This value is multiplied by the total number of pixels of the resulting OFMAP obtaining

$$\text{Pool}_{time} \approx D^2_{out(pool)} \times (W_{x(pool)} \times W_{y(pool)}) \times t_{ck} = D^2_{out(pool)} \times (W^2_{x(pool)}) \times t_{ck} \tag{18}$$

where $D^2_{out(pool)}$ is the pooled OFMAP size. The worst-case filter dimension is set to W_x^2 for both convolutional and pooling.

5.1.2. Convolutional Layer

At the beginning of the convolution algorithm, the binary inputs are precharged inside the Binary Input RF and **K** matrix is computed in the meanwhile, meaning that for each input set are required W_x^2 clock cycles. Since an entire OFMAP has a number of pixels equals to $D^2_{out(conv)}$, the total number of cycles required in this step are $D^2_{out(conv)} \times W_x^2$ clock cycles. After that, convolution is performed: considering Figure 7, an entire convolutional window is computed when all the XNOR outputs have been scanned. The number of XNOR gates is equal to the Binary Input RF's word length, which is W_x^2. By multiplying the time required by a convolutional window computation with the total number of convolutional windows $D^2_{out(conv)}$, we get the total convolution time which is $\text{Convolution}_{time,OOM}$. The last contribution set is the BatchNorm, that can applied after each convolutional window and α computation together with results' storing. Each of them takes only 1 clock cycle. We can derive the equation for the convolutional layer execution time with 1 input/output feature map as follows.

$$\text{Convolution}_{time,OOM} \approx \left(\overbrace{D^2_{out(conv)} \times W_x^2}^{\text{Store inputs \& }\mathbf{K}\text{ computation}} + \overbrace{D^2_{out(conv)} \times (W_x^2+1)}^{\text{Convolution \& BatchNorm}} + \overbrace{(1+1)}^{\alpha\text{ - Store results}} \right) \times t_{ck} \tag{19}$$

When multiple output channels are considered, the convolution windows computation has to be repeated for each of the OFMAP:

$$\begin{aligned} \text{Convolution}_{time,OOM} \approx & \overbrace{D^2_{out(conv)} \times W_x^2}^{\text{Store inputs \& }\mathbf{K}\text{ computation}} \times t_{ck} + \\ & + C_{out} \times \left(\overbrace{D^2_{out(conv)} \times (W_x^2+1)}^{\text{Convolution \& BatchNorm}} + \overbrace{(1+1)}^{\alpha\text{ - Store results}} \right) \times t_{ck} \end{aligned} \tag{20}$$

the last situation is the multiple input/output channels case. Since the convolution operation is parallelized, the convolutional windows coming from each XNOR-Pop Unit is added in a serial fashion.

This means that to achieve the final convolution value, each contribution has to be added together before executing the BatchNorm. The final Convolution$_{time}$ expression is the following.

$$\text{Convolution}_{time,OOM} \approx \overbrace{D^2_{out(conv)} \times W_x^2}^{\text{Store inputs \& }\mathbf{K}\text{ computation}} \times t_{ck} + \\ + C_{out} \times \left(\overbrace{D^2_{out(conv)} \times (W_x^2 + 1 + C_{in})}^{\text{Convolution \& BatchNorm multiple } C_{in}} + \overbrace{(1+1)}^{\alpha\text{ - Store results}} \right) \times t_{ck} \tag{21}$$

5.1.3. FC Layer

For the FC computation, we have to consider the scheduling explained in Section 4.1.3. As the convolution case, the algorithm starts precharging the inputs inside the array, taking $D_{out(FC)}$ clock cycles, where $D_{out(FC)}$ is the total number of output neurons. Since the dimension of the Binary Input RF is Memory size$_x$, only Memory size$_x$ input neurons are considered per time, so as performed for the convolutional layer, the time required for a FC output is equal to $D_{out(FC)} \times$ Memory size$_x$ that has to be added to the previous contribution. FC results have to be stored, and this can be made by scanning the content of Store temp register (depicted in Figure 12), taking $D_{out(FC)}$ clock cycles. The execution time for a single step of the FC scheduling is given by:

$$\text{FC}_{time,OOM} \approx \left(\overbrace{D_{out(FC)}}^{\text{Store inputs}} + \overbrace{D_{out(FC)} \times \text{Memory size}_x}^{\text{FC output computation}} + \overbrace{D_{out(FC)}}^{\text{Store temp scanning}} \right) \times t_{ck} \tag{22}$$

this partial result has to be repeated by the total number of iterations (n_{iter}) required to calculate the FC layer. The final FC execution time expression is:

$$\text{FC}_{time,OOM} \approx \left[n_{iter} \times \left(\overbrace{D_{out(FC)}}^{\text{Store inputs}} + \overbrace{D_{out(FC)} \times \text{Memory size}_x}^{\text{FC output computation}} \right) + \overbrace{D_{out(FC)}}^{\text{Store temp scanning}} \right] \times t_{ck} \tag{23}$$

5.2. LiM Execution Time

Similarly to the OOM case, Pooling, Convolution and FC execution times are provided and explained. Since Pooling layer is the same in both cases, it is not analyzed in this part.

5.2.1. Convolutional Layer

As already done in OOM, the array has to be precharged taking $D^2_{out(conv)}$ clock cycles. After that, all the XNOR gates inside the XNOR part work together at the same time, and the Interface Decoder, which is depicted in Figure 13, takes one by one each XNOR result and provide it to the ones counter. When all XNORs' output have been scanned after W_x^2 clock cycles, the ones counter results are stored inside the LiM ones counter reported in Figure 15. At this point, all the LiM ones counter values must be fetched for each input channel, requiring $C_{in} \times D^2_{out}$ clock cycles to perform the residual part of the algorithm. The final formula for the LiM convolution execution time is

$$\text{Convolution}_{time,LiM} \approx \overbrace{D^2_{out(conv)} \times W_x^2}^{\text{Store inputs \& }\mathbf{K}\text{ computation}} \times t_{ck} + \\ + C_{out} \times \left[\overbrace{W_x^2 + D^2_{out(conv)} \times (1 + \#C_{in})}^{\text{Convolution \& BatchNorm multiple } C_{in}} + \overbrace{(1+1)}^{\alpha\text{ - Store results}} \right] \times t_{ck} \tag{24}$$

5.2.2. FC Layer

Similarly to the OOM case, we have scheduled the algorithm to reduce the complexity. After $D_{out(FC)}$ clock cycles required to store the values inside the LiM array, an entire FC step is computed in Memory size$_x$ clock cycles and the final results are scanned from the LiM ones counter in $D_{out(FC)}$ cycles. In LiM architecture, the Store temp register file is not required since the pop-count values are already stored in the LiM part. By iterating the entire algorithm n_{iter} times, we get the final FC execution time:

$$\mathrm{FC}_{time,LiM} \approx \left[n_{iter} \times \left(\overbrace{D_{out(FC)}}^{\text{Store inputs}} + \overbrace{\text{Memory size}_x}^{\text{FC output computation}} \right) + \overbrace{D_{out(FC)}}^{\text{Store temp scanning}} \right] \times t_{ck} \quad (25)$$

5.3. Comparison Results

The results obtained by performing the ratio between OOM/LiM execution times are now provided. The previous part, and in particular Sections 5.1 and 5.2 take into account an approximate computation of the execution time, since the overheads of idle/dummy states were not considered for sake of simplicity. In this part, we show the real estimations that consider all the contributions. Considering Figure 17, it is possible to see how delay ratio (obtained as execution time OOM/execution time LiM) changes in different cases. A series of sweeps were made, considering the most important variables, in particular $\#C_{in}, \#C_{out}, W_x, D_{in}, D_{out(fc)}, n_{iter}$. On the vertical axes, there is Delay ratio in all plots. Some of the estimations were performed considering the convolution timing equations reported in Equations (21) and (24). These plots are are labelled with "Convolution computation" flag in Figure 17. The remaining one consider FC delay expressions reported in Equations (23) and (25).

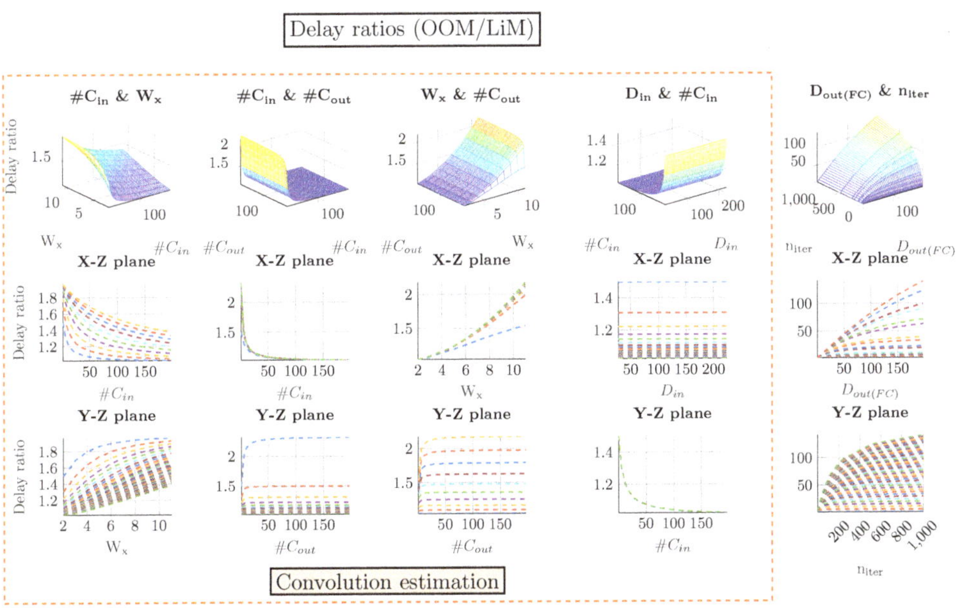

Figure 17. Delay ratio obtained as OOM/LiM for different parameters, in order to see how the two architectures behave for different cases.

- Delay ratio vs $\#C_{in}$ & W_x: the Delay ratio with respect to $\#C_{in}$ has a decreasing trend because, as shown in Figure 15, the Interface Decoder, the multiplexers placed after the LiM ones counter and the serial accumulation of the values of each channel represent a bottleneck for LiM architecture.

As a result a higher execution time for higher values of $\#C_{in}$ is observed. In general, for high values of W_x, the Delay ratio increases, because of the parallelization in LiM architecture.

- Delay ratio vs $\#C_{in}$ & $\#C_{out}$: the trend for $\#C_{in}$ is the same as the previous case. For high values of $\#C_{out}$, we can expect a very good Delay ratio efficiency, because LiM already has the values stored inside the array and it is sufficient to change the weights set by simply selecting it, following the same principle of the OOM case depicted in Figure 7.
- Delay ratio vs W_x & $\#C_{out}$: in general, by increasing both $\#C_{out}$ and W_x we have a higher Delay ratio. By looking to X-Z plane, it is possible to see that for higher $\#C_{out}$ the curve becomes steeper. It is a very good trend for very deep NNs, because usually output channels and filter sizes are high.
- Delay ratio vs D_{in} & $\#C_{in}$: D_{in} is the IFMAP size, which indirectly determines the OFMAP size as reported in Equation (10). High values of D_{in} imply much more complex NN but the Delay ratio remains almost constant, showing that LiM architecture latency is not degraded by the IFMAP's complexity.
- Delay ratio vs $D_{out(FC)}$ & n_{iter}: this last plot set reports an FC layer estimation. In this case the formula for the FC execution time of OOM and LiM is considered. As it can be seen, a higher $D_{out(FC)}$ could be beneficial for a LiM architecture, which has a small increasing trend, because the LiM array performs all the computations in parallel, so there is no need to fetch each data from the memory, compute the result and store inside the Store temp register file as in the OOM case. The predominant variable is n_{iter}, because by looking at Figure 7, the OOM architecture has the important drawback that everytime an FC step terminates, the entire Binary Input RF has to be scanned to perform the FC computation, requiring $n_{iter} \times D_{out(FC)} \times \text{Memory size}_x$ clock cycles. If the number of output neurons is huge ($D_{out(FC)}$),$n_{iter} \times D_{out(FC)} \times \text{Memory size}_x$ becomes very large compared to the LiM case.

From these considerations, it is evident that LiM architecture introduces a gain in terms of execution time, because by increasing the level of parallelism in the architecture, multiple operations can be performed at the same time. The LiM bottlenecks are the Interface Decoder and the multiplexers depicted in Figure 15, that introduce both higher delay and power consumption, but they are required to interface the design blocks.

6. Perfomance Evaluation

In this part, the evaluation steps will be explained. In this work the memories were implemented as register files and each memory cell is a flip flop, so the results obtained are an overestimation (especially for the LiM case). The real performance values can be obtained with a more precise memory model. The performance evaluation is made of three parts:

1. For both OOM and LiM implementations, two NN models were chosen and used as cases of study. These models were implemented, trained and validated by Keras framework [32] and a Matlab script respectively. Then, the architectures were synthesized with Synopsys Design Compiler with 45 nm CMOS technology @ 1.1 V, providing the values of power, area, Critical Path Delay (CPD), execution time and energy consumption. Regarding the power consumption, two kind of estimations are provided: the first is very straight forward and consists of a report power from Synopsys with worst case scenario of switching activity equals to 1 in all the nodes. The second, a post place&route power estimation with Cadence Innovus, using backannotation with `.vcd` file provided by Modelsim, in order to evaluate the effect of both switching activity and interconnections.
2. Parametric sweeps are performed in order to evaluate the trend of the performance parameters in different cases. Power, Area, CPD and Energy ratios are computed between the OOM and LiM values, that are particularly useful to determine the main contributions of both architectures. To perform such procedure, a series of scripts are used to perform several synthesis processes with Synopsys Design Compiler and, everytime a synthesis ends, the performance values are stored in external files. Also in this case, the technology used is 45 nm CMOS @ 1.1 V.

3. An analysis of the differences between our LiM, where memory elements are flip flops, and a LiM circuit with a custom memory is performed. In [8], a very similar XNOR-Net implementation has been implemented with a CAM memory-based XNOR-Pop procedure. Some useful results are provided, since authors have implemented a modified memory array with 65 nm CMOS technology. For this reason, a synthesis with 65 nm CMOS technology @ 1.0 V is performed, trying to use the same metrics as [8] to evaluate how a more real memory model can influence the results obtained.

6.1. Two NN Models Examined

6.1.1. Fashion-MNIST CNN Results

The first NN model is able to classify with an accuracy of 81% a Fashion-MNIST image [33], which is a greyscale picture of 28 × 28 pixels that can belong to one of 10 different categories such as T-shits, trousers, pullovers, dresses, coats, sandals, shirts, sneakers, bags and ankle boots. The NN model is reported in Table 1. The parameters listed in Table 1 give an indication on the dimensions of the hardware implementations, such as the dimensions Memory $size_x$ and Memory $size_y$ of the Binary Input RF/LiM arrays in order to perform an ad-hoc synthesis optimized for that NN model. Binary Input RF, LiM XNOR part and LiM ones counter have Memory $size_y = 24 \times 24 = 576$ rows with a bitwidth of Memory $size_x = 32$ bits to host both convolution and FC algorithms. Since there are a maximum number of input channels equals to 6 as reported in Table 1, the XNOR-Pop Unit for both OOM and LiM has been replicated 6 times.

Table 1. Fashion-MNIST CNN under test parameters.

Layer Number	Type	IFMAP Size	Kernel Size	C_{in}	C_{out}	Stride
1	Convolutional	28 × 28 × 1	5 × 5	1	6	1
2	Max Pooling	24 × 24 × 6	2 × 2	6	6	1
3	Convolutional	12 × 12 × 6	5 × 5	6	6	1
4	Max Pooling	8 × 8 × 6	2 × 2	6	6	1
5	FC	96	-	-	-	-
6	BatchNorm	-	-	-	-	-
7	FC	120	-	-	-	-
8	BatchNorm	-	-	-	-	-
9	FC	84	-	-	-	-
10	BatchNorm	-	-	-	-	-
11	FC	10	-	-	-	-
12	BatchNorm	-	-	-	-	-
13	ReLU	-	-	-	-	-

The number of bits used to perform the extra calculations required (such as multiplications, BatchNorm etc) are 18 expressed in fixed point format. After performing the synthesis with Synopsys Design Compiler, the results obtained for area, CPD and power are reported in Table 2.

Table 2. Synthesis results for the Fashion-MNIST CNN model.

Type	Area (mm^2)	Power (mW)	CPD (ns)
LiM	1.68	254.50	4.11
OOM	1.10	193.30	4.14

A preliminary analysis of the results listed in Table 2, highlights a higher area and power consumption in LiM with respect to the OOM alternative, since the LiM implementation is highly parallellized and, consequently, a higher number of logic elements are required. The CPD is slightly higher in the OOM case because of a more complicated FC scheduling handling circuit (depicted in Figure 12), which requires the Store temp register file. From these results, a simple comparison between power, area and CPD is not enough to determine which is the best architecture between the two proposed ones. For this reason we estimate also the execution time and the energy estimation obtained as Power × Execution time, as shown in Table 3.

Table 3. Power-Execution time-Energy results for Fashion-MNIST CNN model.

Type	Power (mW)	Execution Time (ms)	Energy (μJ)
LiM	254.50	0.21	53.44
OOM	193.30	0.92	178.41

From Table 3, we can derive two very important values which are the Energy ratio and Delay ratio as follows

$$\text{Delay ratio} = \frac{\text{Execution time}_{OOM}}{\text{Execution time}_{LiM}} = \frac{0.92\,\text{ms}}{0.21\,\text{ms}} \simeq 4.38 \tag{26}$$

$$\text{Energy ratio} = \frac{\text{Energy}_{OOM}}{\text{Energy}_{LiM}} = \frac{178.41\,\mu\text{J}}{53.44\,\mu\text{J}} \simeq 3.34 \tag{27}$$

Energy and delay ratio give very important indications on LiM strong points: it consumes less energy and it's faster, although has a higher power value. From an energetic point of view, LiM architecture is more efficient for that particular NN model. In Figure 18 and Table 4 are reported the results obtained from a post place&route estimation with `.vcd` backannotation. Considering both switching activity and interconnections, the resulting power of the LiM architecture is increased by ∼ 22%, bringing a lower energy ratio, but still promoting LiM as an energy efficient architecture.

$$\text{Energy ratio}_{\text{post place and route}} = \frac{130.91\,\mu\text{J}}{68.9\,\mu\text{J}} \simeq 1.9 \tag{28}$$

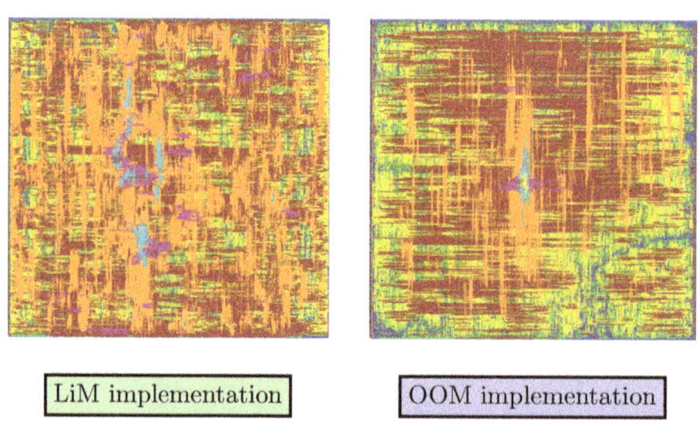

Figure 18. Snapshot of the chips obtained after a post place and route procedure for Fashion-MNIST CNN.

Table 4. Post place and route estimation of Fashion-MNIST CNN implementation.

Type	Area (mm^2)	Power (mW)	CPD (ns)	Execution Time (ms)	Energy (μJ)
LiM	1.70	328.3	4.11	0.21	68.9
OOM	1.07	142.3	4.14	0.92	130.91

6.2. MNIST-MLP Network

Another NN model is evaluated, in order to verify the behavior of both LiM and OOM architectures with different computational models. The realized MLP network is made of a set of FC layers organized as 784-196-196-10 neurons and it is able to achieve up to ∼90% of accuracy on MNIST dataset. Further details on MLP structure are presented in Table 5. Dropout layers indicated in Table 5 are useful in training procedure, since they prevent the network from overfitting by simply "turning off" neurons with a given probability [34]. As already done in Section 6.1.1, the results that will be presented are the ones obtained by the synthesis and the estimated value of energy, based on the execution time. The chosen dimensions of the implementation are Memory size$_x$ = 14, #C_{in} = 1 while the memory arrays have 196 rows because the maximum number of output neurons ($D_{out(FC)}$) is 196.

Table 5. MNIST MLP under test parameters.

Layer Number	Type	IFMAP Size	Kernel Size	C_{in}	C_{out}	Stride
1	Dropout	28 × 28 × 1	-	-	-	-
2	FC	784	-	-	-	-
3	BatchNorm	196	-	-	-	-
4	ReLU	196	-	-	-	-
5	Dropout	196	-	-	-	-
6	FC	196	-	-	-	-
7	BatchNorm	196	-	-	-	-
8	ReLU	196	-	-	-	-
9	Dropout	196	-	-	-	-
10	FC	196	-	-	-	-
11	BatchNorm	10	-	-	-	-

In Table 6, the architectures have a similar power consumption, because in OOM it is required a Store temp register file that has 196 rows. The hardware complexity of the LiM implementation is not so different from the OOM's one, because the memory arrays have a very small size of 196 × 14. Since it is an MLP network, n_{iter} parameter becomes very important, because it gives an indication on how many times the FC scheduling has to be executed for each layer: n_{iter} changes for each FC layer and can be obtained as $n_{iter} = D_{in(FC)}/\text{Memory size}_x$. From the energy and execution time results in Table 6, it is evident that OOM is not competitive with respect to the LiM version. This is due to the much more inefficient FC handling, since the whole Binary Input RF has to be scanned, while the LiM version performs all the calculations directly inside the array.

$$\text{Delay ratio} = \frac{1.62\,\text{ms}}{0.132\,\text{ms}} \simeq 12.27 \tag{29}$$

$$\text{Energy ratio} = \frac{23.20\,\mu\text{J}}{1.99\,\mu\text{J}} \simeq 11.7 \tag{30}$$

After performing post place and route estimation, the results obtained are reported in Figure 19 and in Table 7.

Table 6. Perfomance parameters of MLP implementation.

Type	Area (mm^2)	Power (mW)	CPD (ns)	Execution Time (ms)	Energy (µJ)
LiM	0.11	15.10	4.22	0.132	1.99
OOM	0.09	14.32	4.32	1.62	23.20

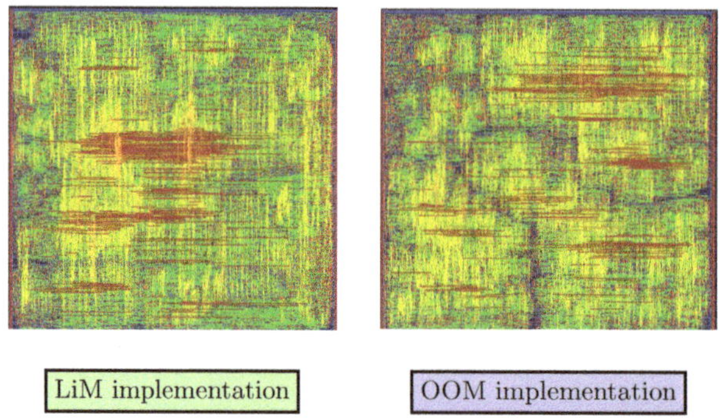

Figure 19. Snapshot of the chips obtained after a post place and route procedure for MLP NN.

Table 7. Post place and route estimation of MLP implementation.

Type	Area (mm^2)	Power (mW)	CPD (ns)	Execution Time (ms)	Energy (μJ)
LiM	0.1033	13.06	4.22	0.132	1.72
OOM	0.086	10.68	4.32	1.62	17.30

In Table 7, the power results are slightly lower than the synthesis ones, since the `.vcd` file and the switching activity information have relevant roles, giving a more precise power estimation, instead of the worst case reported in Table 6. The energy ratio results to be equal to $\sim 10\times$ compared to the previous one equal to $\sim 11.7\times$ provided by the synthesis.

6.3. Parametric Sweeps

The meaningful parameters of the designs such as $\#C_{in}$ and memory arrays dimensions (Memory size$_x$,Memory size$_y$) were varied to determine the differences between the two architectures in terms of performance. Two parameters are chosen per time and a sweep is executed on them, while the remaining are kept constant. For sake of clarity, from now on the following substitution is considered:

$$\begin{cases} H = \text{Memory size}_y^2 \\ W_a = \text{Memory size}_x \end{cases} \tag{31}$$

In Figures 20 and 21 are depicted the Power, Area and CPD for different values of $\#C_{in}$, W_a and $\sqrt{H}$. By increasing the W_a, power and area increase almost quadratically since W_a directly influences the bitwidth of the memories. Also, the trends depending by $\sqrt{H}$ behave quadratically, meaning that a the memory complexities influence a lot the performance of both architectures. In general, the power and area for LiM case are slightly higher than the OOM ones, since the total number of logic elements required by the LiM implementation is greater than OOM. CPD remains almost the same, even for more complex implementations. To better understand the differences of the obtained parameters for both architectures, a ratio was computed for all the cases: the results obtained are reported in Figure 22, where in general for an increasing size of $\#C_{in}$, W_a and $\sqrt{H}$ the power and area ratios decreases, confirming the bigger grade of complexity of the LiM. Another useful estimation can be performed on the energy ratios for the various cases.

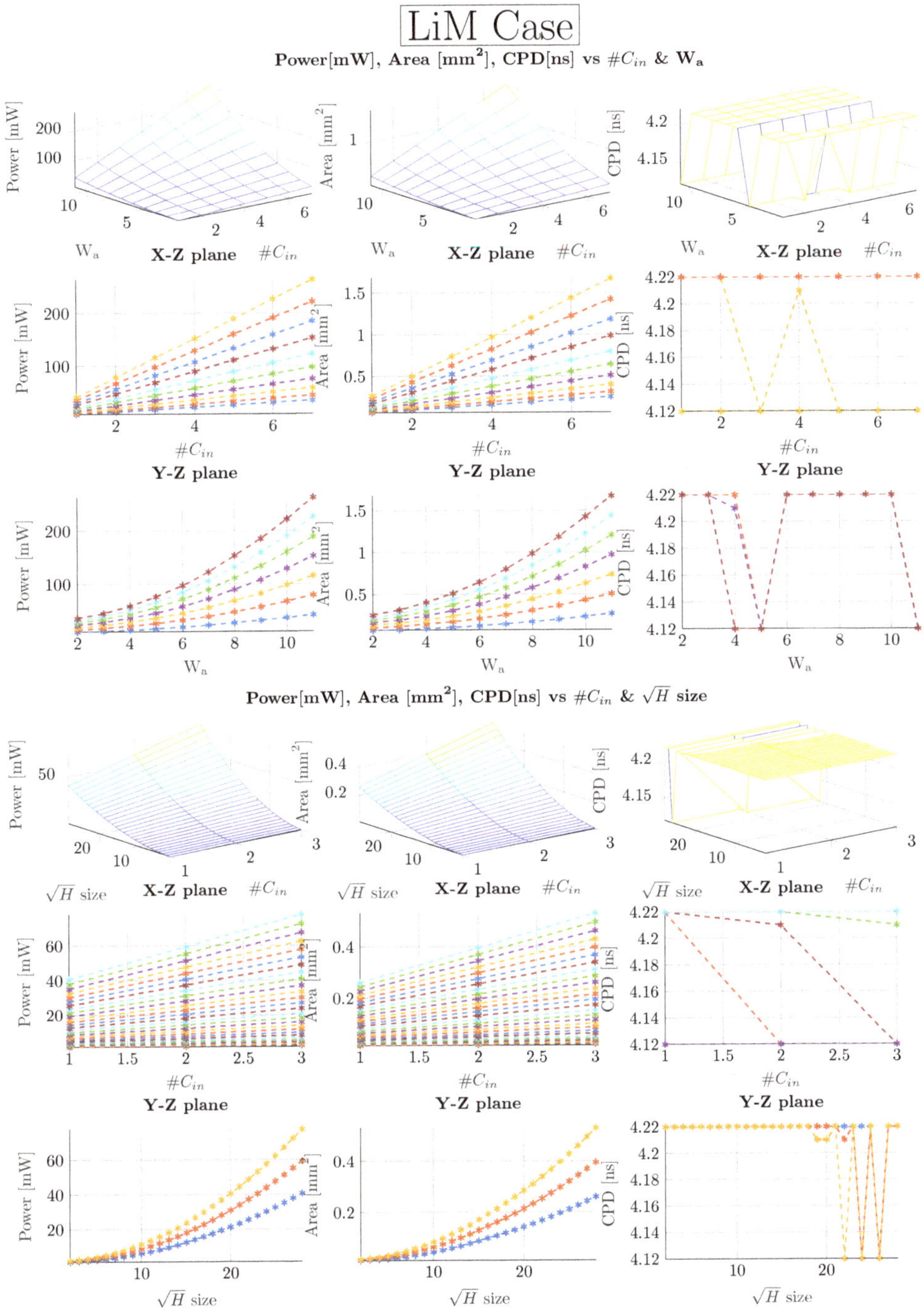

Figure 20. Power, Area and CPD results for different values of $\#C_{in}$, W_a and $\sqrt{H}$ considering LiM implementation.

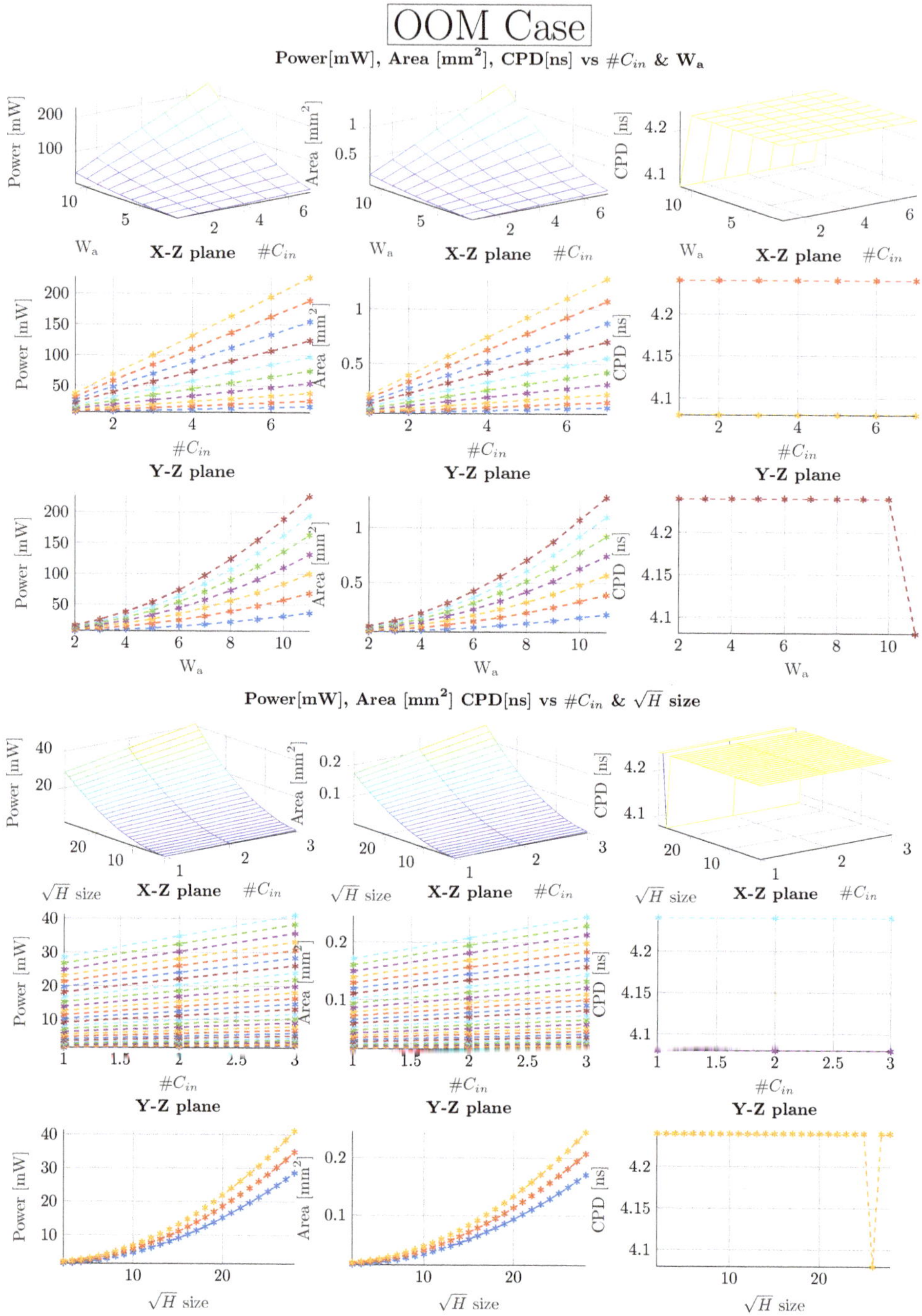

Figure 21. Power, Area and CPD results for different values of $\#C_{in}$, W_a and $\sqrt{H}$ considering OOM implementation.

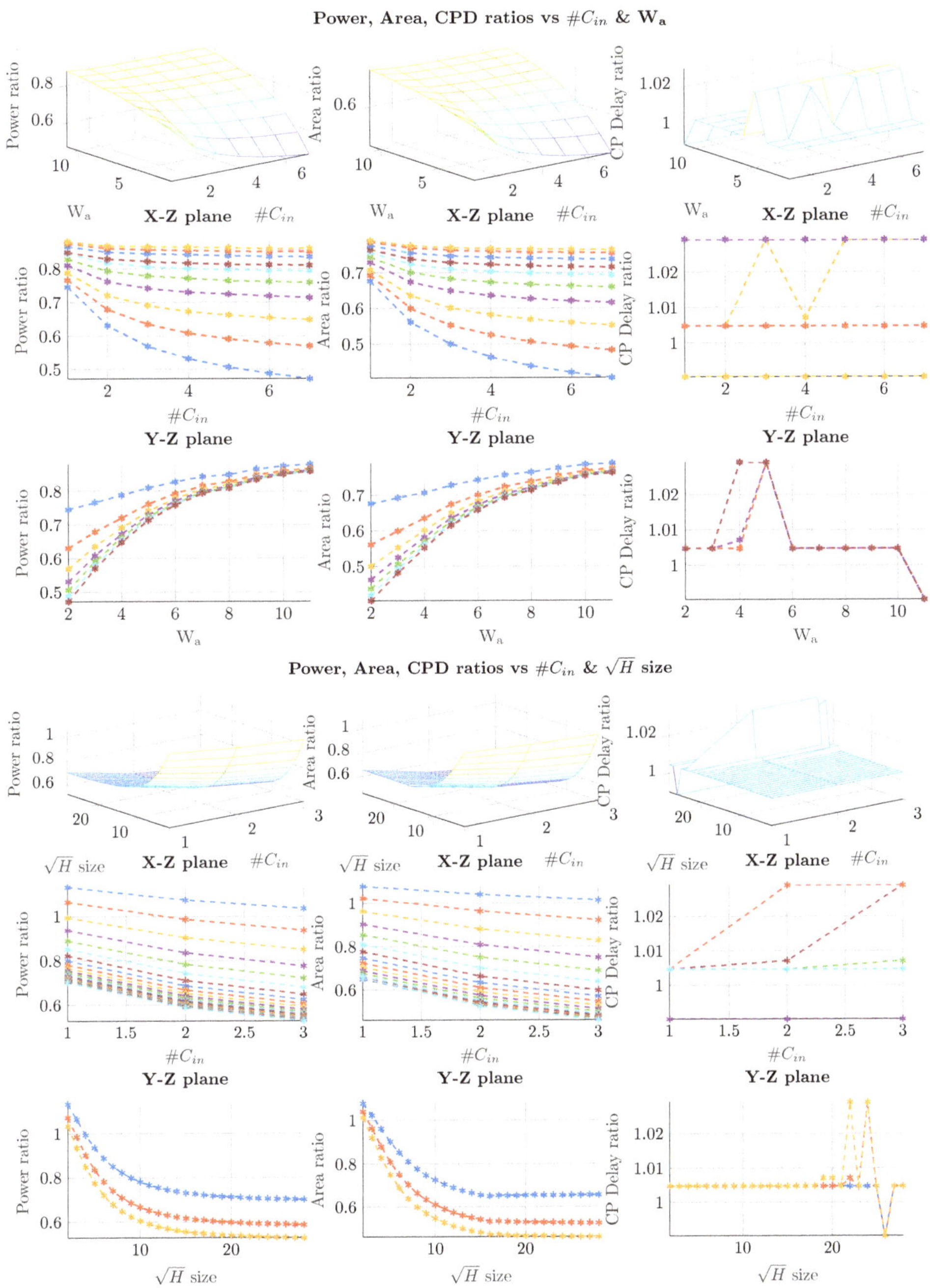

Figure 22. Power, Area and CPD ratios with respect to #C_{in}, W_a and $\sqrt{H}$.

Those values are obtained as $\text{Energy}_{OOM}/\text{Energy}_{LiM}$, as shown in Figure 23, decrease for higher memory dimensions, since power of the LiM architecture starts to assume a predominant contribution in the energy equation. It is important to keep in mind that these are very pessimistic estimations, and they can be improved by employing more realistic memory cells. The pessimistic case, which is reported in Figure 23 in $\#C_{in}$ & $\sqrt{H}$ size plot, is to have a very long ($\sqrt{H}$ big) and narrow (W_a very small) memory structure, which is replicated a lot of times ($\#C_{in}$ big): these set of conditions describes an improbable situation, because the driving force for a memory design is to have a regular squared shape array. The last energy estimation reported in Figure 23 flagged by FC $\#C_{in}$ & $\sqrt{H}$ size, takes into account an FC algorithm mapped on the implementations considering the worst case of big $\sqrt{H}$ and $\#C_{in}$. By varying $\sqrt{H}$, the trend for the energy ratio is increasing, meaning that the more complex is the FC algorithm the lower is the energy for the LiM implementation compared to OOM one.

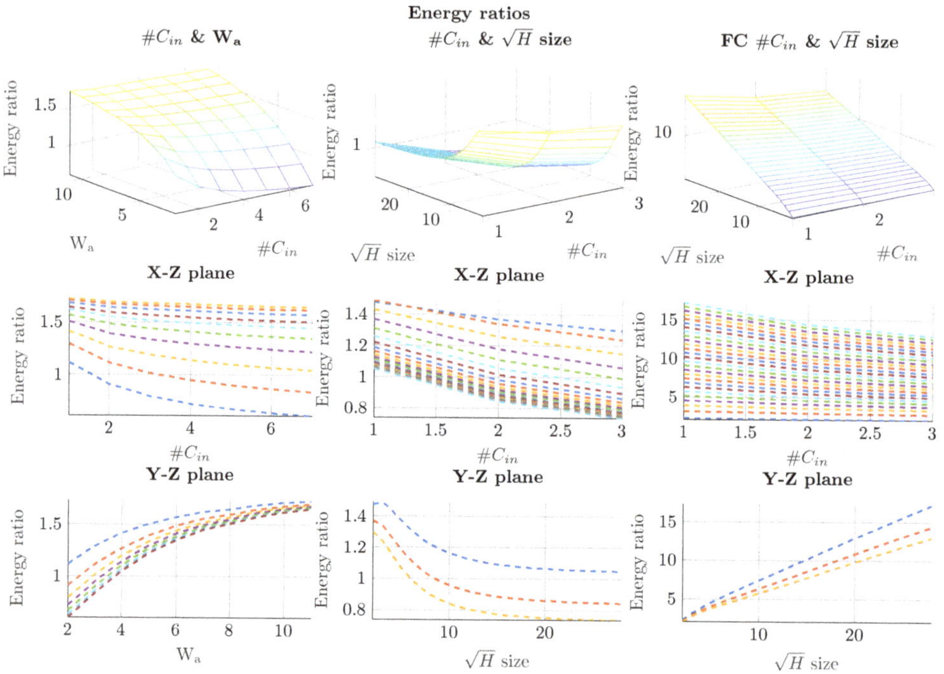

Figure 23. Energy ratio values obtained by varying $\#C_{in}$, W_a and $\sqrt{H}$.

6.3.1. Qualitative Estimation

To give a definitive answer on which architecture performs better, a qualitative estimation is performed, considering the mean values of all the cases explained before.

A ratio obtained as OOM/LiM between each parameter is proposed, which clarify the main points of both implementations. As shown in Figure 24, the values of area and power ratios are below 1, meaning that in general the LiM architecture behaves worse than OOM for the motivations explained before. On the other hand, execution time and energy ratios are equal to $\sim 6\times$ and $\sim 4\times$ respectively, implying that a very good improvement can be achieved by the LiM implementation on these two quantities. These trends confirm our expectations on LiM and further improvements can be achieved by having a more precise LiM array model.

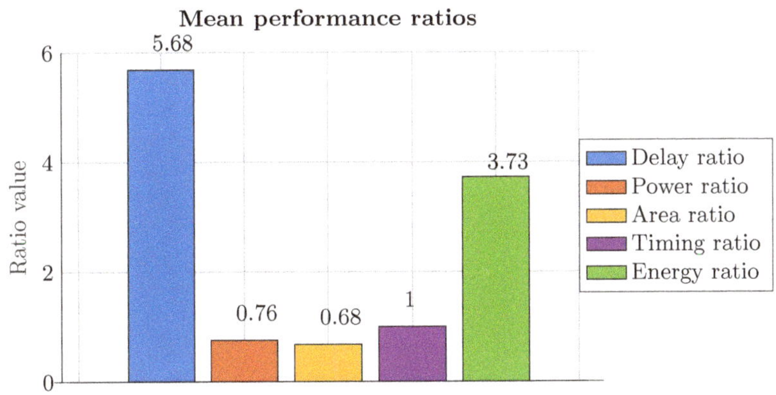

Figure 24. Mean performance ratios obtained as an average of all the cases analyzed from the previously discussed results.

6.3.2. LiM Array Estimation: Impact on Perfomance

In order to estimate the performance of the LiM array, several synthesis estimations were performed with different arrays dimensions. Taking into account the system's structure depicted in Figure 15, the LiM values of power and area are compared with the ones obtained from the same process applied only to the SurroundingLogic unit, in order to understand what are the main performance contributions. In Figure 25 are shown the performance values obtained by sweeping both $\sqrt{H}$, W_a, while $\#C_{in}$ is kept equal to 1, in order to estimate how the array sizes impact the overall performance. As it is possible to see, area and power increases almost quadratically, because of a more complex LiM structure. In Figure 26 it is reported an estimation of the SurroundingLogic unit by varying the same parameters as in Figure 25. The CPD bottleneck is located in the SurroundingLogic unit rather than LiM parts, because of the multipliers/adders employed to perform the final convolution result. Higher values of W_a implies a constant power/area, since there is no correlation between the LiM Memory size$_x$ and the complexity of the SurroundingLogic unit. By increasing $\sqrt{H}$, power and area increase because of the higher complexity required, for example a bigger dimension of the **K** register file (Figure 12). By comparing the performance in terms of power obtained in Figures 25 and 26, it is possible to see that the highest contribution comes from LiM parts, as shown in the breakdown plot depicted in Figure 27. The percentage values are obtained following a rough approach, starting with computing the total power/area, given by the sum of the results obtained in Figures 25 and 26 and by dividing the LiM power/area by the total ones. As it is possible to see, for bigger arrays, LiM parts will assume a predominant contribution on the power/area performance. This behavior recalls the need of employing a more accurate LiM model, instead of the discussed one based on flip flops and static logic gates.

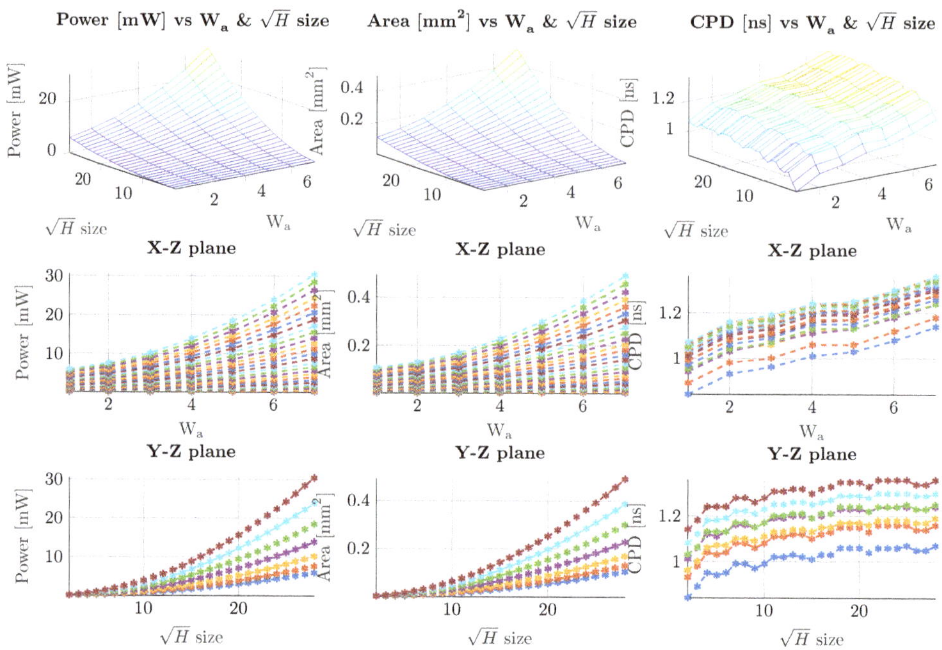

Figure 25. LiM performance estimations by varying W_a and $\sqrt{H}$ sizes. #C_{in} is kept equal to 1.

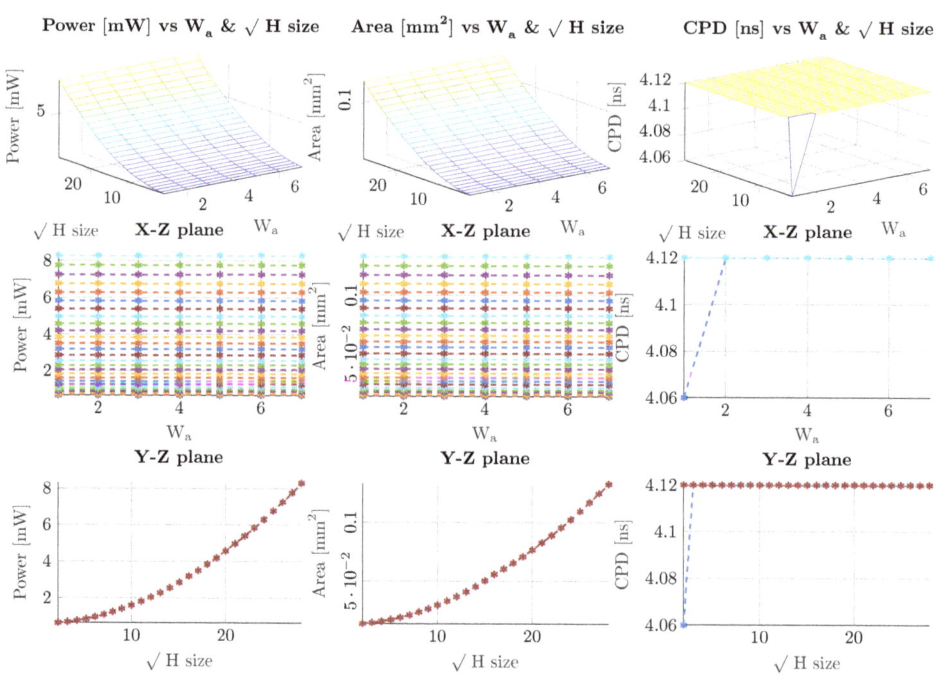

Figure 26. SurroundingLogic unit performance estimations by varying W_a and $\sqrt{H}$ sizes. #C_{in} is kept equal to 1.

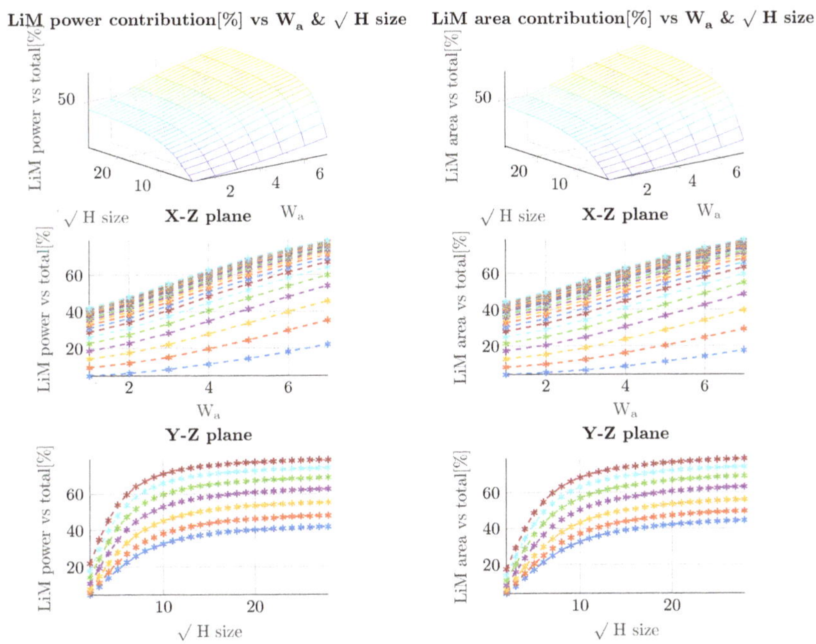

Figure 27. Power and area breakdown of LiM parts.

6.4. A More Detailed LiM Model

Reference [8] proposes a very similar approach, but it performs a Content Addressable Memory (CAM)-based XNOR-Pop procedure, implementing the second convolutional layer of LeNet5 NN model [1], which is depicted in Figure 3. Five arrays are realized and their dimensions are 30×10. They have been implemented with 65nm CMOS technology: the performance results are reported in Table 8. To have a fair comparison with [8], the same conditions have been applied to our LiM design: only the XNOR-Pop part, reported in Figure 15, is synthesized with 65nm technology with a dimension of 30×10 for LiM XNOR part array. To obtain the energy estimation, we started from the power result given by Synopsys and we have mapped the second convolutional layer of LeNet5 CNN, obtaining the corresponding execution time called $\text{Convolution}_{time,\text{II-LeNet5}}$ using the more precise version of Equation (24).

$$\text{Convolution}_{time,\text{II-LeNet5}} = 15852 \times t_{ck} \tag{32}$$

The power obtained by Synopsys is for only 1 LiM array, so the this value has to be multiplied by five:

$$\text{Power}_{\text{5-arrays}} = 0.2473\,\text{mW} \times 5 \approx 1.24\,\text{mW} \tag{33}$$

From the synthesis, CPD for the LiM array is equal to 1.91 ns, so the total energy is:

$$\text{Energy}_{II-LeNet5} = \text{Power}_{\text{5-arrays}} \times \text{Convolution}_{time,\text{II-LeNet5}} \approx 38\,\text{nJ} \tag{34}$$

We can perform a comparison between our less LiM model based on flip flops with the case described in Table 8: the energy ratio between our work and the reference one is about 4.22 while the Bank Area ratio is almost equal to 4.92. This means that, if we design a custom memory, instead of relying on flip flops the performance of our architetcure can be greatly improved. But even considering this fact, the results here presented highlight that LiM architetcures have a huge advantage over traditional Von Neumann circuit, in terms of energy and overall execution speed.

Table 8. CAM-based XNOR-Pop [8] and our LiM architectures performance parameters comparison.

Design	Technology	Bank Size	# of Banks	Bank Area (μm^2)	Energy (nJ)
[8]	65 nm	30 × 10	5	2456.6	∼9
This work (LiM)	65 nm	30 × 10	5	12090.6	∼38

7. Conclusions and Future Works

In this work LiM and OOM architectures have been designed to demonstrate if a logic-in-memory approach is effectively better than a Von Neumann one in designing architectures for memory-intensive applications. From the results here highlighted, LiM design obtains remarkable results in terms of energy dissipation, because of a higher degree of parallel execution of the algorithm. Since the memory part of our designs was synthesized with Synopsys, the results that we obtained are overestimated, meaning that the energy can be significantly smaller with a proper memory design. We can conclude therefore that Logic-In-Memory architectures are worth it. Even considering the increased complexity of the memory design, they provide significant advantages over Von-Neumann architectures.

As a future work we are designing custom memories, based both and CMOS and eventually on emerging technologies, to further improve our analysis.

Author Contributions: Conceptualization, A.C, M.V. and G.T.; methodology, A.C.; software, A.C.; validation, A.C.; formal analysis, A.C.; investigation, A.C.; resources, A.C.; data curation, A.C.; writing—original draft preparation, A.C.; writing—review and editing, G.T. and M.V.; visualization, M.V.; supervision, M.V.; project administration, M.V. and G.T. All authors have read and agreed to the published version of the manuscript.

Funding: This research received no external funding.

Conflicts of Interest: The authors declare no conflict of interest.

Abbreviations

The following abbreviations are used in this manuscript:

LiM	Logic-in-Memory
OOM	Out-Of-Memory
RF	Register File

References

1. LeNet-5, Convolutional Neural Networks. Available online: http://yann.lecun.com/exdb/lenet (accessed on 10 January 2020).
2. Krizhevsky, A.; Sutskever, I.; Hinton, G.E. Imagenet classification with deep convolutional neural networks. In *Advances in Neural Information Processing Systems;* Neural Information Processing Systems Foundation, Inc.: San Diego, CA, USA, 2012; pp. 1097–1105.
3. Szegedy, C.; Liu, W.; Jia, Y.; Sermanet, P.; Reed, S.; Anguelov, D.; Erhan, D.; Vanhoucke, V.; Rabinovich, A. Going deeper with convolutions. In Proceedings of the IEEE Conference on Computer Vision and Pattern Recognition, Boston, MA, USA, 7–12 June 2015; pp. 1–9.
4. Simonyan, K.; Zisserman, A. Very deep convolutional networks for large-scale image recognition. *arXiv* **2014**, arXiv:1409.1556.
5. He, K.; Zhang, X.; Ren, S.; Sun, J. Deep residual learning for image recognition. In Proceedings of the IEEE Conference on Computer Vision and Pattern Recognition, Las Vegas, NV, USA, 27–30 June 2016; pp. 770–778.
6. Courbariaux, M.; Bengio, Y.; David, J.P. Binaryconnect: Training deep neural networks with binary weights during propagations. In *Advances in Neural Information Processing Systems;* Neural Information Processing Systems Foundation, Inc.: San Diego, CA, USA, 2015; pp.3123–3131.
7. Rastegari, M.; Ordonez, V.; Redmon, J.; Farhadi, A. Xnor-net: Imagenet classification using binary convolutional neural networks. In *European Conference on Computer Vision;* Springer: Berlin, Germany, 2016; pp. 525–542.

8. Choi, W.; Jeong, K.; Choi, K.; Lee, K.; Park, J. Content Addressable Memory Based Binarized Neural Network Accelerator Using Time-domain Signal Processing. In *Proceedings of the 55th Annual Design Automation Conference (DAC '18)*; ACM: New York, NY, USA, 2018; pp. 138:1–138:6, doi:10.1145/3195970.3196014. [CrossRef]
9. Santoro, G.; Turvani, G.; Graziano, M. New Logic-In-Memory Paradigms: An Architectural and Technological Perspective. *Micromachines* **2019**, *10*, 368. [CrossRef] [PubMed]
10. Akin, B.; Franchetti, F.; Hoe, J.C. Data reorganization in memory using 3D-stacked DRAM. *ACM SIGARCH Comput. Architect. News* **2015**, *43*, 131–143, doi:10.1145/2749469.2750397. [CrossRef]
11. Durlam, M.; Naji, P.; DeHerrera, M.; Tehrani, S.; Kerszykowski, G.; Kyler, K. Nonvolatile RAM based on magnetic tunnel junction elements. In Proceedings of the 2000 IEEE International Solid-State Circuits Conference, San Francisco, CA, USA, 9 February 2000; pp. 130–131, doi:10.1109/ISSCC.2000.839718. [CrossRef]
12. Rakin, A.S.; Angizi, S.; He, Z.; Fan, D. Pim-tgan: A processing-in-memory accelerator for ternary generative adversarial networks. In Proceedings of the 2018 IEEE 36th International Conference on Computer Design (ICCD), Orlando, FL, USA, 7–10 October 2018; pp. 266–273.
13. Roohi, A.; Angizi, S.; Fan, D.; DeMara, R.F. Processing-In-Memory Acceleration of Convolutional Neural Networks for Energy-Efficiency, and Power-Intermittency Resilience. In Proceedings of the 20th International Symposium on Quality Electronic Design (ISQED), Santa Clara, CA, USA, 6–7 March 2019; pp. 8–13.
14. Wang, H.; Yan, X. Overview of Resistive Random Access Memory (RRAM): Materials, Filament Mechanisms, Performance Optimization, and Prospects. *Phys. Status Solidi (RRL) Rapid Res. Lett.* **2019**, *13*, 1900073, doi:10.1002/pssr.201900073. [CrossRef]
15. Krestinskaya, O.; James, A.P. Binary weighted memristive analog deep neural network for near-sensor edge processing. In Proceedings of the 2018 IEEE 18th International Conference on Nanotechnology (IEEE-NANO), Cork, Ireland, 23–26 July 2018; pp. 1–4.
16. Eshraghian, J.K.; Kang, S.M.; Baek, S.; Orchard, G.; Iu, H.H.C.; Lei, W. Analog weights in ReRAM DNN accelerators. In Proceedings of the 2019 IEEE International Conference on Artificial Intelligence Circuits and Systems (AICAS), Hsinchu, Taiwan, 18–20 March 2019; pp. 267–271.
17. Lee, J.; Eshraghian, J.K.; Cho, K.; Eshraghian, K. Adaptive precision cnn accelerator using radix-x parallel connected memristor crossbars. *arXiv* **2019**, arXiv:1906.09395.
18. Roohi, A.; Sheikhfaal, S.; Angizi, S.; Fan, D.; DeMara, R.F. ApGAN: Approximate GAN for Robust Low Energy Learning from Imprecise Components. *IEEE Trans. Comput.* **2019**. [CrossRef]
19. Agatonovic-Kustrin, S.; Beresford, R. Basic concepts of artificial neural network (ANN) modeling and its application in pharmaceutical research. *J. Pharm. Biomed. Anal.* **2000**, *22*, 717–727. [CrossRef]
20. Nwankpa, C.; Ijomah, W.; Gachagan, A.; Marshall, S. Activation functions: Comparison of trends in practice and research for deep learning. *arXiv* **2018**, arXiv:1811.03378.
21. Wang, Y.; Lin, J.; Wang, Z. An Energy-Efficient Architecture for Binary Weight Convolutional Neural Networks. *IEEE Trans. Very Large Scale Integr. (VLSI) Syst.* **2018**, *26*, 280–293, doi:10.1109/TVLSI.2017.2767624. [CrossRef]
22. Scherer, D.; Müller, A.; Behnke, S. Evaluation of pooling operations in convolutional architectures for object recognition. In *International Conference on Artificial Neural Networks*; Springer: Berlin, Germany, 2010; pp. 92–101.
23. Ioffe, S.; Szegedy, C. Batch normalization: Accelerating deep network training by reducing internal covariate shift. *arXiv* **2015**, arXiv:1502.03167.
24. Sari, E.; Belbahri, M.; Nia, V.P. How Does Batch Normalization Help Binary Training? Available online: http://xxx.lanl.gov/abs/1909.09139 (accessed on 20 December 2019).
25. Whatmough, P.N.; Lee, S.K.; Wei, G.; Brooks, D. Sub-uJ deep neural networks for embedded applications. In Proceedings of the 2017 51st Asilomar Conference on Signals, Systems, and Computers, Pacific Grove, CA, USA, 29 October–1 November 2017; pp. 1912–1915, doi:10.1109/ACSSC.2017.8335697. [CrossRef]
26. Pan, Y.; Ouyang, P.; Zhao, Y.; Kang, W.; Yin, S.; Zhang, Y.; Zhao, W.; Wei, S. A Multilevel Cell STT-MRAM-Based Computing In-Memory Accelerator for Binary Convolutional Neural Network. *IEEE Trans. Magnet.* **2018**, *54*, 1–5, doi:10.1109/TMAG.2018.2848625. [CrossRef]

27. Fan, D.; Angizi, S. Energy Efficient In-Memory Binary Deep Neural Network Accelerator with Dual-Mode SOT-MRAM. In Proceedings of the 2017 IEEE International Conference on Computer Design (ICCD), Boston, MA, USA, 5–8 November 2017; pp. 609–612. doi:10.1109/ICCD.2017.107. [CrossRef]
28. Yonekawa, H.; Sato, S.; Nakahara, H.; Ando, K.; Ueyoshi, K.; Hirose, K.; Orimo, K.; Takamaeda-Yamazaki, S.; Ikebe, M.; Asai, T.; et al. In-memory area-efficient signal streaming processor design for binary neural networks. In Proceedings of the 2017 IEEE 60th International Midwest Symposium on Circuits and Systems (MWSCAS), Boston, MA, USA, 6–9 August 2017; pp. 116–119, doi:10.1109/MWSCAS.2017.8052874. [CrossRef]
29. Jiang, L.; Kim, M.; Wen, W.; Wang, D. XNOR-POP: A processing-in-memory architecture for binary Convolutional Neural Networks in Wide-IO2 DRAMs. In Proceedings of the 2017 IEEE/ACM International Symposium on Low Power Electronics and Design (ISLPED), Taipei, Taiwan, 24–26 July 2017; pp. 1–6, doi:10.1109/ISLPED.2017.8009163. [CrossRef]
30. Sun, X.; Yin, S.; Peng, X.; Liu, R.; Seo, J.; Yu, S. XNOR-RRAM: A scalable and parallel resistive synaptic architecture for binary neural networks. In Proceedings of the 2018 Design, Automation Test in Europe Conference Exhibition (DATE), Dresden, Germany, 19–23 March 2018; pp. 1423–1428, doi:10.23919/DATE.2018.8342235. [CrossRef]
31. Wang, W.; Li, Y.; Wang, M.; Wang, L.; Liu, Q.; Banerjee, W.; Li, L.; Liu, M. A hardware neural network for handwritten digits recognition using binary RRAM as synaptic weight element. In Proceedings of the 2016 IEEE Silicon Nanoelectronics Workshop (SNW), Dresden, Germany, 19–23 March 2016; pp. 50–51, doi:10.1109/SNW.2016.7577980. [CrossRef]
32. Keras. Available online: https://github.com/fchollet/keras (accessed on 20 December 2019).
33. Xiao, H.; Rasul, K.; Vollgraf, R. Fashion-MNIST: A Novel Image Dataset for Benchmarking Machine Learning Algorithms. *arXiv* **2017**, arXiv:1708.07747.
34. Srivastava, N.; Hinton, G.; Krizhevsky, A.; Sutskever, I.; Salakhutdinov, R. Dropout: A Simple Way to Prevent Neural Networks from Overfitting. *J. Mach. Learn. Res.* **2014**, *15*, 1929–1958.

Journal of
Low Power Electronics and Applications

MDPI

Review

Rediscovering Majority Logic in the Post-CMOS Era: A Perspective from In-Memory Computing

John Reuben

Chair of Computer Science 3—Computer Architecture, Friedrich-Alexander-Universität Erlangen-Nürnberg (FAU), 91058 Erlangen, Germany; johnreuben.prabahar@fau.de

Received: 5 August 2020; Accepted: 2 September 2020; Published: 4 September 2020

Abstract: As we approach the end of Moore's law, many alternative devices are being explored to satisfy the performance requirements of modern integrated circuits. At the same time, the movement of data between processing and memory units in contemporary computing systems ('von Neumann bottleneck' or 'memory wall') necessitates a paradigm shift in the way data is processed. Emerging resistance switching memories (memristors) show promising signs to overcome the 'memory wall' by enabling computation in the memory array. Majority logic is a type of Boolean logic which has been found to be an efficient logic primitive due to its expressive power. In this review, the efficiency of majority logic is analyzed from the perspective of in-memory computing. Recently reported methods to implement majority gate in Resistive RAM array are reviewed and compared. Conventional CMOS implementation accommodated heterogeneity of logic gates (NAND, NOR, XOR) while in-memory implementation usually accommodates homogeneity of gates (only IMPLY or only NAND or only MAJORITY). In view of this, memristive logic families which can implement MAJORITY gate and NOT (to make it functionally complete) are to be favored for in-memory computing. One-bit full adders implemented in memory array using different logic primitives are compared and the efficiency of majority-based implementation is underscored. To investigate if the efficiency of majority-based implementation extends to n-bit adders, eight-bit adders implemented in memory array using different logic primitives are compared. Parallel-prefix adders implemented in majority logic can reduce latency of in-memory adders by 50–70% when compared to IMPLY, NAND, NOR and other similar logic primitives.

Keywords: memristor; memristive logic; Non-Volatile Memory (NVM); Resistive RAM; in-memory computing; majority logic; adder; Boolean logic; parallel-prefix adder

---

1. Introduction

Extraordinary innovation in the field of Integrated circuits is the last 50 years was based on Moore's law scaling and predominantly the Complementary Metal Oxide Semiconductor (CMOS) technology. Whether we have reached the end of Moore's law or approaching it in the near future (an issue being debated), it is evident that some signs are clear. The processor clock frequency, a key measure of performance has plateaued [1], the regular doubling of integration density has slowed down in 14 nm and 10 nm CMOS [2] and 2D lithography has reached its limits [3]. Beyond-CMOS research has been underway in the last decade to find an alternative device which is better than CMOS in its characteristics. This includes CMOS-like devices (tunnel FET, GaN TFET, Graphene ribbon pn junction, Ferroelectric FET) [4], quantum-dot cellular automata (QCA), nanomagnet logic, resistance-switching devices (Resistive RAM, Phase Change Memory, conductive bridge RAM), spin-based devices, and plasmonic-based devices [5]. Although some of these post-CMOS devices possessed valuable features like low-voltage operation and non-volatility, recent bench-marking efforts seem to suggest that none of these devices could outperform CMOS in the most critical aspects of computing (energy, latency and area) [4,6]. Hence it is envisaged that post-CMOS devices will augment and enhance CMOS-based computational fabrics and will not completely replace CMOS technology.

Majority logic, a type of Boolean logic, is defined to be true if more than half of the n inputs are true, where n is odd. Hence, a majority gate is a democratic gate and, it can be expressed in terms of Boolean AND/OR as $MAJ(a,b,c) = a.b + b.c + a.c$, where a, b, c are Boolean variables. Although majority logic was known since 1960, there has been a rediscovery in using it for computation in many post-CMOS devices. A majority gate based on spin waves [7], Quantum-Dot cellular automata [8], nano magnetic logic [9], and Single Electron Tunneling [10] have been demonstrated and in some of these technologies, it is more efficient to implement a majority gate [11] than other other logic primitives (NAND, NOR, XOR). Recent research [6,12–14] has confirmed that majority logic is to be preferred not only because a particular nanotechnology can realize it, but also because of its ability to implement arithmetic-intensive circuits with less gates, i.e., in a compact manner. For arithmetic intensive benchmarks, it has been proved that Majority-Invert Graphs (MIGs) can achieve up to 33% reduction in logical depth compared to And-Invert Graphs (AIGs) produced by Berkeley's ABC synthesis tool [12]. Such findings from research in logic synthesis implies that circuits implemented using majority logic will be better regardless of the post-CMOS device used. In this review, we limit our discussion to how majority logic could be implemented using RRAM technology since in-memory computing is the focus of this review. A review of how a majority gate could be implemented using other post-CMOS devices is presented in [15,16].

The movement of data between processing and memory units is the major cause for the degraded performance of contemporary computing systems, often referred to as the 'von Neumann bottleneck' or 'memory wall' [17,18]. 'Computation energy' is dominated by 'data movement energy' since the energy for memory access grows exponentially along the memory hierarchy (from cache to off-chip DRAM). There has been an ongoing effort (for 10-15 years) to combat the memory wall by bringing the processor and memory unit closer to each other. Resistive RAMs are two terminal devices (usually a Metal-Insulator-Metal structure [19]) capable of storing data as resistance. Although RRAM (memristor) was initially experimented as a non-volatile memory technology, it was later discovered that certain Boolean logic operations (IMPLY logic [20,21] and NOR [22] were the first logic gates that were explored) can be implemented in the memory array. Boolean gates were implemented by modifying the structure of the memory array or modifying the peripheral circuitry or a combination of these. In-memory computing (also called 'processing-in-memory') refers to any effort to process data at the residence of data (i.e., in the memory array) without moving it out to a separate processing unit. 'Processing/computing' could mean a wide variety of operations from arithmetic operations to cognitive tasks like machine learning and pattern recognition [23]. In this review, the focus is on arithmetic operations and how majority logic can enable efficient in-memory computing.

The rest of this review is structured as follows. In Section 2, we first give a brief overview on 'memristive logic', the methodology of designing logic circuits using memristors. This is followed by a discussion on how majority gate is implemented in RRAM array in Section 3. Three possible ways are discussed. In Section 4, we analyse the latency of in-memory one-bit adder using different logic primitives and highlight the latency reduction obtained by majority logic. To investigate if majority logic can be efficient for n-bit adders, 8-bit adders implemented using different logic primitives and different types (ripple carry, carry look-ahead, parallel-prefix) are analysed and compared, followed by conclusion in Section 6

2. Memristive Logic

A short introduction to memristors and different array configurations of such non-volatile memories is appropriate before the introduction of memristive logic. Memristors are a class of emerging Non-Volatile Memories (NVMs) which store data as resistance. Under voltage/current stress, the resistance can be switched between a Low Resistance State (LRS) and a High Resistance State (HRS). The word 'memristor' is used because such a device is basically a 'resistor' with a 'memory'. Depending on what causes the change in resistance, a memristor can be classified as follows: Resistive Random Access Memory (RRAM) where the change in resistance is due to the formation and rupture of a conductive filament [24]; Phase Change Memory (PCM) where the change in resistance is due to the amorphous or crystalline state of the chalcogenide phase-change material; Spin Transfer Torque-Magnetic RAM (STT-MRAM) where the change in resistance is due to the magnetic polarization. To construct a memory array using such devices, two configurations are common: 1Transistor-1 Resistor (1T–1R) and 1Selector-1 Resistor (1S–1R), as illustrated in Figure 1a. The 1T–1R configuration uses a transistor as an access device for each memory cell, allowing one to access a particular cell without interfering with its neighbours in the array [25,26]. The 1S–1R configuration uses a two-terminal device called a 'selector' which has a diode-like characteristic. The selector is assembled in series with the memristive device. Different types of selectors have been experimentally demonstrated in [27–30]. The 1S–1R is area-efficient, but suffers from sneak–path problem because it is not possible to program (read or write to a cell) a cell without interfering with its neighbours [22].

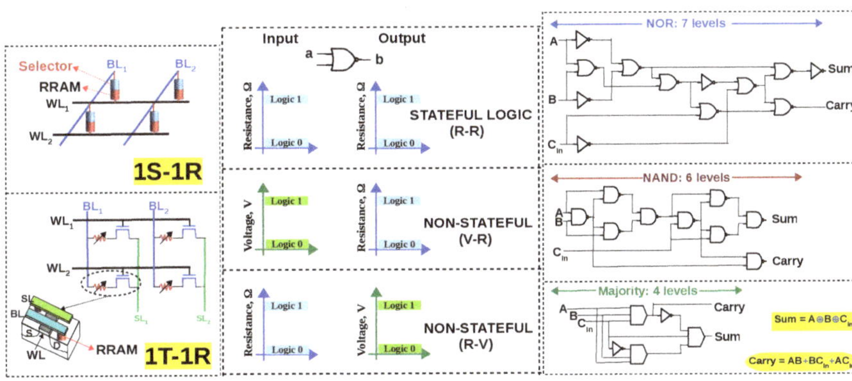

(a) (b) (c)

Figure 1. (**a**) 1S–1R and 1T–1R configuration of memristive memory array (**b**) If resistance is the only state variable, a memristive logic is said to be stateful. If voltage is also used in addition to resistance, it is said to be non-stateful (**c**) 1-bit full adder in terms of NOR gates [31], NAND gates [32,33] and majority gates [34]; Majority logic achieves less logical depth than NAND/NOR for 1-bit full adder.

Memristive logic is the art of designing logic circuits using memristors [17,18]. Conventionally, arithmetic circuits have been implemented using logic gates built from CMOS transistors. In contrast, a memristive logic family formulates a 'functionally complete' Boolean logic using a memristive device (RRAM/PCM/STT-MRAM) as the primary switching device (CMOS circuitry may also be used, but in a peripheral manner). For example, NOR is 'functionally complete' since any Boolean logic can be expressed in terms of NOR gates. Therefore, if a NOR gate can be designed using memristive devices, any Boolean logic can be implemented using memristive devices. Furthermore, most researchers try to make their logic gates executable in an array configuration so that they can be exploited for in-memory computing. NAND, IMPLY+FALSE [35] and Majority+NOT [12] are also functionally complete. From the perspective of the state variable used for computation, memristive logic family can be classified as either stateful or non-stateful. A memristive logic family is said to be stateful if the Boolean variable is represented only as the internal state of the memristor (i.e., its resistance) and computation is performed by manipulating this state [36]. If voltage is also used in addition to resistance, the logic family is said to be non-stateful (Figure 1b). Some logic families are classified on this criteria in [18].

A characteristic of memristive logic families is that, with certain modifications to the conventional memory, a particular logic primitive can be implemented and, other logic primitives have to be realized in terms of that logic primitive. For example, in the NOR-based memristive family (MAGIC [31]), all other gates (AND, OR, XOR) have to be expressed in terms of NOR gates and then mapped to the memory array. It must be noted that even the NOR logic primitive is implemented with modifications to the peripheral circuitry of the conventional memory array, namely the row decoder (modified to bias the rows at 'isolation voltage' to prevent unintended NOR operation in those rows) and the WRITE circuitry (modified to apply the MAGIC execution voltage which is twice the WRITE voltage). Similarly, in the NAND-based logic family reported in [37], XOR gate is implemented as a sequence of four NAND operations. This implies that if the fundamental logic primitive of a memristive logic family is weak, all in-memory computation performed using that logic family will be in-efficient (requiring long sequences of operations). To illustrate, Figure 1c depicts a 1-bit full adder expressed in terms of a particular logic primitive (NOR/NAND/Majority), as required for in-memory implementation. For a 1-bit adder, majority logic (together with NOT gates) can achieve 33–43% reduction in logic levels compared to NAND/NOR, while for bigger circuits, this percentage may vary. Research in logic synthesis suggests that circuits synthesized in terms of majority and NOT gates (Majority-Invert-Graphs) can achieve up to 33% reduction in logical depth compared to And-Invert-Graphs (AIGs) for arithmetic intensive circuits [12]. It must be emphasized that for any memristive logic, the number of cycles/steps to execute a circuit in-memory will be larger than the number of logic levels, i.e, n levels of Boolean logic will require $n + x$ cycles in-memory, where x depends on the memristive logic family and its capability to execute gates in parallel. Therefore, it is evident that to reduce the latency of in-memory computing, the synthesized logic must be latency optimized (before mapping to CMOS or a post-CMOS device). Stronger logic primitives like majority can minimize latency and the purpose of this review is to highlight the efficiency of memristive logic family with majority as the fundamental logic primitive (complemented with NOT since majority as a sole logic primitive is functionally incomplete).

3. In-Memory Majority Logic

In literature, there are two viable ways in which a majority gate is implemented in Resistive RAM array. Both are non-stateful logic families. Following the naming convention introduced in [38] ('input state variable-output state variable' logic), a non-stateful logic family can be V–R logic (input state variable is voltage and the output is resistance) or R–V logic (input state variable is resistance and output is voltage), as illustrated in Figure 1b. In this section, the principle of implementing a in-memory majority gate in V–R and R–V logic is reviewed and the advantages and disadvantages are analysed. In addition to the aforementioned methods, a in-memory minority gate (inverse of majority gate) is also theoretically proposed in [39]. The minority gate is realized by exploiting voltage division between three RRAMs (which store the inputs) and an output RRAM. However, the correct functioning of such a gate is not guaranteed since recent research has shown that variability is intrinsic to RRAM technology and cannot be completely eradicated [40,41]. In the presence of variations (in RRAM's switching voltages and resistive states), such a minority gate is not feasible in RRAM array, and hence it is not discussed in detail in this review.

3.1. V–R Majority Logic

In [42–44], majority gate is implemented in RRAM array (1S–1R) by applying two inputs of the majority gate as voltages at WL and BL of the array (the third input being the initial state of the RRAM) and the output is the new non-volatile state of the device. Hence this way of implementing majority can be called V–R logic, though in the strict sense, it should be VandR–R logic since the third input is resistance (initial state of the RRAM). However, it can be justified to be simply called V–R logic since the output (switching of resistance) is triggered on the applications of voltages. The fourth column of Table 1 depicts $M_3(A, B, C)$, the 3-input majority function of the first three columns. Note that $M_3(A, B, C) = AB + BC + AC$. To understand how a Resistive RAM cell can implement the majority function, consider a situation in which the Boolean variable C of Table 1 is the initial state of a memory cell (following the convention used in this field, logic 0 is HRS and logic 1 is LRS). Let us assume that the RRAM cell holding C has a symmetric switching characteristic, i.e., its internal resistance value changes from HRS to LRS when a voltage V_{SET} is applied across its terminals and from LRS to HRS when -V_{SET} is applied. As in the CMOS realm, logic 1 is a high voltage, which we will fix as V_{SET}, and logic 0 corresponds to ground.

Table 1. Establishing the link between the majority function and Resistive RAM.

A	B	C	$M_3(A,B,C)$	$\overline{B}$	$M_3(A,\overline{B},C)$	$RM_3(A,B,C)$
0	0	0	0	1	0	0
0	0	1	0	1	1	1
0	1	0	0	0	0	0
0	1	1	1	0	0	0
1	0	0	0	1	1	1
1	0	1	1	1	1	1
1	1	0	1	0	0	0
1	1	1	1	0	1	1

If A and B are applied across the two terminals of the RRAM cell, its state will either switch or remain the same in accordance with the initial state. Figure 2 illustrates the different combinations of (A, B, C) on a RRAM cell. When A is logic 1 and B is logic 0, the applied voltage across the RRAM cell is V_{SET}, triggering a transition from HRS to LRS and vice versa. When both A and B are (0,0) or (1,1), the state of the memristor will not change. This specific behavior can be captured as a new functionally complete Boolean function, called 'Resistive Majority', $RM_3(A,B,C)$, which describes the new nonvolatile state of the cell as a function of an initial internal state C and the voltages A and B applied across the terminals of the device. Note that $RM_3(A,B,C) = M_3(A,\overline{B},C)$, as listed in Table 1. Complex functions can be easily expressed and manipulated as RM_3 operators using Majority-Inverter Graphs (MIG), a recently introduced logic manipulation structure consisting of three input majority nodes and regular/complemented edges [12]. In [18], the authors elaborate how an eight-bit adder is expressed in MIGs and then mapped to the memristive memory array using the aforementioned resistive majority function.

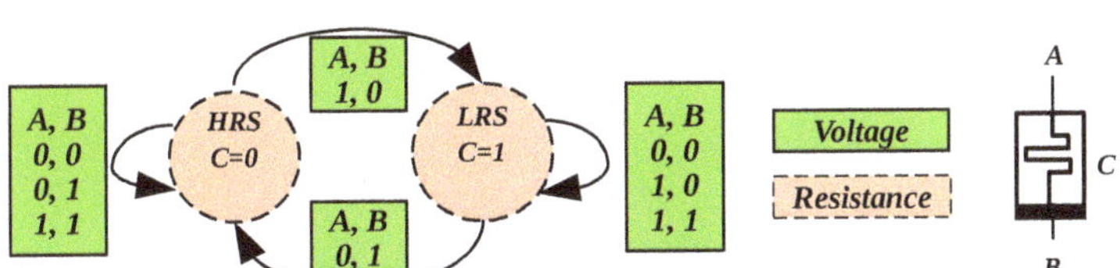

Figure 2. Illustration of V–R majority logic. Arrow indicates the state transition, which depends on the initial state of the RRAM cell C and the voltage applied across its terminals (A, B); dotted lines indicate the state variable of C, which is resistance, while A and B are voltages [18].

3.2. R–V Majority Logic

In [45,46], a majority gate is implemented while reading from a 1T–1R array, i.e., the inputs of the majority gate are the resistances of the cells and the output is sensed as a voltage, a R–V logic. Consider an array of RRAM cells arranged in a 1T-1R configuration, as depicted in Figure 3. Each cell can be individually read/written into by activating the corresponding wordline (WL) and applying appropriate voltage across the cell (BL and SL). Now, if three rows are activated simultaneously during read operation (Rows 1 to 3 in Figure 3a, the resistances in column 1 are in parallel (neglecting the parasitic resistance of BL and SL). The effective resistance between BL and SL will therefore be $R_{eff} = (R_A + r_{DS})||(R_B + r_{DS})||(R_C + r_{DS}) \approx (R_A||R_B||R_C)$, if the drain-to-source resistance of transistor (r_{DS}) is small compared to LRS. A Sense Amplifier (SA) which can accurately sense the effective resistance implements a 'in-memory' majority gate. Table 2 lists the truth table of a 3-input majority gate ($M_3(A, B, C)$) and the effective resistance for all the eight possibilities. If we assume a LRS and HRS of 10 kΩ and 133 kΩ, respectively (IHP's RRAM), the crucial aspect of the proposed gate is to be able to differentiate between R_{eff}^{001} (two LRS and one HRS) and R_{eff}^{110} (two HRS and one LRS). In other words, resistance $\leq$ 4.8 kΩ must be sensed as '0' and resistance $\geq$ 8.7 kΩ must be sensed as '1' (shaded grey in Table 2). If we call the resistance to be differentiated as sensing window (8.7 kΩ − 4.8 kΩ = 3.9 kΩ), any sense amplifier which can differentiate this sensing window can be used to implement the majority gate. A current-mode SA is used in [45] and a time-based SA is used in [46] to verify the correct functioning of majority gate, even in the presence of reasonable RRAM variations. It must be noted unlike NAND and NOR, majority as a logic primitive is not functionally complete. However, it forms a functionally complete logic when used together with NOT, i.e., any Boolean logic can be expressed in terms of majority and NOT gates [12]. Therefore, a NOT gate is implemented by latching the inverted output of the SA, as illustrated in Figure 3b.

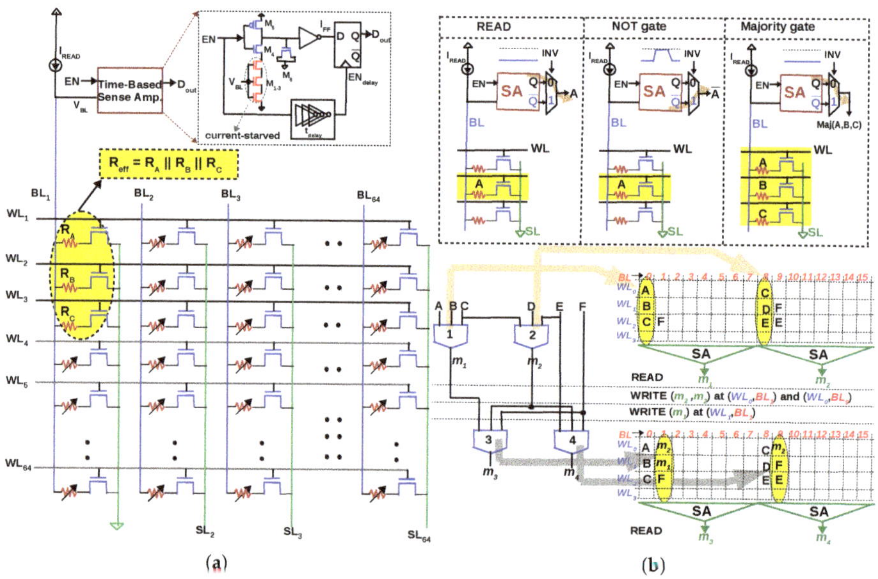

Figure 3. (**a**) In-memory majority gate proposed in [45,46]: When three rows are activated (WL_{1-3}) simultaneously in a 1T-1R array, the three resistances R_A, R_B, R_C will be in parallel (Inputs of the majority gate A, B, C are represented as resistances R_A, R_B, R_C). An 'in-memory' majority gate can be implemented by accurately sensing the effective resistance R_{eff} during READ. (**b**) NOT operation implemented with a 2:1 multiplexer at the output of the SA. With majority and NOT gate implemented as READ, multiple levels of logic can be executed by writing the data back to the array, simplifying computing to READ and WRITE operations in memory. Multiple majority gates can be executed in parallel in the memory array, thereby reducing latency of in-memory computation.

Table 2. Precisely sensing R_{eff} results in majority: Logic '0' is LRS (10 kΩ) and logic '1' is HRS (133.3 kΩ). Sense amplifier distinguishes between rows shaded grey and those that are not.

A	B	C	$M_3(A,B,C)$	R_{eff}	R_{eff}
0	0	0	**0**	$\frac{LRS}{3}$	3.3 kΩ
0	0	1	**0**	$\frac{HRS \cdot LRS}{LRS+2 \cdot HRS}$	4.8 kΩ
0	1	0	**0**	$\frac{HRS \cdot LRS}{LRS+2 \cdot HRS}$	4.8 kΩ
0	1	1	**1**	$\frac{HRS \cdot LRS}{HRS+2 \cdot LRS}$	8.7 kΩ
1	0	0	**0**	$\frac{HRS \cdot LRS}{LRS+2 \cdot HRS}$	4.8 kΩ
1	0	1	**1**	$\frac{HRS \cdot LRS}{HRS+2 \cdot LRS}$	8.7 kΩ
1	1	0	**1**	$\frac{HRS \cdot LRS}{HRS+2 \cdot LRS}$	8.7 kΩ
1	1	1	**1**	$\frac{HRS}{3}$	44.4 kΩ

A comparison between V–R logic and R–V logic is presented pictorially in Figure 4. In the V–R implementation [42–44] in memory, the inputs of the majority gate are applied as voltages at WL/BL. This manner of computation complicates the row/column decoders of the memory array, which were conventionally used to select rows/columns. Thus the peripheral circuitry will get complicated, i.e., the row/column decoders have to be significantly modified to do row selection (during memory operation) and apply inputs (during majority operation). In contrast, in the R–V implementation [45,46], the row/column decoders retain their functionality as in a conventional memory, with a minor modification (the row decoder must be enhanced to select three rows during majority operation, which can be achieved by interleaving decoders [46]). Furthermore, the R–V implementation [45,46] is conducive for parallel-processing since multiple gates can be mapped to the same set of rows, as illustrated in in Figure 4. This will aid the implementation of in-memory parallel-prefix adders (Section 5) and ternary computing [47].

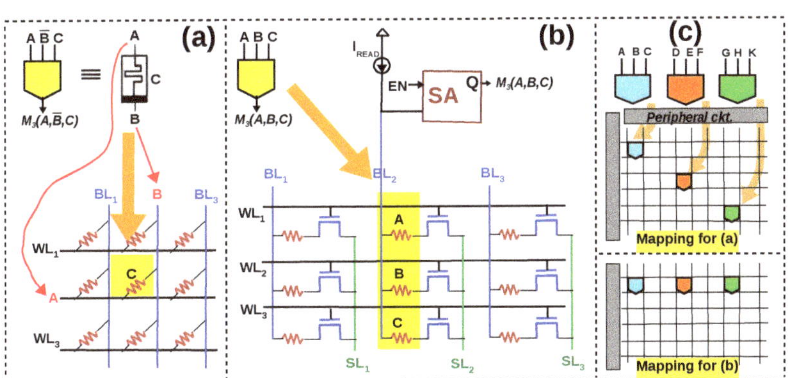

Figure 4. (**a**) V–R majority gate [42–44] (**b**) R–V majority gate [45,46] (**c**) When multiple gates have to be executed in parallel, the majority gates of [42–44] have to be mapped diagonally because two gates cannot be executed in the same row/column.

4. In-Memory One-Bit Full Adders Using Different Logic Primitives

As stated, in-memory addition is achieved by a sequence of Boolean logic operations executed in memory. To compute in memory, the circuit must first be expressed in terms of the logic gates the particular memristive logic family implements. A one-bit full adder in memristive logic family based on NOR [48], NAND [49] and MAJORITY [45] is compared in Figure 5. It is evident that the number of steps (memory cycles) to compute in memory is larger than the number of logic levels. When mapped to the memory array, n levels of logic will require $n + x$ cycles, where x depends on the characteristics of the memristive logic family. This includes attributes like statefulness, capability to executes gates in parallel etc. In a non-stateful logic family, the output of the gate may be a voltage and it may be needed as resistance for the next level of logic, requiring an additional WRITE operation. In a stateful logic family, the output of the gate needs to be aligned with the inputs of following gate (next logic level), requiring an additional WRITE operation. In this manner, the interconnecting wires between logic levels contribute to additional cycles in memory. Furthermore, a memristive logic family should have the capability to execute multiple gates simultaneously. Consequently, multiple gates in a logic level can be mapped to the memory array in a single cycle. If the memristive logic family does not support the simultaneous execution of multiple gates, x will increase. Thus the parallel-friendliness of the logic family is also an important characteristic to minimize latency.

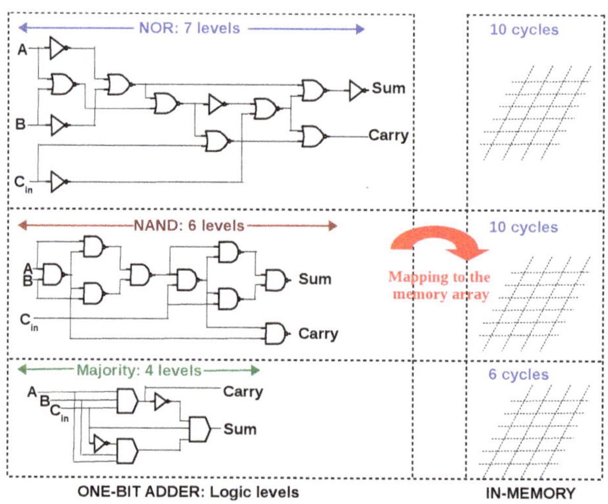

Figure 5. n levels of Boolean logic will require $n + x$ cycles in-memory, where x depends on the memristive logic family. It must be noted that the number of cycles required (10 cycles for NOR, NAND and 6 cycles for MAJORITY) is already optimized by executing multiple gates in parallel (see the mapping for NOR [31], NAND [49] and MAJORITY [45]).

To evaluate the effectiveness of majority logic for in-memory computing, one-bit adders using different logic primitives are analysed from literature. Table 3 lists the latency of one-bit adders. IMPLY logic primitive was the most researched logic primitive because of it's stateful nature. IMPLY was explored in different array configurations (1S–1R, 1T–1R) and the full adder, expressed in terms of XOR and AND gates was implemented as sequence of IMPLY operations. However, all the adders using IMPLY primitive have a latency of at least 13 cycles, implying a weak primitive. As summarized in Table 3, the number of steps to compute in an array, reduces from IMPLY to NAND/NOR logic primitive, and, further from NAND/NOR to MAJORITY, proving the strength of majority as a logic primitive.

Table 3. Latency of in-memory one-bit full adders.

Primitive	Structure	Latency	Ref
IMPLY	1D–1R	43 steps	[50]
IMPLY	1R	35 steps	[32]
IMPLY	1R	27 steps	[51]
IMPLY	1R	23 steps	[52]
IMPLY(semi-parallel)	1T–1R	17 steps	[53]
IMPLY	1T–1R	13 steps	[54]
ORNOR	1T–1R	17 steps	[55]
NOR	1S–1R	10 steps	[48]
NAND	1S–1R	10 steps	[49]
XOR+NAND (unipolar memristors)	1S–1R	8 steps	[56]
MAJORITY+NOT	1T–1R	6 steps	[45]

5. In-Memory Eight-Bit Adders Using Different Logic Primitives

Will the reduced latency obtained by majority logic for 1-bit full adder translate to n-bit adders? In this section, eight-bit adders using different logic primitives are analysed and compared to answer this question. From Figure 5, it is evident that to minimize in-memory latency, the number of logic levels which is mapped to the memory array must be minimized. Parallel-prefix (PP) adders are a family of adders originally proposed to overcome the latency incurred by the rippling of carry in ripple carry adders. Such adders have the capability to minimize the latency to O(log n), for n-bit adders. PP adders are conventionally expressed as "propagate" ($a_i \oplus b_i$) and "generate" terms ($a_i.b_i$). Hence, they are implemented as AND, OR and XOR gates. As already stated, a memristive logic family cannot implement such a heterogeneity of gates. As illustrated in Figure 6, the XOR gate has to be implemented as NAND gates [37,49], increasing the logic levels to 12. Such an eight-bit PP adder (Sklansky) is expressed in OR/AND logic primitive and implemented in the memory array in 37 cycles [57]. Using majority logic, an 8-bit PP adder is implemented in memory in [46]. Since majority gate is the basic building block for many emerging nanotechnologies, prior works [13,14] have formulated such PP adders in majority logic. The majority-based eight-bit adder depicted in Figure 7 is derived from [13,14]. For an eight-bit adder, the logical depth is six levels of majority gates and one level of NOT gates, and at most eight gates are needed simultaneously in each level. Since multiple majority gates can be executed in parallel (Figure 4), they can be mapped to the array in 19 cycles, as elaborated in [46].

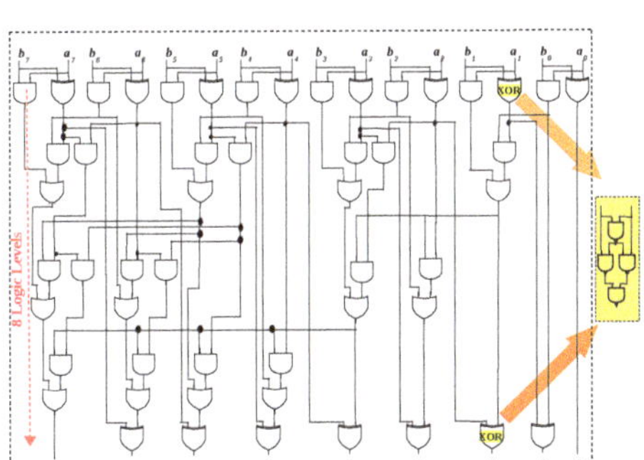

Figure 6. An eight-bit parallel-prefix adder (Ladner-Fischer) has 8 logic levels of AND, OR and XOR gates. If the logic family cannot execute XOR gate, it must be expressed as NAND gates, increasing the logic levels to 12.

A detailed comparison of the latency of 8-bit in-memory adders based on different logic primitives and the corresponding adder configuration is presented in Table 4. Since IMPLY logic incurred highest latency for 1-bit addition, the trend continues for 8-bit addition which is to be expected. In ripple carry configuration, IMPLY logic based adders incur a latency of at least 54 steps and parallel-prefix configuration could reduce it to 25 steps. It may be safe to conclude that for the same logic primitive, parallel-prefix configurations results in lower latency, although the mapping of the parallel-prefix adder to the memory array is not clearly elaborated in [58]. Regarding NOR, a carry look-ahead configuration incurs 48 steps while a computerised algorithm is used to map 8-bit NOR-based adder to the memory array in 38 steps. OR/AND-based logic primitive could achieve a latency of 37 steps in parallel-prefix configuration. An eight-bit parallel-prefix adder in majority logic could achieve a latency of 19 steps [46]. Finally, a XOR-based adder [59] could achieve a latency of 16 steps even in ripple carry configuration, but it must be emphasized that [59] used multiple arrays since multiple XOR gates could not be executed simultaneously in the same array. To conclude, the latency minimization achieved by majority logic for 1-bit addition does extend to 8-bit addition. Majority logic used in synergy with parallel-prefix configuration is one of the best performing in-memory adders. Finally, any comparison among in-memory adders is not complete without considering energy consumption and area of the memory array and the peripheral circuitry needed to implement the logic operations in memory. Such a holistic comparison is beyond the scope of this work. However, latency can be a good measure of performance if the individual logic operations are achieved in an energy efficient manner and sneak-path energy leakage is avoided (in 1S–1R configuration). Note that there are other works implementing adders using memristors along with CMOS in a non-array configuration. However, such works are not included in the comparisons performed in this work since they cannot be exploited for in-memory computing.

Table 4. Latency of in-memory 8-bit adders.

Primitive	Array	Adder Type	Latency	Comment/Ref
IMPLY	1S-1R	Ripple carry	58	Each step is IMPLY operation [35]
IMPLY+OR	1S-1R	Ripple Carry	54	Each step is IMPLY/OR/NOR operation [60]
IMPLY	–	Parallel-prefix	25	Each step is IMPLY operation [58]
NOR/NOT	1T-1R	Look-Ahead	48	Each step has one or more NOR/NOT operations [61]
NOR	1S-1R	algorithm	38	Each step has one or more NOR operations [18]
OR/AND	1S-1R	Parallel-prefix	37	Each step has one or more OR/AND operation [57]
ORNOR	1S-1R	Parallel-clocking	31	Each step has one or more ORNOR/IMPLY operation [55]
MAJORITY+NOT	1T-1R	Parallel-prefix	19	Each step is Majority/NOT or WRITE [46]
XOR	1T-1R	Ripple carry	16 *	Each step is XOR [59]

* XOR gate proposed in [59] is not parallel-friendly and consequently multiple gates cannot be executed in parallel in the array (to circumvent this, multiple arrays have been used in [59]). Furthermore, XOR is not functionally complete and has to be used in conjunction with other gates to implement other arithmetic circuits. In contrast, majority+NOT is functionally complete.

Latency is a big hurdle for mainstream adoption of in-memory arithmetic. As noted in Tables 3 and 4, in-memory adders require tens of steps for addition operations. Even if a single step takes 5 *ns* (RRAMs can switch in a few *ns*), this would be much larger than the latency incurred in CMOS technology (32-bit addition operation can be performed in 4 *ns* in CMOS technology [62]). However, in in-memory arithmetic, the energy and latency (hundreds of *ns*) for data movement is avoided (the numbers to be added have to be moved from DRAM memory to processor in conventional approach). Therefore, in-memory arithmetic can still be beneficial, provided the latency to compute in memory is minimized. The power of majority logic lies in reducing this latency to compute in memory array.

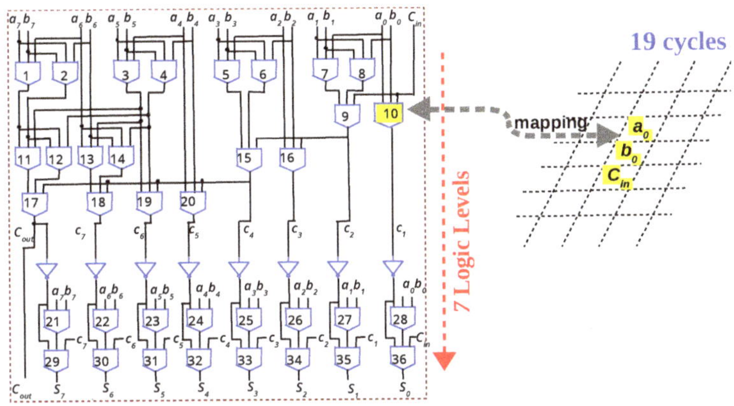

Figure 7. Eight-bit parallel-prefix adder (Ladner-Fischer) expressed as 7 levels of Majority+NOT gates. By executing multiple gates in parallel, the adder can be implemented in memory in 19 cycles, as elaborated in [46].

6. Conclusions

Majority logic did not become the dominant logic to compute in CMOS technology because it was more efficient to implement NAND/NOR gate than a majority gate (12 transistors for an inverted majority gate compared to 6 transistors for NAND3/NOR3). However, in many emerging post-CMOS devices, a majority gate can be implemented efficiently and therefore, majority logic needs to be re-evaluated for its computing efficiency. This review attempted to investigate the efficiency of majority logic from the perspective of in-memory computing. When the logic levels are minimized and mapped to the memory array using a memristive logic family (which can implement an in-memory majority gate), it leads to a latency optimized in-memory adder. Unlike CMOS implementation which accommodated a heterogeneity of logic gates, in-memory computing favours a homogeneous implementation of logic gates because peripheral circuitry of the array needs to be enhanced with capability to execute a particular logic primitive (different logic primitives necessitate different modifications to the peripheral circuitry). Therefore, majority-based memristive logic may be all the more preferred since they can implement any logic succinctly when used together with NOT gates. Comparisons with different logic primitives revealed that majority logic incurs least latency for 1-bit adders. For n-bit adders, majority logic has the potential to achieve a latency reduction of 70% and 50% when compared to IMPLY and NAND/NOR logic primitives, if implemented in a parallel-prefix configuration in the memory array. Minimizing latency also aids in lowering the power consumption since the array will be powered for a shorter time. Latency is a significant disadvantage in in-memory addition and the power of majority logic lies in reducing this latency. Therefore, majority logic and its advantages needs to be rediscovered in the era of in-memory computing.

Funding: This research received no external funding.

Conflicts of Interest: The author declares no conflict of interest.

References

1. Theis, T.N.; Wong, H.P. The End of Moore's Law: A New Beginning for Information Technology. *Comput. Sci. Eng.* **2017**, *19*, 41–50. [CrossRef]
2. Bohr, M.T.; Young, I.A. CMOS Scaling Trends and Beyond. *IEEE Micro* **2017**, *37*, 20–29. [CrossRef]
3. Shalf, J.M.; Leland, R. Computing beyond Moore's Law. *Computer* **2015**, *48*, 14–23. [CrossRef]

4. Nikonov, D.E.; Young, I.A. Benchmarking of Beyond-CMOS Exploratory Devices for Logic Integrated Circuits. *IEEE J. Explor. Solid State Comput. Devices Circuits* **2015**, *1*, 3–11. [CrossRef]
5. Testa, E.; Soeken, M.; Amar, L.G.; De Micheli, G. Logic Synthesis for Established and Emerging Computing. *Proc. IEEE* **2019**, *107*, 165–184. [CrossRef]
6. Young, I.A.; Nikonov, D.E. Principles and trends in quantum nano-electronics and nano-magnetics for beyond-CMOS computing. In Proceedings of the 2017 47th European Solid-State Device Research Conference (ESSDERC), Leuven, Belgium, 11–14 September 2017; pp. 1–5.
7. Ciubotaru, F.; Talmelli, G.; Devolder, T.; Zografos, O.; Heyns, M.; Adelmann, C.; Radu, I.P. First experimental demonstration of a scalable linear majority gate based on spin waves. In Proceedings of the 2018 IEEE International Electron Devices Meeting (IEDM), San Francisco, CA, USA, 1–5 December 2018; pp. 36.1.1–36.1.4. [CrossRef]
8. Imre, A.; Csaba, G.; Ji, L.; Orlov, A.; Bernstein, G.H.; Porod, W. Majority Logic Gate for Magnetic Quantum-Dot Cellular Automata. *Science* **2006**, *311*, 205–208. [CrossRef] [PubMed]
9. Breitkreutz, S.; Kiermaier, J.; Eichwald, I.; Ju, X.; Csaba, G.; Schmitt-Landsiedel, D.; Becherer, M. Majority Gate for Nanomagnetic Logic With Perpendicular Magnetic Anisotropy. *IEEE Trans. Magn.* **2012**, *48*, 4336–4339. [CrossRef]
10. Oya, T.; Asai, T.; Fukui, T.; Amemiya, Y. A Majority-Logic Nanodevice Using a Balanced Pair of Single-Electron Boxes. *J. Nanosci. Nanotechnol.* **2002**, *2*, 333–342. [CrossRef]
11. Amarú, L.; Gaillardon, P.; De Micheli, G. Majority-based synthesis for nanotechnologies. In Proceedings of the 2016 21st Asia and South Pacific Design Automation Conference (ASP-DAC), Macau, China, 25–28 January 2016; pp. 499–502. [CrossRef]
12. Amarú, L.; Gaillardon, P.E.; Micheli, G.D. Majority-Inverter Graph: A New Paradigm for Logic Optimization. *IEEE Trans. Comput. Aided Des. Integr. Circuits Syst.* **2016**, *35*, 806–819. [CrossRef]
13. Jaberipur, G.; Parhami, B.; Abedi, D. Adapting Computer Arithmetic Structures to Sustainable Supercomputing in Low-Power, Majority-Logic Nanotechnologies. *IEEE Trans. Sustain. Comput.* **2018**, *3*, 262–273. [CrossRef]
14. Pudi, V.; Sridharan, K.; Lombardi, F. Majority Logic Formulations for Parallel Adder Designs at Reduced Delay and Circuit Complexity. *IEEE Trans. Comput.* **2017**, *66*, 1824–1830. [CrossRef]
15. Amarú, L.; Gaillardon, P.; Mitra, S.; De Micheli, G. New Logic Synthesis as Nanotechnology Enabler. *Proc. IEEE* **2015**, *103*, 2168–2195. [CrossRef]
16. Parhami, B.; Abedi, D.; Jaberipur, G. Majority-Logic, its applications, and atomic-scale embodiments. *Comput. Electr. Eng.* **2020**, *83*, 106562. [CrossRef]
17. Reuben, J.; Ben-Hur, R.; Wald, N.; Talati, N.; Ali, A.; Gaillardon, P.E.; Kvatinsky, S. Memristive Logic: A Framework for Evaluation and Comparison. In Proceedings of the Power And Timing Modeling, Optimization and Simulation (PATMOS), Thessaloniki, Greece, 25–27 September 2017; pp. 1–8.
18. Reuben, J.; Talati, N.; Wald, N.; Ben-Hur, R.; Ali, A.H.; Gaillardon, P.E.; Kvatinsky, S. A Taxonomy and Evaluation Framework for Memristive Logic. In *Handbook of Memristor Networks*; Chua, L., Sirakoulis, G.C., Adamatzky, A., Eds.; Springer International Publishing: Cham, Switzerland, 2019; pp. 1065–1099. [CrossRef]
19. Simmons, J.G.; Verderber, R.R. New conduction and reversible memory phenomena in thin insulating films. *Proc. R. Soc. Lond. Ser. A Math. Phys. Sci.* **1967**, *301*, 77–102. [CrossRef]
20. Borghetti, J.; Snider, G.S.; Kuekes, P.J.; Yang, J.J.; Stewart, D.R.; Williams, R.S. 'Memristive' switches enable 'stateful' logic operations via material implication. *Nature* **2010**, *464*, 873–876. [CrossRef]
21. Zhou, F.; Guckert, L.; Chang, Y.F.; Swartzlander, E.E.; Lee, J. Bidirectional voltage biased implication operations using SiOx based unipolar memristors. *Appl. Phys. Lett.* **2015**, *107*, 183501. [CrossRef]
22. Talati, N.; Ben-Hur, R.; Wald, N.; Haj-Ali, A.; Reuben, J.; Kvatinsky, S. mMPU—A Real Processing-in-Memory Architecture to Combat the von Neumann Bottleneck. In *Applications of Emerging Memory Technology: Beyond Storage*; Suri, M., Ed.; Springer: Singapore, 2020; pp. 191–213. [CrossRef]
23. Rahimi Azghadi, M.; Chen, Y.C.; Eshraghian, J.K.; Chen, J.; Lin, C.Y.; Amirsoleimani, A.; Mehonic, A.; Kenyon, A.J.; Fowler, B.; Lee, J.C.; et al. Complementary Metal-Oxide Semiconductor and Memristive Hardware for Neuromorphic Computing. *Adv. Intell. Syst.* **2020**, *2*, 1900189. [CrossRef]
24. Chang, K.C.; Chang, T.C.; Tsai, T.M.; Zhang, R.; Hung, Y.C.; Syu, Y.E.; Chang, Y.F.; Chen, M.C.; Chu, T.J.; Chen, H.L.; et al. Physical and chemical mechanisms in oxide-based resistance random access memory. *Nanoscale Res. Lett.* **2015**, *10*. [CrossRef]

25. Reuben, J.; Fey, D.; Wenger, C. A Modeling Methodology for Resistive RAM Based on Stanford-PKU Model With Extended Multilevel Capability. *IEEE Trans. Nanotechnol.* **2019**, *18*, 647–656. [CrossRef]
26. Golonzka, O.; Arslan, U.; Bai, P.; Bohr, M.; Baykan, O.; Chang, Y.; Chaudhari, A.; Chen, A.; Clarke, J.; Connor, C.; et al. Non-Volatile RRAM Embedded into 22FFL FinFET Technology. In Proceedings of the 2019 Symposium on VLSI Technology, Kyoto, Japan, 9–14 June 2019; pp. T230–T231. [CrossRef]
27. Hsieh, C.C.; Chang, Y.F.; Chen, Y.C.; Shahrjerdi, D.; Banerjee, S.K. Highly Non-linear and Reliable Amorphous Silicon Based Back-to-Back Schottky Diode as Selector Device for Large Scale RRAM Arrays. *ECS J. Solid State Sci. Technol.* **2017**, *6*, N143–N147. [CrossRef]
28. Lin, C.Y.; Chen, P.H.; Chang, T.C.; Chang, K.C.; Zhang, S.D.; Tsai, T.M.; Pan, C.H.; Chen, M.C.; Su, Y.T.; Tseng, Y.T.; et al. Attaining resistive switching characteristics and selector properties by varying forming polarities in a single HfO_2-based RRAM device with a vanadium electrode. *Nanoscale* **2017**, *9*, 8586–8590. [CrossRef] [PubMed]
29. Kim, S.; Lin, C.Y.; Kim, M.H.; Kim, T.H.; Kim, H.; Chen, Y.C.; Chang, Y.F.; Park, B.G. Dual Functions of V/SiOx/AlOy/p++Si Device as Selector and Memory. *Nanoscale Res. Lett.* **2018**, *13*. [CrossRef]
30. Chen, C.; Lin, C.; Chen, P.; Chang, T.; Shih, C.; Tseng, Y.; Zheng, H.; Chen, Y.; Chang, Y.; Lin, C.; et al. The Demonstration of Increased Selectivity During Experimental Measurement in Filament-Type Vanadium Oxide-Based Selector. *IEEE Trans. Electr. Devices* **2018**, *65*, 4622–4627. [CrossRef]
31. Ben-Hur, R.; Ronen, R.; Haj-Ali, A.; Bhattacharjee, D.; Eliahu, A.; Peled, N.; Kvatinsky, S. SIMPLER MAGIC: Synthesis and Mapping of In-Memory Logic Executed in a Single Row to Improve Throughput. *IEEE Trans. Comput. Aided Des. Integr. Circuits Syst.* **2019**. [CrossRef]
32. Adam, G.C.; Hoskins, B.D.; Prezioso, M.; Strukov, D.B. Optimized stateful material implication logic for three- dimensional data manipulation. *Nano Res.* **2016**, *9*, 3914–3923. [CrossRef]
33. Kumar, A.P.; Aditya, B.; Sony, G.; Prasanna, C.; Satish, A. Estimation of power and delay in CMOS circuits using LCT. *Indones. J. Electr. Eng. Comput. Sci.* **2019**, *14*, 990–998.
34. Rumi, Z.; Walus, K.; Wei, W.; Jullien, G.A. A method of majority logic reduction for quantum cellular automata. *IEEE Trans. Nanotechnol.* **2004**, *3*, 443–450. [CrossRef]
35. Kvatinsky, S.; Satat, G.; Wald, N.; Friedman, E.G.; Kolodny, A.; Weiser, U.C. Memristor-Based Material Implication (IMPLY) Logic: Design Principles and Methodologies. *IEEE Trans. Very Larg. Scale Integr. (VLSI) Syst.* **2014**, *22*, 2054–2066. [CrossRef]
36. Lehtonen, E.; Poikonen, J.H.; Laiho, M. Memristive Stateful Logic. In *Handbook of Memristor Networks*; Chua, L., Sirakoulis, G.C., Adamatzky, A., Eds.; Springer International Publishing: Cham, Switzerland, 2019; pp. 1101–1121, doi:10.1007/978-3-319-76375-0_38. [CrossRef]
37. Shen, W.; Huang, P.; Fan, M.; Han, R.; Zhou, Z.; Gao, B.; Wu, H.; Qian, H.; Liu, L.; Liu, X.; et al. Stateful Logic Operations in One-Transistor-One- Resistor Resistive Random Access Memory Array. *IEEE Electr. Device Lett.* **2019**, *40*, 1538–1541. [CrossRef]
38. Ielmini, D.; Wong, H.S.P. In-memory computing with resistive switching devices. *Nat. Electr.* **2018**, *1*, 333–343. [CrossRef]
39. Gupta, S.; Imani, M.; Rosing, T. FELIX: Fast and Energy-efficient Logic in Memory. In Proceedings of the International Conference on Computer-Aided Design (ICCAD '18), San Diego, CA, USA, 5–8 November 2018; pp. 55:1–55:7. [CrossRef]
40. Reuben, J.; Fey, D. A Time-based Sensing Scheme for Multi-level Cell (MLC) Resistive RAM. In Proceedings of the 2019 IEEE Nordic Circuits and Systems Conference (NORCAS): NORCHIP and International Symposium of System-on-Chip (SoC), Helsinki, Finland, 29–30 October 2019; pp. 1–6. [CrossRef]
41. Reuben, J.; Biglari, M.; Fey, D. Incorporating Variability of Resistive RAM in Circuit Simulations Using the Stanford–PKU Model. *IEEE Trans. Nanotechnol.* **2020**, *19*, 508–518. [CrossRef]
42. Gaillardon, P.; Amaru, L.; Siemon, A.; Linn, E.; Waser, R.; Chattopadhyay, A.; De Micheli, G. The Programmable Logic-in-Memory (PLiM) computer. In Proceedings of the 2016 Design, Automation Test in Europe Conference Exhibition (DATE), Dresden, Germany, 14–18 March 2016; pp. 427–432.
43. Shirinzadeh, S.; Soeken, M.; Gaillardon, P.; Drechsler, R. Logic Synthesis for RRAM-Based In-Memory Computing. *IEEE Trans. Comput. Aided Des. Integr. Circuits Syst.* **2018**, *37*, 1422–1435. [CrossRef]
44. Bhattacharjee, D.; Easwaran, A.; Chattopadhyay, A. Area-constrained technology mapping for in-memory computing using ReRAM devices. In Proceedings of the 2017 22nd Asia and South Pacific Design Automation Conference (ASP-DAC), Chiba, Japan, 16–19 January 2017; pp. 69–74. [CrossRef]

45. Reuben, J. Binary Addition in Resistance Switching Memory Array by Sensing Majority. *Micromachines* **2020**, *11*, 496. [CrossRef]
46. Reuben, J.; Pechmann, S. A Parallel-friendly Majority Gate to Accelerate In-memory Computation. In Proceedings of the 2020 IEEE 31st International Conference on Application-Specific Systems, Architectures and Processors (ASAP), Manchester, UK, 6–8 July 2020; pp. 93–100.
47. Fey, D.; Reuben, J. Direct state transfer in MLC based memristive ReRAM devices for ternary computing. In Proceedings of the 2020 European Conference on Circuit Theory and Design (ECCTD), Sofia, Bulgaria, 7–10 September 2020; pp. 1–5.
48. Hur, R.B.; Wald, N.; Talati, N.; Kvatinsky, S. SIMPLE MAGIC: Synthesis and In-memory Mapping of Logic Execution for Memristor-aided Logic. In Proceedings of the 36th International Conference on Computer-Aided Design (ICCAD '17), Irvine, CA, USA, 13–16 Novenber 2017; pp. 225–232.
49. Huang, P.; Kang, J.; Zhao, Y.; Chen, S.; Han, R.; Zhou, Z.; Chen, Z.; Ma, W.; Li, M.; Liu, L.; et al. Reconfigurable Nonvolatile Logic Operations in Resistance Switching Crossbar Array for Large-Scale Circuits. *Adv. Mater.* **2016**, *28*, 9758–9764. [CrossRef] [PubMed]
50. Chang, Y.; Zhou, F.; Fowler, B.W.; Chen, Y.; Hsieh, C.; Guckert, L.; Swartzlander, E.E.; Lee, J.C. Memcomputing (Memristor + Computing) in Intrinsic SiOx-Based Resistive Switching Memory: Arithmetic Operations for Logic Applications. *IEEE Trans. Electr. Devices* **2017**, *64*, 2977–2983. [CrossRef]
51. Cheng, L.; Zhang, M.Y.; Li, Y.; Zhou, Y.X.; Wang, Z.R.; Hu, S.Y.; Long, S.B.; Liu, M.; Miao, X.S. Reprogrammable logic in memristive crossbar for in-memory computing. *J. Phys. D Appl. Phys.* **2017**, *50*, 505102. [CrossRef]
52. Teimoory, M.; Amirsoleimani, A.; Shamsi, J.; Ahmadi, A.; Alirezaee, S.; Ahmadi, M. Optimized implementation of memristor-based full adder by material implication logic. In Proceedings of the 2014 21st IEEE International Conference on Electronics, Circuits and Systems (ICECS), Marseille, France, 7–10 December 2014; pp. 562–565.
53. Rohani, S.G.; Taherinejad, N.; Radakovits, D. A Semiparallel Full-Adder in IMPLY Logic. *IEEE Trans. Very Larg. Scale Integr. (VLSI) Syst.* **2019**; 28, 297–301. [CrossRef]
54. Kim, K.M.; Williams, R.S. A Family of Stateful Memristor Gates for Complete Cascading Logic. *IEEE Trans. Circuits Syst. I Regul. Pap.* **2019**, *66*, 4348–4355. [CrossRef]
55. Siemon, A.; Drabinski, R.; Schultis, M.J.; Hu, X.; Linn, E.; Heittmann, A.; Waser, R.; Querlioz, D.; Menzel, S.; Friedman, J.S. Stateful Three-Input Logic with Memristive Switches. *Sci. Rep.* **2019**, *9*, 14618. [CrossRef]
56. Xu, L.; Yuan, R.; Zhu, Z.; Liu, K.; Jing, Z.; Cai, Y.; Wang, Y.; Yang, Y.; Huang, R. Memristor-Based Efficient In-Memory Logic for Cryptologic and Arithmetic Applications. *Adv. Mater. Technol.* **2019**, *4*, 1900212. [CrossRef]
57. Siemon, A.; Menzel, S.; Bhattacharjee, D.; Waser, R.; Chattopadhyay, A.; Linn, E. Sklansky tree adder realization in 1S1R resistive switching memory architecture. *Eur. Phys. J. Spec. Top.* **2019**, *228*, 2269–2285. [CrossRef]
58. Revanna, N.; Swartzlander, E.E. Memristor based adder circuit design. In Proceedings of the 2016 50th Asilomar Conference on Signals, Systems and Computers, Pacific Grove, CA, USA, 6–9 November 2016; pp. 162–166.
59. Wang, Z.; Li, Y.; Su, Y.; Zhou, Y.; Cheng, L.; Chang, T.; Xue, K.; Sze, S.M.; Miao, X. Efficient Implementation of Boolean and Full-Adder Functions with 1T1R RRAMs for Beyond Von Neumann In-Memory Computing. *IEEE Trans. Electr. Devices* **2018**, *65*, 4659–4666. [CrossRef]
60. Cheng, L.; Li, Y.; Yin, K.S.; Hu, S.Y.; Su, Y.T.; Jin, M.M.; Wang, Z.R.; Chang, T.C.; Miao, X.S. Functional Demonstration of a Memristive Arithmetic Logic Unit (MemALU) for In-Memory Computing. *Adv. Funct. Mater.* **2019**, *29*, 1905660. [CrossRef]

61. Kim, Y.S.; Son, M.W.; Song, H.; Park, J.; An, J.; Jeon, J.B.; Kim, G.Y.; Son, S.; Kim, K.M. Stateful In-Memory Logic System and Its Practical Implementation in a TaOx-Based Bipolar-Type Memristive Crossbar Array. *Adv. Intell. Syst.* **2020**, *2*, 1900156. [CrossRef]
62. Xiao, T.P.; Bennett, C.H.; Hu, X.; Feinberg, B.; Jacobs-Gedrim, R.; Agarwal, S.; Brunhaver, J.S.; Friedman, J.S.; Incorvia, J.A.C.; Marinella, M.J. Energy and Performance Benchmarking of a Domain Wall-Magnetic Tunnel Junction Multibit Adder. *IEEE J. Explor. Solid State Comput. Devices Circuits* **2019**, *5*, 188–196. [CrossRef]

Journal of
Low Power Electronics and Applications

MDPI

Tutorial

Graph Coloring via Locally-Active Memristor Oscillatory Networks

Alon Ascoli [1,*], Martin Weiher [1], Melanie Herzig [2], Stefan Slesazeck [2], Thomas Mikolajick [2,3] and Ronald Tetzlaff [1]

[1] Chair of Fundamentals of Electrical Engineering, Institute of Circuits and Systems, Faculty of Electrical and Computer Engineering, Technische Universität Dresden, 01062 Dresden, Germany; martin.weiher@tu-dresden.de (M.W.); ronald.tetzlaff@tu-dresden.de (R.T.)
[2] Nano-Electronic Materials Laboratory (NaMLab) gGmbH, 01187 Dresden, Germany; melanie.herzig@namlab.com (M.H.); stefan.slesazeck@namlab.com (S.S.); thomas.mikolajick@namlab.com (T.M.)
[3] Institute für Halbleiter-und Mikrosystemtechnik, Technische Universität Dresden, 01062 Dresden, Germany
* Correspondence: alon.ascoli@tu-dresden.de

Abstract: This manuscript provides a comprehensive tutorial on the operating principles of a bio-inspired Cellular Nonlinear Network, leveraging the local activity of NbO_x memristors to apply a spike-based computing paradigm, which is expected to deliver such a separation between the steady-state phases of its capacitively-coupled oscillators, relative to a reference cell, as to unveal the classification of the nodes of the associated graphs into the least number of groups, according to the rules of a non-deterministic polynomial-hard combinatorial optimization problem, known as vertex coloring. Besides providing the theoretical foundations of the bio-inspired signal-processing paradigm, implemented by the proposed Memristor Oscillatory Network, and presenting pedagogical examples, illustrating how the phase dynamics of the memristive computing engine enables to solve the graph coloring problem, the paper further presents strategies to compensate for an imbalance in the number of couplings per oscillator, to counteract the intrinsic variability observed in the electrical behaviours of memristor samples from the same batch, and to prevent the impasse appearing when the array attains a steady-state corresponding to a local minimum of the optimization goal. The proposed Memristor Cellular Nonlinear Network, endowed with ad hoc circuitry for the implementation of these control strategies, is found to classify the vertices of a wide set of graphs in a number of color groups lower than the cardinality of the set of colors identified by traditional either software or hardware competitor systems. Given that, under nominal operating conditions, a biological system, such as the brain, is naturally capable to optimise energy consumption in problem-solving activities, the capability of locally-active memristor nanotechnologies to enable the circuit implementation of bio-inspired signal processing paradigms is expected to pave the way toward electronics with higher time and energy efficiency than state-of-the-art purely-CMOS hardware.

Keywords: graph coloring; cellular nonlinear networks; memristor oscillatory networks; locally-active memristors; control theory

Citation: Ascoli, A.; Weiher, M.; Herzig, M.; Slesazeck, S.; Mikolajick, T.; Tetzlaff, R. Graph Coloring via Locally-Active Memristor Oscillatory Networks. *J. Low Power Electron. Appl.* **2022**, *12*, 22. https://doi.org/10.3390/jlpea12020022

Academic Editors: Alexander Serb and Adnan Mehonic

Received: 22 January 2022
Accepted: 2 April 2022
Published: 18 April 2022

1. Introduction

Memristor technologies promise to revolutionise the world of electronics in the years to come, allowing to boost the performance of integrated circuits beyond the Moore era. Theoretically introduced in 1971 by L. Chua [1], memristors are essentially resistances with state- and input-dependent programmable capability [2]. Despite the first association between Chua's theory and the experimental observation of fingerprints of memristive behaviour at the nanoscale was made by R.S. Williams and his team at Hewlett Packard Labs in 2008 [3], the appearance of memory resistance switching effects in miniaturized physical structures were reported in numerous occasions throughout the past two centuries [4], constituting the object of extensive and intensive investigations for the development of novel solid-state memories first in the 1960s [5]. In fact, the main application of memristors [6]—certainly the most profitable one from a business perspective point of view—is the design of memories with higher retention, lower power consumption, and larger density as compared to state-of-the-art data storage units [7]. Another major field of application regards the development of innovative neuromorphic systems, which resemble biological entities more closely than traditional artificial networks ([8–11]). A closely-related branch of research takes inspiration from the high levels of organization and energy efficiency of biological systems to develop mem-computing machines ([12–14]), which extend the functionalities of traditional cellular architectures [15]. Another bio-inspired research direction aims to the circuit implementation of in-memory computing paradigms through high-capacity memristive memories, stacked in 3D crossbar arrangement above underlying CMOS circuitry [16], and employed, alternatively, to store data or to execute processing tasks ([17–20]), which reveals the high potential of novel hardware platforms of this kind to resolve the von Neumann bottleneck, limiting the maximum operating speed of traditional computing engines, in the near future. Additionally, the unique combined capability of non-volatile resistance switching memories to sense data ([21,22]), learn how to recognize patterns [23], process information [24], and store multiple states [25] within a single tiny physical volume, and the availability of nanoscale locally-active [26] volatile memristors, which may amplify the small signal upon suitable polarization ([27,28]), open up yet-unexplored opportunities for the Internet-of-Things (IoT) industry, which urgently calls for the development of miniaturized, low-power, light-weight, portable, and smart technical systems, which, within a very short time frame, are able to acquire a large amount of information from the environment, to extract features of interest from noisy data so as to solve specific optimization problems, and to store or transmit to a prescribed user the most relevant results of the computation.

One of the most challenging tasks for computing machines, based upon the classical von Neumann architecture, is the solution of combinatorial optimisation problems belonging to the non-deterministic polynomial (NP)-hard complexity class[1]. Nowadays there is a huge interest in developing novel inexpensive low-power high-speed hardware solutions capable to solve NP-hard problems more efficiently than traditional computers, given that high-performance technical systems of this kind may find application in various industry sectors, e.g., for traffic management, airline scheduling, gene sequencing, and electronic chip wiring.

A recent Nature Electronics publication [29] proposed the use of a memristive crossbar array—refer to Figure 1—for accelerating the vector-matrix multiplications (VMMs) at the basis of the update rule of an iterative machine learning algorithm, which is credited to Hopfield, and enables to solve combinatorial problems, within an overall disruptive analogue hardware architecture, leveraging and controlling its numerous inherent noise sources to allow a power-efficient derivation of the optimal solutions.

[1] The time it takes for a von Neumann computing machine to find the optimal solution to a NP-hard problem, which involves n elements, scales exponentially with n.

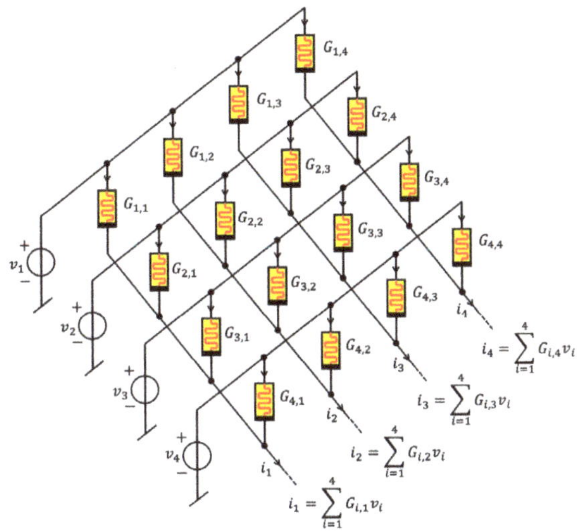

Figure 1. In-memory computing in a $N \times N$ memristive crossbar array (here $N = 4$). In-memory computing in an $N \times N$ memristive crossbar array (here, $N = 4$). Naturally obeying Kirchhoff's Current Law (KCL), the bio-inspired network enables a time-efficient computation of VMMs. With $j \in \{1, \ldots, N\}$, the current flowing down the jth column of the array is simply given by $i_j = \sum_{i=1}^{N} G_{i,j} \cdot v_i$, where $G_{i,j}$ denotes the conductance of the memory resistive switch located at the intersection between the conductive nanowires stretching along row i and column j [29]. The computation of the currents at the outputs of the crossbar columns assumes that the bottom terminals of all the memristors—refer to the thick black horizontal segments in their circuit-theoretic symbols—are at virtual ground.

Cellular Nonlinear Networks (CNNs) with locally-active volatile memristors [30] constitute a powerful engine for the implementation of bio-inspired spike-based computing paradigms. This manuscript, inspired to a recent publication [31], is devoted to explain in a pedagogical form how these bio-inspired memristor cellular arrays[2], may be adopted for solving a complex NP-hard problem known as vertex coloring. While page limitation prevented a complete description of the theory at the basis of the spike-based computing paradigm implemented by the proposed cellular arrays[3], all details are pedagogically reported in this tutorial. In regard to the structure of the manuscript, Section 2 provides a brief description of the memristor model adopted for the study. Section 3 introduces the memristive computing engine for solving the vertex coloring problem, including a discussion of its operating principles, and the specification of strategies aimed to compensate for non-idealities, including an imbalance in the number of couplings per oscillator, and the memristor device-to-device variability. Section 4 presents a rigorous iterative procedure for coloring a graph via the network phase dynamics. Importantly, control paradigms to resolve local minima-based impasse conditions [33] are proposed in Section 5, which further compares the performance of the proposed spike-based computing engine, endowed with ad hoc circuitry to implement such strategies, with the solutions of state-of-the-art software and hardware competitor systems. Finally, conclusions are drawn in Section 6.

2 CNNs with non-volatile memristors ([12–14]) may pave the way toward the development of advanced visual-sensor processors [32] with high spatial resolution and intrinsic memory capability.

3 Importantly, depending upon the graph under focus, the proposed Memristor CNN (M-CNN) may feature either local or non-local capacitive couplings. The solution of the vertex coloring problem through the proposed M-CNNs depends upon the phase differences among the oscillations developing in the constitutive units of the array at steady state. In order to highlight the steady-state oscillatory behaviours of the cells during operation, the bio-inspired arrays are also referred to as Memristor Oscillatory Networks (MONs) in the remainder of the manuscript.

2. A Physics-Based Model for the Threshold Switching Dynamics of a Nano-Scale Locally-Active Memristor Device Stack

In a past study [34,35] we employed physics laws to construct state evolution and memductance functions of a NbO_x memristor, fabricated at NaMLab, after inferring the physical mechanisms, which underlie its nonlinear dynamics, from the outcome of experimental measurements, and the insights gained through theoretic investigations of a mathematical model, derived previously on the basis of Chua's Unfolding Theorem [28]. A thorough analysis of the proposed physics-based model [34] revealed that *the Mott insulator-to-metal transition does not constitute the key physical mechanism at the origin of the threshold switching process, that each of our NbO_x-based memristors undergoes under the application of a generic quasi-static voltage stimulus between its two terminals.* Conversely, a temperature-activated trap-assisted Poole-Frenkel conduction mechanism underlies the abrupt turn-on dynamics of the volatile memristor. Importantly, shortly after our discovery, a further proof of evidence for the validity of our conjecture was provided by engineers from Hewlett Packard (HP) Labs [36]. Remarkably, despite it was originally proposed for the NbO microstructure, the physics-based model in [34] was found to fit rather well also experimental data extracted from nanoscale variants of the NbO_x threshold switching resistance from NaMLab after minor adaptations [37,38].

As illustrated in Figure 2(a), ([30,31]), the equivalent circuit model of the memristor $\mathcal{M}$ consists of the series combination between a linear resistor R_c, capturing the action of the top electrode resistance, and a parallel one-port, formed by a core memristor $\tilde{\mathcal{M}}$, and a nonlinear resistor $\mathcal{R}$, which accounts for the parasitics inherent to the NbO_x nanostructure, and is responsible for the manifestation of leakage current effects. Several studies [34,35] have revealed that the state of the core memristor is well captured by its body temperature T. In fact, as anticipated earlier, threshold switching effects in the nano-device originate from runaway Joule self-heating governed by Poole-Frenkel electrical conduction mechanisms. Taking this into account for the formulation of a state-dependent Ohm's law, with $\tilde{v}_m$ ($\tilde{i}_m$) representing the voltage (current) of $\tilde{\mathcal{M}}$, and choosing Newton's law of cooling to dictate the time evolution of the state, the DAE set, governing the static and dynamic behaviour of the core memristor may be expressed as

$$\frac{dT}{dt} = g(T, \tilde{v}_m) \triangleq \frac{1}{C_{th}} \cdot \tilde{i}_m \cdot \tilde{v}_m - \frac{\Gamma_{th}}{C_{th}} \cdot (T - T_{amb}), \tag{1}$$

$$\tilde{i}_m = G(T, \tilde{v}_m) \cdot \tilde{v}_m \triangleq \frac{1}{R_{01}} \cdot \exp\left(-\frac{a_{01} - a_{11} \cdot \sqrt{|\tilde{v}_m|}}{T}\right) \cdot \tilde{v}_m, \tag{2}$$

where C_{th} (Γ_{th}) stands for the effective thermal capacitance (conductance) of the core device $\tilde{\mathcal{M}}$, T_{amb} denotes the ambient temperature, R_{01}, a_{01} and a_{11} are constants[4], while v_m (i_m) symbolises the voltage (current) falling across (flowing though) the memristor $\mathcal{M}$, whose circuit-theoretic symbol is shown in Figure 2(b). Remarkably, the DAE set (1)–(2) of the core memristor falls into the voltage-controlled extended memristor family from Chua's classification[5] [40]. The constitutive relationship $f(v_{\mathcal{R}}, i_{\mathcal{R}}) = 0$ of the nonlinear resistor $\mathcal{R}$ is given in implicit form as

$$v_{\mathcal{R}} = R_{02} \cdot i_{\mathcal{R}} \cdot \exp\left(\frac{a_{02} - a_{12} \cdot \sqrt{|v_{\mathcal{R}}|}}{T_{amb}}\right). \tag{3}$$

[4] The reader is invited to consult [34,35] for details on the association between the real parameters R_{01}, a_{01} and a_{11} and the physical properties of the core nanostructure.

[5] It is instructive to observe that there exist memristor physical realisations, whose models feature an even more general input- and state- dependent Ohm law than what is admissible for extended memristors. For these two-terminal devices, including the TiO_2 memristor from HP Labs [39], the mathematical description includes an implicit Ohm law of the form $h(x, v_m, i_m) = 0$, with x, v_m, and i_m denoting the device state, voltage, and current, respectively.

revealing that a variant of the Poole-Frenkel law explains current transport phenomena in the parasitic resistor as well[6]. The nominal parameter values for the core memristor and nonlinear resistor models were obtained by fitting the underlying equations to experimental data extracted from nano-device samples (see [37,38] for details on the NbO_x nanostructure fabrication process). Importantly, since the non-negligible intrinsic spread in dynamic behaviour from sample to sample may impair the capability of a computing engine based on memristive hardware to perform a predefined data processing task as desired, this non-ideality may not be neglected in circuit design considerations. Particularly, in the analysis to follow, where a Memristor Oscillatory Network (MON) is adopted to find optimal solutions to graph coloring problems [31], it will be accounted through the replacement of specific parameters in the NbO_x nano-scale threshold switch physics model, specifically Γ_{th}, R_{01}, a_{01}, a_{11}, R_C, R_{02}, and a_{12}, with corresponding ones, i.e., in turn, $\Gamma_{th,\alpha}$, $R_{01,\alpha}$, $a_{01,\alpha}$, $a_{11,\alpha}$, $R_{C,\alpha}$, $R_{02,\alpha}$, and $a_{12,\alpha}$, that are controlled via a real variable α, which is set randomly to a distinct value chosen from a uniform distribution across the closed range $[0,1]$ for each nanostructure individually, prior that the simulation of the ODE, modelling the array associated to a pre-specified graph, is commenced.

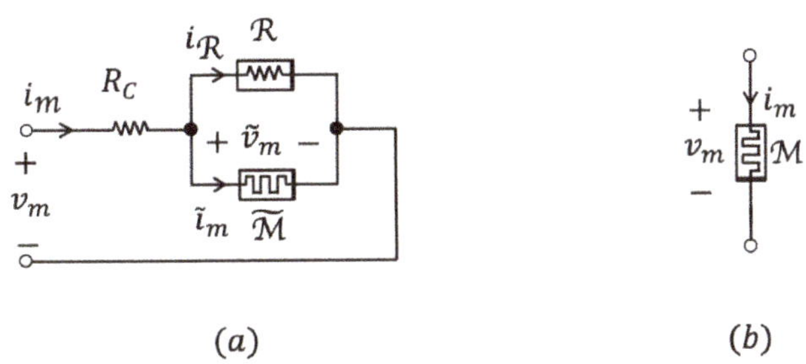

Figure 2. (**a**) Equivalent circuit of the physical model of a NbO_x nanoscale memristor $\mathcal{M}$ from NaMLab. The linear resistor R_C and the nonlinear resistor $\mathcal{R}$ respectively account for the effects of electrode contact resistance and parasitics. (**b**) Memristor circuit-theoretic symbol.

Table 1 reports the parameter setting of the nano-scale memristor physics-based model modulated according to the device-to-device variability estimated statistically beforehand through the analysis of current-voltage characteristics of a large number of samples under a common quasi-static stimulation [31].

Table 1. Parameter setting for NbO_x nanoscale threshold switch from NaMLab. The effects of the memristor-to-memristor variability are accounted through the assignment of a distinct value, chosen randomly within the closed set $[0,1]$ to the variable α, controlling specific coefficients of Equations (1), (2), and (3).

C_{th} / J · K^{-1}	$\Gamma_{th,\alpha}$ / W · K^{-1}	T_{amb} / K	$R_{01,\alpha}$ / Ω	$a_{01,\alpha}$ / K
$1 \cdot 10^{-14}$	$1.889 \cdot 10^{-6} \cdot 1.064^{\alpha}$	293	$3.047 \cdot 0.831^{\alpha}$	$3620 \cdot 1.061^{\alpha}$
$a_{11,\alpha}$ / K· V^{-1}	$R_{c,\alpha}$ / Ω	$R_{02,\alpha}$ / Ω	a_{02} / K	$a_{12,\alpha}$ / K· V$^{-1/2}$
$820.4 \cdot 1.137^{\alpha}$	$173.8 \cdot 1.092^{\alpha}$	$565 \cdot 1.377^{\alpha}$	1000	$168.8 \cdot 1.083^{\alpha}$

3. Memristive Computing Engine for Solving the Vertex Coloring Problem

One of the most popular NP-hard problems is *graph or vertex coloring*. Given an undirected graph, consisting of a certain arrangement of edge-coupled vertices, the aim of the problem is to assign a color to each vertex, satisfying the constraint, which dictates

[6] The mathematical description of the nonlinear resistor, formulated in Equation (3), is in fact equivalent to the model of a NbO_x memristor, as originally presented in [34,35], which reveals the correspondence of the real parameters R_{02}, a_{02} and a_{12} to physical properties of the nanostructure, in the limit when changes occurring in the device state, defined as its body temperature, are negligible.

that a given color should be shared between as many vertices as possible so long as no edge connects any two of them. The lowest number of color groups, the vertices of the unconnected graph may be classified into, is called *chromatic number*.

As revealed back in 1988, graph coloring may be achieved by harnessing synchronisation mechanisms in arrays of of coupled oscillatory cells [41]. A more recent work [42] showed that the assignment of colors to vertices of a graph may be naturally implemented through the analysis of the steady-state phase shifts between capacitively-coupled relaxation oscillators exploiting negative differential resistance (NDR) effects in vanadium dioxide (VO_2) nano-structures. Inspired from this research, we have recently investigated the capability of an oscillatory network, leveraging locally-active dynamics in NbO_x memristors, and featuring capacitive couplings, to identify the minimum possible number of colors assignable to the vertices of an associated undirected graph via phase dynamics.

In order to color a given graph of N vertices or nodes, a unique number in the set $\{0, 1, \ldots, N-1\}$ is first attributed to each vertex. A one-to-one association is then established between the vertices (edges) of the graph, and the oscillators (coupling capacitors, each of capacitance C_C) of the associated network. The oscillator 0, corresponding to the node 0, assumes a critical role in the solution of the graph coloring problem, and is called *reference cell*. A general indication, regarding the selection of a suitable reference oscillator among the N possible candidates, will be given shortly. As an example, Figure 3(a) and (b) show a 6-node ring and the associated MON, respectively.

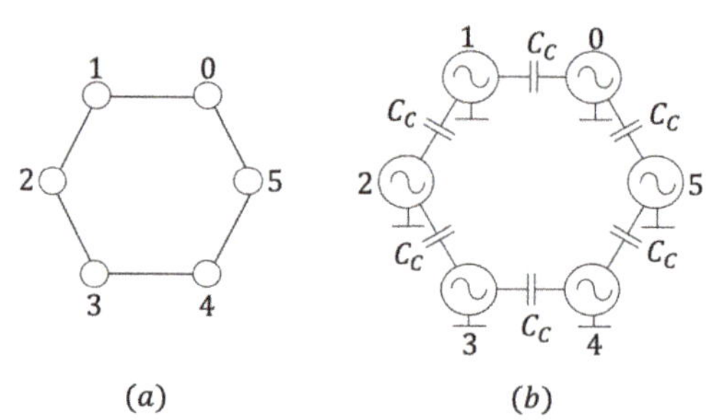

Figure 3. (**a**) A 6-node ring-shaped undirected graph (**b**) Associated MON. The oscillator i of the network corresponds to the vertex i of the graph ($i \in \{0, 1, 2, 3, 4, 5\}$).

With reference to Figure 4, plots (a) and (b) show the oscillator circuit and its symbol, respectively. Each oscillatory cell is composed of the parallel connection between a NbO_x memristor $\mathcal{M}$, a bias circuit, consisting of the series combination of a DC voltage source V_S with a series resistor R_S, allowing to polarize [28] the resistance switching memory within the locally-active region of its DC current-voltage locus, and a capacitor C.

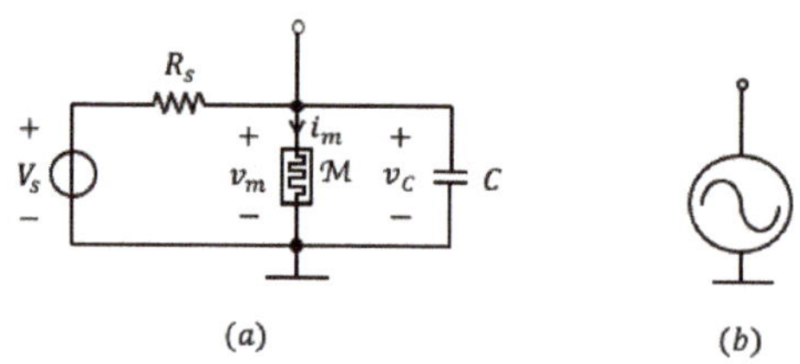

Figure 4. Memristive oscillatory cell (**a**) and its symbol (**b**).

On the basis of the NbO_x nano-device physics-based model, expressed by Equations (1), (2), and (3), under the variability-aware parameter setting of Table 1, we carried out a deep numerical investigation of the capability of an array of capacitively-coupled memristive oscillatory cells to classify the vertices of a pre-defined undirected graph in the least number of groups.

While the memristor model parameters, accounting for the inherent fluctuations in static and dynamic properties among distinct nano-device samples, were already reported in Table 1, only the values assigned to the physical quantities of the non-memristive circuit elements in the proposed MON are provided in Table 2.

Table 2. Parameter setting for the non-memristive circuit elements in the oscillator of Figure 4(a).

V_S/ V	R_S/ Ω	C/ F	C_C/ F
2.5	5525	$10 \cdot 10^{-9}$	$0.2 \cdot 10^{-9}$

3.1. Operating Principles of the Capacitively-Coupled Networks

The capacitive nature of the couplings in the network is responsible for pulling the phases of physically-connected oscillators far apart one from the other at steady-state. This repelling mechanism may be exploited to colour the vertices of the associated graph. The phases of uncoupled oscillators tend to form clusters, which may be interpreted as color groups for the corresponding vertices. The larger is the separation between phase clusters, the simpler is the classification of the nodes of the graph into color groups.

To illustrate this concept, let us consider a simple undirected graph, composed of one edge, which couples two nodes, as shown in Figure 5(a). The chromatic number of this graph is obviously equal to 2. The corresponding oscillatory network is depicted in plot (b) of the same figure. Simulating the circuit with nominal parameter setting[7], the time waveforms of the voltages across the capacitors or those of the currents through the memristors within the circuits of the two capacitively-coupled memristive oscillators are expected to feature a steady-state phase shift of about 180°. Upon the emergence of anti-phase synchronisation between the two oscillators of this simple network, it would be natural to assign one color to vertex 0 and another one to vertex 1 of the associated graph of Figure 5(a).

As a further example, Figure 5(c) visualises another undirected graph with chromatic number equal to 2. The respective oscillatory array is shown in plot (d) of the same figure. Simulating this network, the phases of oscillators 1 and 2 are expected to cluster together, and to shift away from the phase of reference oscillator 0 as much as possible, approaching a relative value of approximately 180° at steady state. With the network exhibiting such a phase pattern at steady state, it would be straightforward to divide the vertices of the associated graph into two color groups, including vertex 0 and vertices 1 and 2, respectively.

Due to a couple of non-idealities the expectations on the phase dynamics of the networks in plots (b) and (d) of Figure 5 are not fulfilled in practice. Details will be provided in the next two sections.

[7] The nominal parameter setting is obtained from Table 1 for $\alpha = 0.5$. As will be shown later, simulating the memristor model under a quasi-DC voltage stimulus and with the variability parameter stepped across its existence domain, the locus observed for $\alpha = 0.5$ appears in the center of the distribution of characteristics emerging in the voltage-current plane.

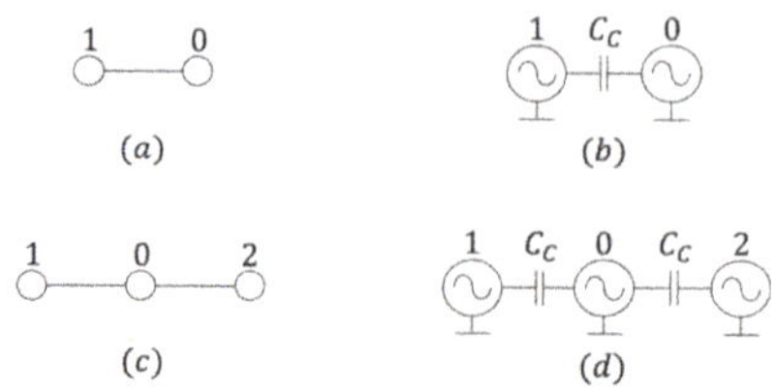

Figure 5. (**a**) A 2-node 1-edge graph. Its chromatic number is 2. (**b**) Oscillatory network corresponding to the graph in (**a**). (**c**) A 3-vertex 2-edge graph. Its chromatic number is once again 2. Interestingly, the number of edges departing from vertex 0 (from either vertex 1 or vertex 2) is 2 (1). (**d**) Oscillatory network corresponding to the graph in (**c**).

Before proceeding, the following remark explains how the autonomous memristive array is initialised, and clarifies how the steady-state phases of the cells are computed when the network converges to an oscillatory solution.

Remark 1. *With V_S and R_S fixed to specific values, as reported in Table 1, all memristors in the network are biased in a common operating point lying along the NDR of the DC I_m–V_m locus of the NbO_x resistive nanoswitch. Defining as $v_{C,i}$ and T_i the states of the second-order cell i of the memristive array ($i \in \{0, \ldots, N-1\}$), the initial conditions for the capacitor voltage and the memristor[8] temperature are set in each oscillator to 0 V, and to the ambient temperature T_{amb}, fixed to 293 K in Table 1, respectively. A random sequence is generated to mismatch, in a non-deterministic way, the time instants, at which the signals generated by the DC voltage sources within the oscillators ramp toward the nominal V_S value over a time span of 1 µs at the beginning of a simulation. If sustained periodic oscillations develop across the network at steady state, for each oscillator $i \in \{0, \ldots, N-1\}$, the phase of the memristor current $i_{m,i}$ relative to the phase of the current $i_{m,0}$ through the NbO_x device in the reference oscillator 0 is then computed, as follows. First, the common period T of the oscillations, observed in the time waveforms of the memristor currents at steady state, is estimated. For each i-value in the set $\{0, \ldots, N-1\}$, the time instant t_i, at which the memristor current $i_{m,i}$ in the cell i attains a given threshold value I_{th}, specifically 0.5 mA, during its ascending phase, within a single common steady-state cycle, is then recorded. The cycle, utilised for these calculations, covers the time span $[t_0, t_0 + T]$, where t_0 marks the time instant, when this threshold crossing event occurs for the memristor current $i_{m,0}$ in the reference cell 0. Next, for each i-value, the temporal span $\Delta t_i \triangleq t_i - t_0$, which separates the instants t_i and t_0, at which the threshold crossing event occurs for the memristor currents in the cells i and 0, respectively, is calculated. Finally, this allows to compute the steady-state phase shift between cells i and 0 via[9] $\varphi_i^{(s)} \triangleq \omega_0 \cdot \Delta t_i$, where $\omega_0 \triangleq \frac{2\pi}{T}$, for each i-value.*

3.2. Compensation for an Imbalance in the Number of Couplings per Oscillator

If an imbalance in the number of edges per node characterises the coupling structure of a given graph, the phases of physically-coupled oscillators in the corresponding memristive oscillatory network may be found to hold only a marginal distance one from the other at steady state. The balanced nature of the graph of Figure 3 (Figure 5(a)) originates from the fact that each of its six (two) nodes is coupled to 2 other nodes (the other node). On the other hand, inspecting the graph of Figure 5(c), vertex 0 is coupled to vertices 1 and 2, while each of vertices 1 and 2 is connected to vertex 0 only. The resulting imbalance in the number of couplings per cell, arising in the associated array of memristive oscillators—refer to Figure 5(d)—prevents the phases of cells 1 and 2 from separating as expected from the

[8] The voltage across (current through) the memristor in the cell i is indicated via $v_{m,i}$ ($i_{m,i}$).

[9] The units of the steady-state relative phase of oscillator i, computed via $\varphi_i^{(s)} = \omega_0 \cdot \Delta t_i$, are radiants. In order to express $\varphi_i^{(s)}$ in degrees, its formula needs to be scaled by the factor $\frac{180°}{\pi}$ ($i \in \{0, \ldots, N-1\}$).

phase of cell 0. This is shown in the phase diagram[10] of Figure 6(a), where the dashed orange and green traces, illustrating the time evolution of the phases of cells 1 and 2 relative to cell 0, respectively, feature a separation as small as 50° from the reference 0° level at steady state. The unbalanced nature of the network of Figure 5(d) results in an imbalance in the capacitive load per oscillator. Looking at the coupling configuration in this array, and recalling the circuit of each oscillator, shown in Figure 4(a), in which, for simplicity, the memristor is replaced with its small-signal equivalent circuit model [43], the application of basic circuit-theoretic principles allows to obtain an expression in the sinusoidal regime for the capacitive impedance $Z_{C_i}(j\omega)$, loading oscillator i for each value of i in the set $\{0,1,2\}$, i.e.,

$$Z_{C_0}(j\omega) = Z_C \parallel (Z_{C_C}(j\omega) + Z_C(j\omega)) \parallel (Z_{C_C}(j\omega) + Z_C(j\omega)), \tag{4}$$
$$Z_{C_1}(j\omega) = Z_C \parallel \{Z_{C_C}(j\omega) + [Z_C(j\omega) \parallel (Z_{C_C}(j\omega) + Z_C(j\omega))]\}, \tag{5}$$
$$Z_{C_2}(j\omega) = Z_C \parallel \{Z_{C_C}(j\omega) + [Z_C(j\omega) \parallel (Z_{C_C}(j\omega) + Z_C(j\omega))]\}, \tag{6}$$

where

$$Z_C(j\omega) = \frac{1}{j\omega C}, \quad \text{and} \tag{7}$$
$$Z_{C_C}(j\omega) = \frac{1}{j\omega C_C}. \tag{8}$$

Defining

$$C_a \parallel C_b \triangleq \frac{C_a \cdot C_b}{C_a + C_b}, \tag{9}$$

the load capacitance C_i of the oscillator $i \in \{0,1,2\}$ in the network of Figure 5(d) may be extracted easily from the $(i+1)^{\text{th}}$ equation in the triplet (4)–(6), yielding

$$C_0 = C + 2 \cdot (C_C \parallel C), \tag{10}$$
$$C_1 = C + C_C \parallel (C + C_C \parallel C) \approx C + C_C \parallel C, \tag{11}$$
$$C_2 = C + C_C \parallel (C + C_C \parallel C) \approx C + C_C \parallel C, \tag{12}$$

where the approximations stem from the inequality $C >> C_C$, yielding $C + C_C \parallel C \approx C$. Analysing Equations (10)–(12) it is clear that, in order to compensate for the imbalance in the capacitive load per oscillator within the network of Figure 5(d), an additional capacitor with capacitance $C_{\text{comp},j} = C_C \parallel C$ should be added in parallel to the capacitor C in the circuit of oscillator j, for each j-value in the set $\{1,2\}$, as shown in Figure 6(b). Simulating the balanced network, the relative phases of cells 1 and 2 cluster together, converging to values close to the expected 180° level at steady state, as may be evinced from plot (a) in the same figure, where the orange and green solid traces reveal the time evolution of φ_1 and φ_2, respectively.

[10] A new graphical tool—which we call *phase diagram*—is introduced in this research study [31] for visualising the phase dynamics of the network. Referring, for example, to the phase diagram of Figure 6(a), a specific trace visualises the time evolution of the phase of oscillator j relative to oscillator 0 ($j \in \{1,\ldots,N-1\}$). Reading the time flow along the radial direction, the angle between the segment, joining the origin to the point, where the trace is found to lie at time t, and the blue horizontal line, denoting the 0°-valued reference level, represents the phase shift $\varphi_j(t)$ of oscillator j with respect to oscillator 0 at time t.

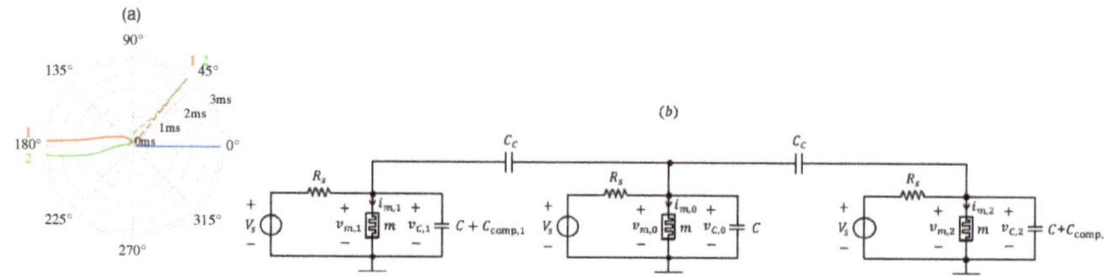

Figure 6. (**a**) Phase diagram visualising the time evolution of the phases of oscillators 1 (in orange) and 2 (in green) relative to oscillator 0, sitting on the 0° phase state throughout the simulation (blue horizontal line) for the original unbalanced network of Figure 5(c) (see the dashed traces) and for the compensated network in plot (**b**) of this figure (refer to the solid traces). In the first (latter) case, the phases of oscillators 1 and 2 are found to cluster together, and to distance themselves from the reference 0° phase, associated to oscillator 0, by approximately 50° (180°) at steady state. (**b**) Complete circuitry of the memristive oscillatory network of Figure 5(d) after compensation for the imbalance in the number of couplings per oscillator. Here $C_{comp,1} = C_{comp,2} = C_C \parallel C$.

With reference to Figure 6(b), it is interesting to observe that the compensating capacitance $C_{comp,j}$ for oscillator $j \in \{1,2\}$ is equal to the product between $C_C \parallel C$ and the difference between the number of connections for oscillator 0, coinciding with the maximum number of connections per oscillator in the unbalanced network of Figure 5(d), and the number of connections for oscillator j. Taking inspiration from this finding, for a general unbalanced network with N oscillators, the compensating capacitance $C_{comp,i}$ for oscillator $i \in \{0,1,\ldots,N-1\}$ is computed via

$$C_{comp,i} = (n_{max} - n_i) \cdot (C_C \parallel C), \quad (13)$$

where n_i is the number of couplings for oscillator i, while

$$n_{max} = \max_{0 \le i < N} \{n_i\}, \quad (14)$$

is the maximum number of couplings per oscillator in the original network. While reducing the coupling capacitance may allow to accelerate the phase dynamics in the memristive computing engine, a number of factors influence its selection. In this regard, to name but a couple of key aspects, first C_C should not be too small, otherwise the capacitive path between any two oscillators, to be paired so as to reproduce the links, joining the vertices of the associated graph, across the proposed cellular medium, would effectively act as an open circuit. Concurrently, C_C may not be so large as to violate the validity of the approximation $C >> C_C$, used to derive the compensating capacitance $C_{comp,i}$ for each oscillator $i \in \{0,1,\ldots,N-1\}$ (refer to Equation (13)), which enables to counteract the non-uniformity in the load capacitance per oscillator across the cellular medium, allowing to keep the natural frequencies[11] of the oscillators close together, which facilitates the convergence of the memristive computing engine to some steady state.

In the remainder of this paper, any unbalanced network will first be compensated, and then simulated for the solution of a given graph coloring problem.

3.3. Compensation for the Memristor Device-to-Device Variability

Simulating the network of Figure 5(b) for the case, where the device-to-device variability is taken into account, the memristor currents in the two oscillatory circuits may be unable to settle on steady-state oscillatory waveforms. Figure 7(a) shows the current-voltage loci obtained by simulating the memristor model Equations (1), (2), and (3) under a quasi-DC voltage stimulus for each value of the variability parameter α in the set

[11] The natural frequency of an oscillator is the inverse of the period of the oscillations developing across its circuitry.

$\{0, 0.2, 0.4, 0.6, 0.8, 1.0\}$. It is worth to pinpoint that the distribution of quasi-static characteristics, visualised in this figure, matches the variability observed in analogous loci measured from 196 device samples. Assuming that the oscillator 0 (1) in the two-cell network hosts a memristor, featuring a quasi-DC i_m–v_m locus lying in the center (on the right end) of the distribution of Figure 7(a), the network is found to fail to converge to steady-state oscillatory dynamics. This issue is essentially due to the significant mismatch between the DC operating points of the resistive nano-switches in the two oscillators. It may be addressed by reprogramming the DC operating point of the memristor in cell 1. This may be achieved [28] by adjusting either the resistance R_S of the series resistor, as done in this research study, or the voltage V_S of the DC source within the circuit of oscillator 1 until an anti-phase synchronisation pattern is found to emerge in the network.

As demonstrated in the phase diagram visualised in plot (b) of Figure 7, stepping the increment $\Delta R_{S,1}$ in the series resistance R_S of oscillator 1 from $0\,\Omega$ up to $151\,\Omega$, the network keeps featuring non-convergent phase dynamics for a while (see the orange trace relative to the case $\Delta R_{S,1} = 50\,\Omega$). Then, from some point onward, the memristor currents in the two oscillators settle on steady-state oscillatory waveforms, sharing the same frequency, but differing in phase by an offset, which progressively approaches the expected 180° level as the series resistance R_S of oscillator 1 gets larger (compare the green and purple traces, obtained for the first and second $\Delta R_{S,1}$-value in the set $\{100, 125\}\,\Omega$, respectively). Finally, the two capacitively-coupled cells attain anti-phase synchronisation (refer to the red trace corresponding to the scenario $\Delta R_{S,1} = 151\,\Omega$).

As another example, Figure 7(c), where the orange and green dashed traces illustrate the time evolution of the phase of oscillators 1 and 2 relative to oscillator 0, respectively, demonstrates that the balanced network of Figure 6(b) fails to converge to a steady-state oscillatory solution when the first, second, and third value in the set $\{0.5, 0, 1\}$ is assigned in turn to the variability parameter α in the model of the memristor in oscillator 0, 1, and 2. A numerical procedure, tuning separately[12], one at a time, the series resistances in oscillators 1 and 2 with the intention to maximise the steady-state phase shifts $\varphi_1^{(s)}$ and $\varphi_2^{(s)}$, determines that decrementing (incrementing) the series resistance of oscillator 1 (2) by $\Delta R_{S,1} = -134\,\Omega$ ($\Delta R_{S,2} = +151\,\Omega$), the network exhibits a steady-state oscillatory pattern characterised by the anti-phase synchronisation between oscillators 1 and 2, on one side, and oscillator 0, on the other side.

In the remainder of this section, in order to compensate for the negative effects that the memristor device-to-device variability has on the performance of a balanced network, the following approach shall be adopted. First, a reference cell, hosting the memristor, to which the random number generator assigns a variability parameter value closest to 0.5 among all NbO_x devices in the array, should be selected. In a hardware implementation of the memristive network, the memristor of the reference oscillator would approximately display the average dynamical behaviour among all the resistive nano-switches employed in the array[13]. This would minimise the subsequent adjustment to be carried out on the bias circuit of each oscillator $j \in \{1, \ldots, N-1\}$ to reprogram the DC operating point of the respective memristor[14]. Then, for each j-value in the set $\{1, \ldots, N-1\}$, the oscillator j is capacitively coupled only to the reference oscillator 0, and the series resistance R_S in its

12 In order to reprogram appropriately the operating point of the memristor in the cell $j \in \{1, 2\}$ of the 3-oscillator network under focus, the cell j itself is capacitively coupled only to the reference cell 0, and, as described earlier, $\Delta R_{S,j}$ is tuned until anti-phase synchronisation emerges in the resulting two-cell network. This procedure is carried out separately for oscillators 1 and 2.

13 It is important to pinpoint that, while the choice of a reference cell for the preliminary compensation of the memristor device-to-device variability should fall for a specific oscillator, as specified here, no rule dictates the selection of a reference cell for the later computation of the relative phase pattern of the array, as discussed in Section 4.

14 It is important to observe that, while taking the proposed device-to-device compensation measure, care need to be taken so as to keep the natural oscillation frequency of each oscillator within a close range. In fact, a wide spread in this parameter, inevitably differing across the cellular medium, due to the R_S tuning procedure, would jeopardize the convergence of the bio-inspired computing engine to some steady state.

DC bias circuit is adjusted through a numerical procedure till the point when its increment or decrement by $\Delta R_{s,j}$ induces a 180° phase shift between the memristor currents of the two cells.

It is important to observe that, in hardware, such R_S tuning procedure needs to be carried out only once, during the computing machine testing phase, directly after its fabrication. In order to simplify the programmability of the series resistor, it could be implemented through a voltage-controlled CMOS transistor forced to operate in the linear region.

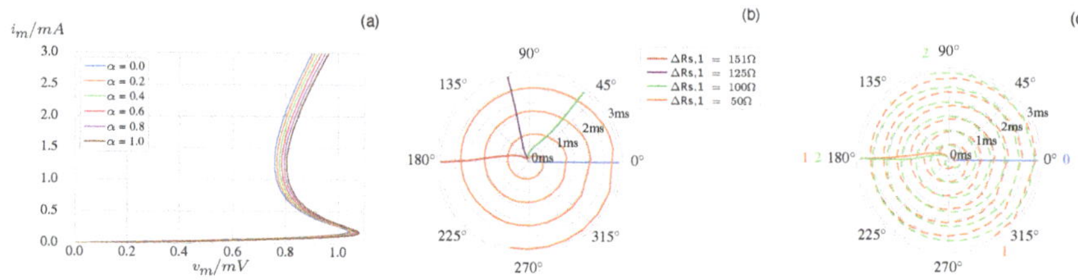

Figure 7. (**a**) Spread in the distribution of quasi-DC memristor current-voltage loci, as emerging from numerical simulations of the model Equations (1), (2), and (3) for all values of the variability parameter α in the set $\{0, 0.2, 0.4, 0.6, 0.8, 1.0\}$. (**b**) Phase diagram showing the time evolution of the phase shift of oscillator 1 relative to the 0°-valued phase of the reference oscillator 0 for each of the values of the series resistance increment $\Delta R_{S,1}$ in the set $\{50, 100, 125, 151\}\,\Omega$ (refer in turn to the orange, green, purple, and red traces). The two capacitively-coupled oscillators achieve anti-phase synchronisation for the largest $\Delta R_{S,1}$-value in this set. (**c**) Phase diagram illustrating the phase dynamics of oscillators 1 (in orange) and 2 (in green) for the balanced network of Figure 6(b) for the case where specific parameters in the model of the memristor in the cells 0, 1, and 2 are respectively controlled by the first, second, and third α-value within the set $\{0.5, 0, 1\}$ (see the dashed traces), and after the negative effects on the network performance associated to the memristor device-to-device variability have been compensated by incrementing (decrementing) the series resistance R_S by $\Delta R_{S,1} = -134\,\Omega$ ($\Delta R_{S,2} = +151\,\Omega$) (refer to the solid lines).

Remark 2. *In view of a future hardware implementation, the parameter setting of the memristive computing engine could be optimized to increase the data processing speed further. However, a comprehensive investigation, aimed to ensure this would not impair the accuracy of the engine calculations, should concurrently be carried out. On one hand, the operating speed of the computing engine should not be so large to prevent the memristor to respond to the stimuli, it experiences over time, which would not allow to exploit thoroughly its rich dynamics for solving the challenging vertex coloring task under focus. On the other hand, the array of capacitively-coupled memristor oscillators is expected to attain a steady state within a sufficiently-short time frame. For all the case studies, investigated in this research work, except for those simulations, in which the proposed computing engine was unable to exit the transient phase, the value assigned to the coupling capacitance C_C enabled the oscillators' phases to converge to steady-state values within tens of milliseconds. Each design parameter may in fact affect some key figure of merit of the bio-inspired network. For example, reducing the capacitance C of each oscillator may induce an acceleration in the phase dynamics. However, it could also reduce the range of admissible memristor NDR bias points, about which the processing unit would undergo limit-cycle oscillations [44], which, as a consequence, would shrink the tuneable range for the series resistance R_S in the device-to-device variability compensation procedure.*

4. A Rigorous Strategy for Coloring a Graph via the Network Phase Dynamics

As explained earlier, the assignment of colors to the nodes of a graph is based upon the relative phases among the oscillators of the associated network at steady state. However, a rigorous strategy, as proposed below, needs to be applied at the end of the network simulation to classify the vertices of the associated graph into color groups, with the intention to use the minimum possible number of colors[15].

As described earlier, the relative phases among the oscillators of a given network are computed at the end of a simulation, on the basis of the time instants at which, within a common steady-state period, the time waveforms of the currents through the memristors cross a threshold value I_{th}, here set to 0.5 mA, during the ascending phase. This calculation allows to order the relative phases in increasing order, with the 0° reference level sitting on the first position of the arrangement, which we refer to as *phase shift ordering* in the remainder of the paper. The phase shift ordering directly translates into a corresponding ranking among the oscillators, with those, which feature a lower steady-state phase relative to the reference cell, sitting higher in the table. Equivalently, the ranking among the oscillators may be interpreted as a ranking among the associated vertices.

Our rigorous strategy to assign colors to the vertices of the associated graph on the basis of the network phase dynamics is based upon the analysis of the oscillator/vertex ranking through an iterative procedure composed of N iterations. The first iteration may be summarised as follows. Initially the first vertex in the table, i.e., reference vertex 0, is inserted in the first color group. Proceeding toward the bottom of the table, for $j \in \{2, \dots, N\}$, vertex at position j in the table is assigned the same color as $(j-1)^{th}$-placed vertex if the graph features no edge between these two vertices, otherwise the jth-ranked vertex is defined as the first element of a new color group. After coloring vertex at row N in the table, a final check needs to be carried out to verify if the vertices in the last color group may be merged with those in the first color group. This may be done if and only if no pair of vertices in these two groups is connected by means of an edge in the undirected graph. In cycle $i \in \{2, \dots, N\}$ of the iterative procedure, the color assignment step is repeated in a similar fashion, analysing progressively the vertices at positions $i, i+1, \dots, N, 1, 2, \dots, i-1$ in the table. With such iterative procedure, at least one of the cycles will allow to determine the minimum number of color groups, identifiable by the network, given the phase shift ordering it outputs at the end of a certain simulation. In other words, indicating the kth color group, which, on the basis of the prediction of the ith cycle of the iterative procedure, is identifiable by the network, as $C_k^{(i)}$ ($k \in \{1, \dots, m_i\}$, where $m_i \in [n, N]$ denotes the total number of colors assigned to the N nodes of the associated graph in the ith iteration, and n represents the chromatic number of the graph itself, the proposed strategy will output the particular group classification obtained from the qth iteration, whereby $m_q = \min_{i=1}^{i=N}\{m_i\}$.

Remark 3. *Remarkably, the initialisation of the oscillatory network, and, particularly, the temporal order of activation of the DC voltage sources in its oscillators, plays a crucial role on the steady-state phase arrangement, and, as a result, on the outcome of the proposed strategy. Consequently, it is possible that the classification of the vertices of a graph into color groups, as estimated via our iterative procedure, is not optimal. Under these circumstances, the network is unable to identify the chromatic number of the associated graph, since it converges to an oscillatory solution corresponding to a local minimum for some optimisation goal associated to the vertex coloring problem.*

Let us denote the *phase shift vector* as $\boldsymbol{\varphi} \triangleq [\varphi_0 = 0^{\circ}, \dots, \varphi_{N-1}]$, where φ_i stands for the phase shift between oscillator i and reference oscillator 0 over the course of a certain simulation of the network ($i \in \{0, \dots, N-1\}$). Given that the network of capacitively-coupled oscillators tends to pull the phases of physically-connected cells far apart one

15 The proposed strategy will determine the minimum possible number of color groups, which, under a given initialisation setting, the network is able to identify as it classifies the nodes of the associated graph. Importantly, as will be clarified later, this minimum number does not necessarily coincide with the chromatic number of graph, since the network may converge to a correct but suboptimal solution. Methods allowing the memristive array to overcome a suboptimal solution so as to approach the optimal one will be presented shortly.

from the other, it is expected that the phase dynamics unfold toward a steady-state pattern maximising a function $F(\boldsymbol{\varphi})$ of the form

$$F(\boldsymbol{\varphi}) \triangleq \frac{1}{2} \cdot \sum_{i=0}^{N-1} \sum_{j=0}^{N-1} a_{i,j} \cdot |\varphi_i - \varphi_j|, \tag{15}$$

where $a_{i,j}$ is the element at row i and column j of the so-called *adjacency matrix* $\mathbf{A}$, which encodes the coupling arrangement within a N-node graph[16]. Transforming the expression for $F(\boldsymbol{\varphi})$ in Equation (15), another function of the phase shift vector, defined as

$$G(\boldsymbol{\varphi}) \triangleq \frac{1}{2} \cdot \sum_{i=0}^{N-1} \sum_{j=0}^{N-1} a_{i,j} \cdot \cos(\varphi_i - \varphi_j), \tag{16}$$

is chosen to describe the *optimisation goal* of the network as it evolves toward a stable solution. It is worth pinpointing that the phase shift vector $\boldsymbol{\varphi}^{(s)} = [\varphi_0^{(s)} = 0^\circ, \ldots, \varphi_{N-1}^{(s)}]$, emerging in a well-behaved network at steady state, minimises the function $G(\boldsymbol{\varphi})$. For a general network, however, the optimization goal function might feature a number of local minima besides the global minimum. If the oscillatory solution, the network converges to, at the end of a certain simulation, corresponds to a local (the global) minimum of $G(\boldsymbol{\varphi})$, the application of the vertex coloring strategy, described earlier, results in a group number higher than (equal to) the chromatic number of the associated graph.

The analysis of an exemplary network, specifically the one shown in Figure 3(b), allows to gain a deeper insight into the phase dynamics of a memristive array in a pair of simulation scenarios, associated to a suboptimal and to the optimal initialisation scenario, respectively. The network under focus is already balanced, since each of its oscillators is coupled to the 2 adjacent cells, thus only the memristor device-to-device variability needs to be neutralised. Figure 8(a) ((c)) visualises the time evolution of the phase dynamics emerging in the network under a suboptimal (the optimal) initialisation setting. In both scenarios the optimisation goal function decreases over time, but, in the first (latter) case, $G(\boldsymbol{\varphi})$ converges asymptotically toward a local (the global) minimum, as depicted in plot (b) ((d)) of the same figure.

[16] If an (no) edge connects vertices i and j, then $a_{i,j} = a_{j,i} = 1(0)$. Note that $\mathbf{A} = \mathbf{A}^T \in \mathbb{R}^{N \times N}$.

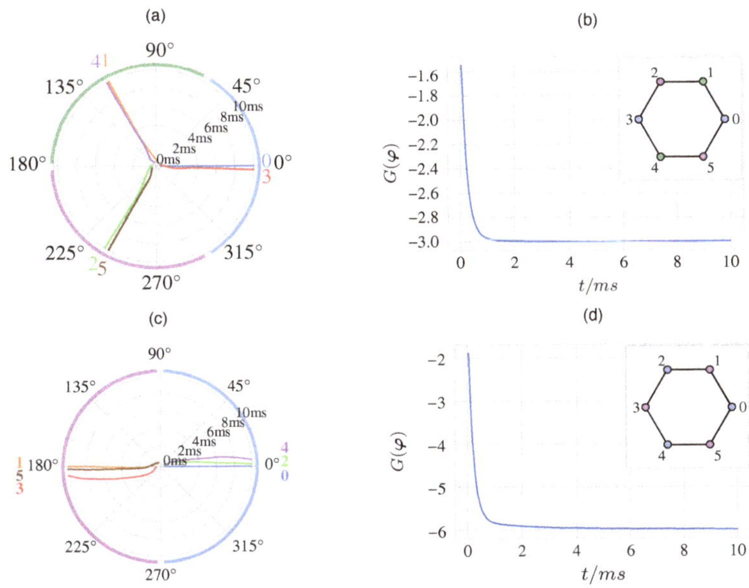

Figure 8. (**a**) ((**c**)) Phase dynamics of the balanced network of Figure 3(b) with compensation for the memristor device-to-device variability and under a suboptimal (the optimal) initialisation setting. (**b**) ((**d**)) Time evolution of the optimisation goal function toward a local (the global) minimum in the simulation scenario illustrated in plot (**a**) ((**c**)). In the first (latter) case the application of the vertex coloring strategy to the respective vertex ranking divides the 6 nodes of the graph of Figure 3(a) into 3 (2) colors. In the first (latter) case the composition of each of the 3 (2) color groups is made clear by the colors assigned in plot (**a**) ((**c**)) to the arcs of the circular sectors, which host the final destinations of the traces associated to phase shifts clustering together, as well as by the colors assigned in the inset of plot (**b**) ((**d**)) to the respective nodes of the graph. Since the chromatic number of the graph is 2, the relative phases among the oscillators of the network converge to a suboptimal (the optimal) pattern in the simulation of plot (**a**) ((**c**)).

Let us gain a deeper insight into the outcome of the proposed graph coloring strategy for the case of the suboptimal simulation, which, as demonstrated in Figure 8(a), outputs the following steady-state phase shift vector:

$$\boldsymbol{\varphi}^{(s)} = [\varphi_0^{(s)}, \varphi_1^{(s)}, \varphi_2^{(s)}, \varphi_3^{(s)}, \varphi_4^{(s)}, \varphi_5^{(s)}]^{\mathrm{T}} = [0^\circ, 118^\circ, 240^\circ, 358^\circ, 120^\circ, 242^\circ]^{\mathrm{T}}. \quad (17)$$

The resulting phase shift ordering, namely $\varphi_0^s - \varphi_1^s - \varphi_4^s - \varphi_2^s - \varphi_5^s - \varphi_3^s$, allows to establish the following vertex ranking[17]:

$$0 - 1 - 4 - 2 - 5 - 3. \quad (18)$$

[17] Since, here, oscillator i is associated to vertex i for each $i \in \{0, \ldots, N-1\}$, the oscillator sequence, corresponding to the phase shift ordering, may be indifferently referred to as oscillator ranking or vertex ranking. As will be clarified later on, this is not always the case, when perturbation actions are performed on the network to enhance its performance.

For each i-value in the set $\{1,\ldots,6\}$, the ith cycle of the proposed iterative graph coloring procedure provides the following group classification of the nodes of the undirected graph of Figure 3(a):

$$\begin{aligned}
C_1^{(1)} &= \{0,3\}, C_2^{(1)} = \{1,4\}, C_3^{(1)} = \{2,5\}, && (19)\\
C_1^{(2)} &= \{1,4\}, C_2^{(2)} = \{2,5\}, C_3^{(2)} = \{3,0\}, && (20)\\
C_1^{(3)} &= \{4,2\}, C_2^{(3)} = \{5,3\}, C_3^{(3)} = \{0\}, C_4^{(3)} = \{1\}, && (21)\\
C_1^{(4)} &= \{2,5\}, C_2^{(4)} = \{3,0\}, C_3^{(4)} = \{1,4\}, && (22)\\
C_1^{(5)} &= \{5,3\}, C_2^{(5)} = \{0\}, C_3^{(5)} = \{1,4\}, C_4^{(5)} = \{2\}, && (23)\\
C_1^{(6)} &= \{3,0\}, C_2^{(6)} = \{1,4\}, C_3^{(6)} = \{2,5\}. && (24)
\end{aligned}$$

In each of cycles 1, 2, 4, and 6, our iterative procedure assigns the vertices of the graph of Figure 3(a) a minimum number of colors, i.e., 3, which, as expected, is higher than the chromatic number of the graph itself, i.e., $n = 2$. In each of these iterations the same three pairs of vertices are grouped together, thus any number in the set $\{1,2,4,6\}$ may be assigned to q, and, as a result, any of Equations (19), (20), (22), and (24) may be taken as outcome of the graph coloring procedure. The suboptimal solution of the graph coloring problem is clearly indicated in Figure 8(a), where 3 different colors are used to mark the arcs of the 3 circular sectors, which in turn host the final destinations of the 2 traces associated to the phase shifts of the oscillator pairs $(0,3)$, $(1,4)$, and $(2,5)$, as well as in the inset of Figure 8(b), where a distinct color is used to fill each node pair in the set $\{(0,3),(1,4),(2,5)\}$.

Let us now analyse comprehensively the working principles of our graph coloring strategy for the case of the optimal simulation, which produces Figure 9(a) for the dynamical behaviour of the memristor currents in the 6 oscillators of the network over the steady-state time interval $t \in [9.9, 10]$ ms. Looking in more detail at the evolution of the memristor currents over the last part of this time interval, i.e., for $t \in [9.97, 10]$ ms, Figure 9(b) shows the common period[18] T of the 6 oscillatory waveforms, and marks the time instants t_0 and t_3, at which $i_{m,0}$ and $i_{m,3}$ cross the threshold value $I_{th} = 0.5$ ms, respectively, allowing to compute the steady-state phase shift $\varphi_3(s)$ between cells 3 and 0, as described in the figure caption. Computing also the relative phase shifts of the other 5 oscillators of the network over the common T-long time interval shown in plot (b) of Figure 9(b), the steady-state phase shift vector $\boldsymbol{\varphi}^{(s)}$ is found to be equal to

$$\boldsymbol{\varphi}^{(s)} = [\varphi_0^{(s)}, \varphi_1^{(s)}, \varphi_2^{(s)}, \varphi_3^{(s)}, \varphi_4^{(s)}, \varphi_5^{(s)}]^{\mathrm{T}} = [0^\circ, 180^\circ, 5^\circ, 195^\circ, 11^\circ, 182^\circ]^{\mathrm{T}}. \quad (25)$$

as indicated in Figure 8(c). The resulting phase shift ordering, namely $\varphi_0^{(s)} - \varphi_2^{(s)} - \varphi_4^{(s)} - \varphi_1^{(s)} - \varphi_5^{(s)} - \varphi_3^{(s)}$, allows to establish the following vertex ranking:

$$0 - 2 - 4 - 1 - 5 - 3. \quad (26)$$

For each i-value in the set $\{1,\ldots,6\}$, the ith cycle of the proposed iterative graph coloring procedure provides the following group classification of the nodes of the undirected graph of Figure 3(a):

[18] in this work the estimation of the common period T of the oscillations developing in a N-cell network, and the associated group ordering of the phase shifts of the cells $1,\ldots,N-1$ relative to the null phase of the reference cell 0 are carried out every cycle throughout the duration of any simulation.

$$\mathcal{C}_1^{(1)} = \{0,2,4\},\ \mathcal{C}_2^{(1)} = \{1,5,3\}, \quad (27)$$
$$\mathcal{C}_1^{(2)} = \{2,4,0\},\ \mathcal{C}_2^{(2)} = \{1,5,3\}, \quad (28)$$
$$\mathcal{C}_1^{(3)} = \{4,1\},\ \mathcal{C}_2^{(3)} = \{5,3\},\ \mathcal{C}_3^{(3)} = \{0,2\}, \quad (29)$$
$$\mathcal{C}_1^{(4)} = \{1,5,3\},\ \mathcal{C}_2^{(4)} = \{0,2,4\}, \quad (30)$$
$$\mathcal{C}_1^{(5)} = \{5,3,1\},\ \mathcal{C}_2^{(5)} = \{0,2,4\}, \quad (31)$$
$$\mathcal{C}_1^{(6)} = \{3,0\},\ \mathcal{C}_2^{(6)} = \{2,4\},\ \mathcal{C}_3^{(6)} = \{1,5\}. \quad (32)$$

In each of cycles 1, 2, 4, and 5, our iterative procedure assigns the vertices of the graph of Figure 3(a) a minimum number of colors, i.e., 2, which, as expected, coincides with the chromatic number of the graph itself, i.e., $n = 2$. In each of these iterations the same two triplets of vertices are grouped together, thus any number in the set $\{1,2,4,5\}$ may be assigned to q, and, as a result, any of Equations (27), (28), (30) and (31) may be taken as outcome of the graph coloring procedure. The optimal solution of the graph coloring problem is clearly indicated in Figure 8(c), where 2 different colors are used to mark the arcs of the 2 circular sectors, which in turn host the final destinations of the 3 traces associated to the phase shifts of the oscillator triplets $(0,2,4)$, and $(1,5,3)$, as well as in the inset of Figure 8(d), where a distinct color is used to fill each node triplet in the set $\{(0,2,4),(1,5,3)\}$.

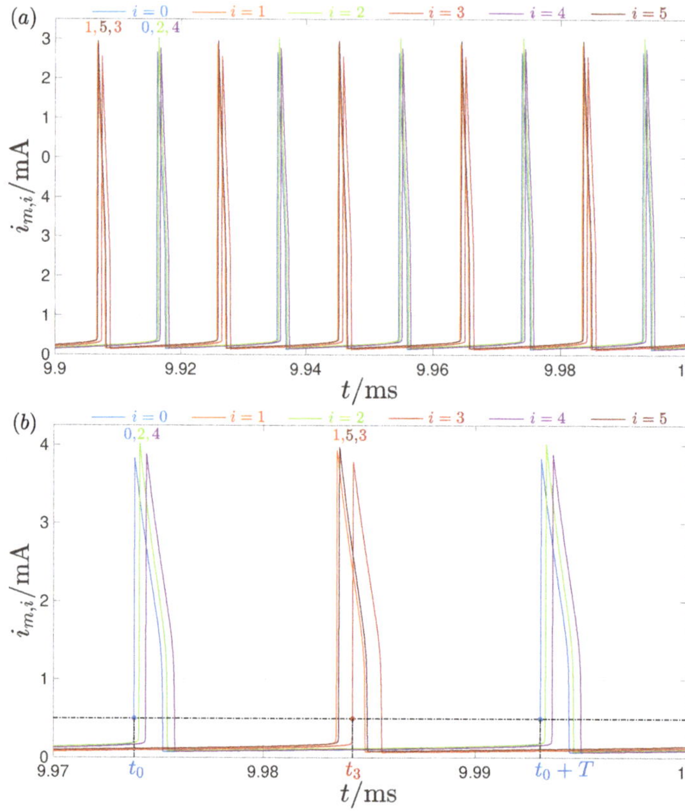

Figure 9. (**a**) Steady-state time evolution of the memristor current $i_{m,i}$ of oscillator $i \in \{0,1,2,3,4,5\}$ over the time interval $[9.9, 10]$ ms for the simulation of the balanced network of Figure 3(b) with compensation for the memristor device-to-device variability and under the optimal initialisation setting (see also Figure 8(c),(d) for more results obtained from this simulation). The differences in the peak values of the waveforms originate from the variability in the static and dynamic properties of the samples, as reproduced by our memristor model. As was already shown in relation to their relative phases in Figure 8(c), the oscillator triplets $(1,5,3)$ and $(0,2,4)$ group together, as respectively indicated over the first and second half of the first observable cycle, where the order of appearance of the nearby peaks of the memristor currents follows in turn the patterns 1-5-3 and 0-2-4. The same color coding map, as established in Figure 8(c) and reused in the inset of Figure 8(d), is adopted here to differentiate between the traces pertaining to distinct oscillators. (**b**) Zoom-in view of the time behaviour of each memristor current in the network across the time span $[9.97, 10]$ ms. The common period T of the oscillatory waveforms is found to be equal to 19.21 µs. As an example, the time instant t_0 (t_3), at which the memristor current $i_{m,0}$ ($i_{m,5}$) of oscillator 0 (3) crosses the threshold $I_{th} = 0.5\,\mathrm{mA}$ in its ascending phase over the first observable cycle, gives 9973.8457µs (9984.2351µs), allowing to compute the steady-state phase of oscillator 3 relative to the reference oscillator 0 via $\varphi_3^{(s)} = \Delta t_3 \cdot \omega_0$, with $\omega_0 = \frac{2\cdot\pi}{T}$. Performing a calculation of this kind for each of the remaining 5 oscillators results in the steady-state phase shift vector $\boldsymbol{\varphi}^{(s)}$ reported in Equation (25).

5. Control Paradigms to Resolve Local Minima-Based Impasse Conditions

In order to overcome the impasse, emerging when a memristive network converges to a stable oscillatory solution associated to a local minimum of the optimisation goal function $G(\boldsymbol{\varphi})$, it is necessary to destabilise the array through an ad-hoc perturbation. We identified two possible strategies allowing to pull the dynamical system out of a local minimum of the respective optimisation problem. In many cases, after recovering from the impasse, the network was found to converge asymptotically toward an oscillatory solution associated the global minimum of the optimisation goal function. The proposed approaches, referred to as *crossover* and *pulse destabilisation* strategies, are presented in the following two sections.

5.1. Crossover Strategy

One of the strategies, allowing the network to overcome an impasse situation, inspired from genetic algorithms [45], is based upon the interchange between the connections of two properly-selected oscillators[19] ([31,33]). This corresponds to the exchange of the associations between the two cells of the network and the corresponding vertices in the relative graph. In order to select the most appropriate pair of oscillators—let us use indices i and j to label them—for the crossover, the following two-step procedure is applied.

1. For each value of k in the set $\{0, \ldots, N-1\}$, the vertex k is removed from the original N-node graph, and the iterative vertex coloring strategy is applied to the resulting graph of $(N-1)$ nodes, using a modified version of the vertex ranking, which is tabulated beforehand, after a simulation of the oscillatory network, under a generic sub-optimal initialisation setting, attains the steady state. Specifically, the label of the vertex k, taken out of the original graph, is removed from the original vertex ranking, resulting in a new table with $N-1$ entries. For each value of k, a $N \times N$ matrix, denoted as $\mathbf{A}^{(k)}$, and obtained from the original adjacency matrix $\mathbf{A}$ by setting to 0 all the elements at row k and at column k, may still be used to define the connectivity of the respective $(N-1)$-node graph. Coloring the $N-1$ vertices of N distinct graphs, at least one of the N problems will be found to admit the best solution, allowing to categorise the $N-1$ nodes of the relative graph through the lowest number of color groups. The particular node k, which, extracted out of the original graph, allows the resulting network to identify the least number of colors according to our iterative vertex coloring procedure, may then be chosen as first vertex i to involve in the crossover[20].
2. Assigning, one at a time, any integer from the set $\{0, \ldots, i-1, i+1, \ldots, N-1\}$ to k, the iterative vertex coloring strategy is then applied to a new vertex ranking, obtained from the original table by interchanging the positions of vertices i and k. Note that the original N-node graph, with connectivity defined by the adjacency matrix $\mathbf{A}$, should be considered in each of the $N-1$ applications of the iterative vertex coloring strategy, since nothing else, except for the correspondence between oscillators and vertices, is affected in a crossover operation[21]. Solving the resulting $N-1$ vertex coloring problems, the solution, assigning the least number of colors to the N nodes of the original graph, will be determined. It may happen that, on the basis of the proposed iterative procedure, for two or more values of k, the exchange between the positions of oscillators k and i in the original vertex ranking results in a common lowest number of color groups for the N nodes of the original graph. In this case, the choice of the second oscillator j to involve in the crossover falls for the particular candidate

19 The application of a crossover to pairs of oscillators implies the necessity to endow the network with reprogrammable connections, e.g. via transistor-based switches, which, however, would add on to the integrated circuit (IC) overhead in a future hardware implementation of the network.

20 In fact, it is highly probable that this node mostly prevents the optimisation measure of Equation (16) from attaining the global minimum, which would provide as solution to the vertex coloring task the chromatic number of the original N-node graph, as desired. In case, for each of two or more values of k, the application of our vertex coloring strategy to the respective $(N-1)$-node graph, obtained by removing the vertex k from the original graph, results in a common lowest number of colors, any of these node i candidates may be finally considered for the crossover

21 The interchange between nodes i and k operated on the original vertex ranking is due to the fact that the application of a crossover between the corresponding oscillators in the network is equivalent to exchanging their associations to the respective pair of vertices in the original graph. The relative phases, inherent to the oscillators, maintain the same ordering, as established originally. As a result, the oscillator ranking remains unaltered, but the mapping from oscillator ranking to vertex ranking is subject to the earlier mentioned node interchange.

k, whose relative phase φ_k features the largest distance from the relative phase φ_i of oscillator i in the steady-state phase shift vector $\boldsymbol{\varphi}$ obtained through the simulation preceding the application of the two-step strategy.

In order to gain a deeper understanding of the proposed two-step procedure, let us apply it to the vertex ranking $0-1-4-2-5-3$, obtained at steady state from a simulation of the balanced network of Figure 3(b) (or, equivalently, of Figure 10(b), the corresponding graph of which is illustrated again in Figure 10(a) for the sake of clarity) with compensation for the memristor device-to-device variability and under the same sub-optimal initialisation setting as in the simulation illustrated in Figure 8(a),(b) (see the caption of Figure 11 for details). According to the first step of the procedure, applying our iterative vertex coloring strategy to the original vertex ranking, i.e., $0-1-4-2-5-3$, after depriving it of the kth node label, with the associated 5-node graph obtained from the original one of Figure 3(a) by breaking all the connections of the kth vertex ($k \in \{0,1,2,3,4,5\}$), the smallest number of colors, the vertices of the 5-node graph are assigned to, is 2 for each k-value in the set $\{0,1\}$, and 3 for each k-value in the set $\{2,3,4,5\}$. The choice for the i cell for the crossover may then fall either on reference oscillator 0 or on oscillator 1. Let us choose the latter cell. In line with the second step of the procedure, exchanging the positions of labels k and $i=1$ in the original vertex ranking $0-1-4-2-5-3$ ($k \in \{0,2,3,4,5\}$), the application of the iterative vertex coloring strategy to the resulting sequence, with respect to the original 6-node graph of Figure 3(a), results in a minimum number of color groups equal to 2 for each k-value in the set $\{0,2\}$, to 3 for each k-value in the set $\{3,4\}$, and to 4 for $k=5$. Now, given that, in the original steady-state phase shift ordering, $\varphi_2^{(s)} - \varphi_0^{(s)} = 122^\circ > \varphi_1^{(s)} - \varphi_0^{(s)} = 118^\circ$, oscillator with label $k=2$ is selected as cell j for the crossover. With reference to Figure 10, where plots (a) and (b) show once again the six-node ring-based graph, and the associated capacitively-coupled array of memristor oscillators, plot (c), redrawn in a different but equivalent form in plot (d), depicts the novel coupling arrangement in the network upon the interchange between the connections of oscillators 1 and 2.

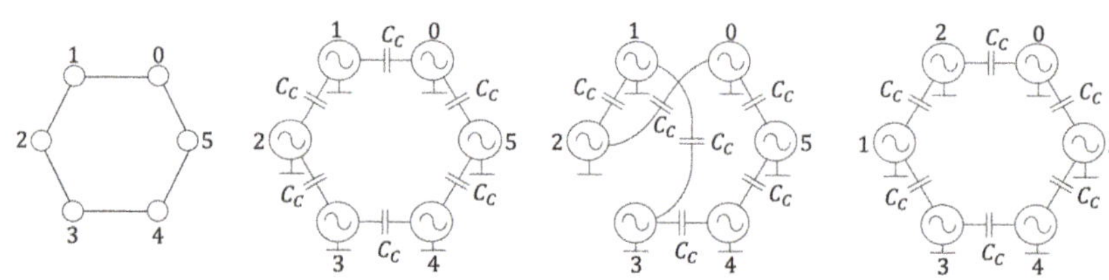

Figure 10. (**a**) Original 6-node ring-based graph. (**b**) ((**c**) or, equivalently, (**d**)) Coupling arrangement in the network associated to the graph in (**a**), before (after) a crossover between cells 1 and 2, which swaps the correspondence between these cells and the respective nodes in the graph.

The numerical results illustrated in Figure 11, where plots (a), (b), and (c) respectively show phase dynamics, time evolution of the optimisation goal function, and temporal trend of the solution of the classification task, respectively, provide evidence for the success of the circuit implementation of the crossover strategy in pulling the dynamical system out of the impasse state, allowing its asymptotic convergence to the solution of the graph coloring problem associated to the global minimum of $G(\boldsymbol{\varphi})$. These results were obtained by simulating the balanced network of Figure 10(b), with compensation for the memristor device-to-device variability, and under the same sub-optimal initialisation setting as in the simulation illustrated in Figure 8(a),(b). With reference to Figure 11, at the end of the first part of the simulation, covering the time span $t \in [0,5)$ ms,the optimisation goal function was found to sit at a local minimum value, specifically -3 (see plot (b)), and the minimum number of colors, which, on the basis of our iterative vertex coloring procedure, may be assigned to the nodes of the graph in Figure 10(a), is 3 (refer to plot (c)). At $t = 5\,\text{ms}$ the connections of oscillators 1 and 2 were interchanged, as shown in plot (c) or, equivalently, in plot (d) of Figure 10. Looking more at Figure 11, the dynamics of the relative phases of the cells resume directly after the crossover, approaching a new stable steady-state pattern, whereby $G(\boldsymbol{\varphi})$ is found to sit at its global minimum level, particularly -6 (see plot (b))), and the network is able to identify the chromatic number of the graph in Figure 10(a), as determined through the proposed iterative vertex coloring procedure (refer to plot (c)).

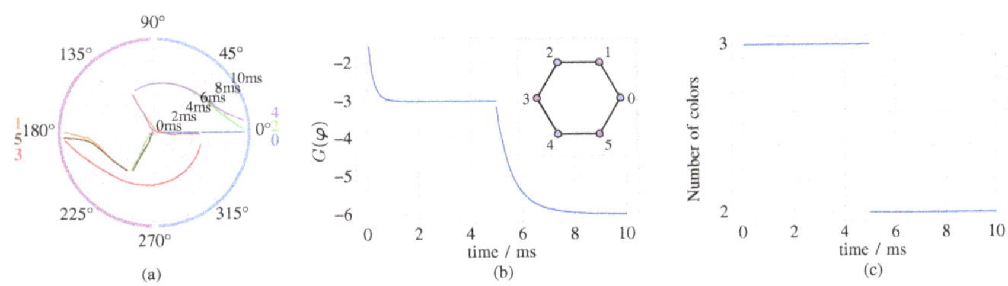

Figure 11. (**a**) Time evolution of the relative phases of the oscillators of the balanced network of Figure 3(b), with compensation for the memristor device-to-device variability, and under the same sub-optimal initialisation setting as in the simulation illustrated in Figure 8(a),(b), for the case where the connections of oscillators 1 and 2 are interchanged at $t = 5\,\text{ms}$. Right before the application of the crossover to the two cells, the network is found to sit on a stable oscillatory solution associated to a local minimum of the optimisation goal function (see also the three phase clusters emerging in plot (**a**) right before the crossover procedure). Despite the phase shift vector, measured much earlier than it was done in the simulation of Figure 8(a),(b), was found to be slightly different from the one reported in Equation (17), specifically $\boldsymbol{\varphi}^{(s)} = [0°, 118°, 238°, 359°, 119°, 240°]^{\mathrm{T}}$, the resulting vertex ordering remains defined by Equation (18). (**b**) Evolution of the optimisation goal function over time. (**c**) Minimum number of color groups assigned through the iterative vertex coloring procedure of Section 4 to the nodes of the graph in Figure 3(a) over time (the procedure is applied once every T-long cycle, with the common period T of the oscillatory waveforms of the currents through the memristors, measured over the time interval $[4.965, 4.984]\,\text{ms}$, found to be equal to $19.24\,\mu\text{s}$). After the crossover these nodes are classified into 2 groups (**c**). The interchange between the couplings of oscillators 1 and 2 was thus found to resolve the impasse, allowing the memristive array to approach the optimal solution associated to the global minimum of $G(\boldsymbol{\varphi})$, and to identify the chromatic number $n = 2$ of the associated graph (see also the two phase clusters emerging in plot (**a**) at the end of the simulation).

As anticipated earlier, the circuit implementation of the crossover strategy crucially requires the availability of reprogrammable connections among the oscillators of the network. This inevitably increases the area overhead of the hardware platform. Furthermore, the circuitry necessary to endow the network with reprogrammable connectivity, may introduce some mismatch between the capacitive loads of the oscillators, which may represent an obstacle toward the convergence of the phase dynamics of the memristive array toward the pattern corresponding to the global minimum of the optimisation problem. For this reason, another strategy for pulling the dynamical system out of a local minimum, which is more amenable to circuit implementation, is presented in the next section.

5.2. Pulse Destabilisation Strategy

Inspired from *simulated annealing algorithms* [46] as well from our latest approach—referred to as *Kick-Fly-Catch* paradigm—for humanoid robot motion control [47,48], the proposed strategy ([31,33]) is based upon supplying energy to the system, while it sits in an impasse state. Particularly, a pulse is applied to an ad-hoc oscillator by offsetting the value V_S of its DC voltage source by an appropriate amount ΔV_S for a given time interval of length T_p. Consequently, the relative phase of the oscillator, sitting at the steady-state level $\varphi^{(s)}$ previous to the stimulation, experiences a sudden shift by $\Delta\varphi$, which is found to depend upon amplitude ΔV_S and length T_p of the voltage pulse. Taking inspiration from the research study presented in [49], extensive numerical simulations revealed that, for a given pulse width[22] T_p, there is an approximately-linear relation between the pulse height ΔV_S and the sudden shift $\Delta\varphi$, that the relative phase of the oscillator experiences upon destabilisation, i.e.,

$$\Delta V_S \approx \frac{V_0 \cdot \Delta\varphi}{180^\circ}, \tag{33}$$

where V_0 was numerically set to -0.23 V for the networks analysed in this research study[23]. Provided the right choice is made regarding the cell to perturb, and the proper amplitude and width are assigned to the voltage pulse stimulus, the resulting resumption of the phase dynamics of the oscillators should ideally allow the optimisation goal function to move out of the local minimum, converging asymptotically toward its global minimum. A two-step procedure, presented below, is proposed here to determine the most suitable cell to perturb and the most appropriate pulse amplitude.

1. The most suitable oscillator $i \in \{0, \ldots, N-1\}$ to target in the pulse destabilisation action is determined in the same way as was done for the selection of the cell $i \in \{0, \ldots, N-1\}$ to involve in the crossover process (see the first step in the procedure aimed to choose the right cell pair (i, j) to involve in the coupling interchange strategy).
2. The second step is aimed to determine the appropriate shift $\Delta\varphi$ to be added to the steady-state relative phase $\varphi_i^{(s)}$ of oscillator i for pulling the network out of the local minimum state, facilitating its convergence to an oscillatory solution, which would ideally correspond to the least number of color groups for the N vertices of the associated graph. To accomplish this task, for each value of k within the set $\{1, \ldots, M-1\}$, with M a predefined positive integer, the offset $\Delta\varphi_k = k \cdot \frac{360^\circ}{M}$ is added to the phase shift $\varphi_i^{(s)}$ of cell i in the steady-state relative phase shift vector $\boldsymbol{\varphi}^{(s)}$ recorded before the application of the pulse destabilisation process, and the iterative graph coloring procedure is applied to the resulting vertex ranking for the original graph. The choice for the most appropriate offset $\Delta\varphi$, within the specified set of k-dependent uniformly-spaced values, goes for the $\Delta\varphi_k$-candidate, which, according to our graph coloring strategy, allows the network to classify the nodes of the associated graph in the lowest number of color groups. If, for two or more k-values, the application of the iterative graph coloring procedure to the vertex ranking, resulting from the phase shift ordering, obtained by adding up the relevant offset $\Delta\varphi_k$ to the phase shift $\varphi_i^{(s)}$ of oscillator i in the steady-state relative phase shift vector $\varphi^{(s)}$, leads to the identification of the same lowest number of colors, the selection goes for the $\Delta\varphi_k$-candidate featuring the largest modulus. Finally, the pulse amplitude of the T_p-long stimulus to be applied to oscillator i is obtained from Equation (33).

In order to gain more insights into the mechanisms underlying this two-step procedure, let us apply it to the vertex ranking $0-1-4-2-5-3$, obtained at steady state from a simulation of the balanced network of Figure 3(b) with compensation for the memristor device-to-device variability and under the same sub-optimal initialisation setting as in the simulation illustrated in Figure 8(a),(b) (see the caption of Figure 12 for details). Given that the first step of the procedure is identical to the first step of the algorithm allowing to

[22] In this work T_p was set to twice the common graph-dependent period T of the oscillatory waveforms of the capacitor voltages and of the memristor currents in the network before the application of the pulse destabilisation paradigm.

[23] We acknowledge, however, that the most suitable formula, expressing the relationship between the amplitude ΔV_S of a destabilising pulse of fixed width T_p and the resulting sudden shift $\Delta\varphi$ in the phase of the perturbed oscillator, may depend upon network properties and parameters. A deeper study, aimed to optimise the shape of the destabilisation stimulus, will be carried out in the future.

select the optimal cell pair (i, j) to involve in the crossover strategy, retrieving the results presented in the previous section, either oscillator from the label set $i \in \{0, 1\}$ may be chosen as target of the pulse destabilisation action. Let us go for the latter one. Following the guidelines established for the second step of the procedure, setting M to 4, for each k-value in the set $\{1, 2, 3\}$, colors are assigned to the nodes of the original graph of Figure 3(a) by applying the proposed iterative methodology to the kth vertex ranking variant derived from the phase shift vector $\boldsymbol{\varphi}^{(s)} = [0°, 118°, 238°, 359°, 119°, 240°]^{\text{T}}$ recorded right before the time instant[24] $t = 5$ ms, at which the pulse perturbation action is commenced by adding up the kth value of $\Delta\varphi_k$ in the set $\{90°, 180°, 270°\}$ to the relative phase $\varphi_1^{(s)}$ of oscillator 1. Analysing the kth vertex ranking within the set $\{0-4-1-2-5-3, 0-4-2-5-1-3, 0-1-4-2-5-3\}$ ($k \in \{1, 2, 3\}$), according to our node coloring paradigm the network identifies a minimum number of color groups equal to the kth number in the set $\{3, 2, 3\}$. Setting k to 2 in the formula for $\Delta\varphi_k$, from Equation (33) the amplitude ΔV_S of the pulse applied to oscillator 1 at $t = 5$ ms for a $T_p = 38.48$ µs-long time interval is set to -0.23 V (see the caption of Figure 12 for details).

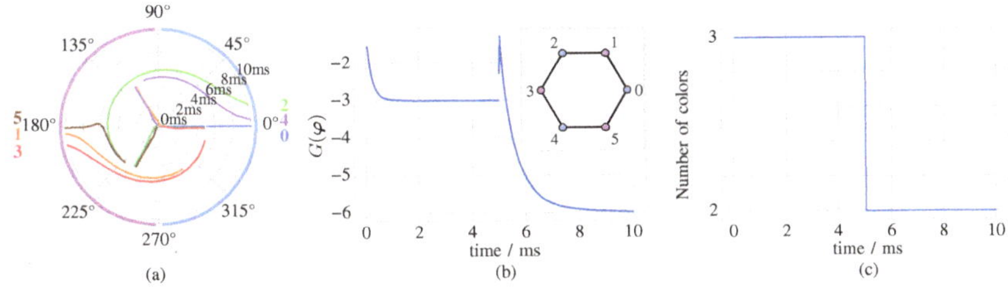

Figure 12. (**a**) Phase dynamics of the balanced network of Figure 3(b), with compensation for the memristor device-to-device variability, and under the same sub-optimal initialisation setting as in the simulation illustrated in Figure 8(a)-(b), for the case where the voltage V_S of the DC source in oscillator 1 is offset by $\Delta V_S = -0.23$ V from the time instant $t = 5$ ms for a temporal window of duration $T_p = 38.48$ µs (the common period T of the oscillatory waveforms of the currents through the memristors, measured over the time interval $[4.965, 4.984]$ ms, was found to be equal to 19.24 µs). As discussed earlier, right before the application of the pulse to cell 1, which triggers a sudden shift in its relative phase by approximately 180°, the network is found to sit on a stable oscillatory solution associated to a local minimum of the optimisation goal function (see also the three phase clusters emerging in plot (**a**) right before the pulse destabilisation action). Despite the phase shift vector, measured much earlier than it was done in the simulation of Figure 8(a),(b), was found to be slightly different from the one reported in Equation (17), specifically $\boldsymbol{\varphi}^{(s)} = [0°, 118°, 238°, 359°, 119°, 240°]^{\text{T}}$, the resulting vertex ordering remains defined by Equation (18). (**b**) Time evolution of the optimisation goal function. (**c**) Minimum number of color groups assigned through the iterative vertex coloring procedure of Section 4 to the nodes of the graph in Figure 3(a) over time (the procedure is applied once every T-long cycle). After the pulse destabilisation 2 colors are assigned to these nodes. The pulse-based perturbation of oscillator 1 was thus found to resolve the impasse, allowing the memristive array to approach the optimal solution associated to the global minimum of $G(\boldsymbol{\varphi})$, and to identify the chromatic number $n = 2$ of the associated graph (see also the two phase clusters emerging in plot (**a**) at the end of the simulation). Despite, at the time instant $t = 5$ ms, when the pulse perturbation commences, $G(\boldsymbol{\varphi})$ undergoes a sudden increase from the local minimum value of -3, it descends steeply straight away, decreasing monotonically toward the global minimum value of -6 thereafter.

The numerical results illustrated in Figure 12, where plots (a), (b), and (c) respectively show phase dynamics, time evolution of the optimisation goal function, and temporal trend

[24] In order to present a fair comparison between the beneficial effects of the crossover and pulse destabilisation control paradigms, we ensured that the simulations in Figures 11 and 12 provided identical results for $t \in [0, 5)$ ms by choosing the same initialisation setting, and assigning a common random set of α-values to the memristors.

of the solution of the classification task, respectively, provide evidence for the success of the circuit implementation of the pulse destabilisation strategy in pulling the dynamical system out of the impasse state, allowing its asymptotic convergence to the solution of the graph coloring problem associated to the global minimum of $G(\boldsymbol{\varphi})$. These results were obtained by simulating the balanced network of Figure 10(b), with compensation for the memristor device-to-device variability, and under the same sub-optimal initialisation setting as in the simulation illustrated in Figure 8(a),(b). With reference to Figure 12, at the end of this first part of the simulation, covering the time interval expressed as $t \in [0, 5)$ ms, the optimisation goal function was found to sit at a local minimum value, specifically -3 (see plot (b)), and the minimum number of colors, which, on the basis of our iterative vertex coloring procedure, may be assigned to the nodes of the graph in Figure 3(a), is 3 (refer to plot (c)). From the time instant $t = 5$ ms and for a $T_p = 38.48$ µs-long time interval, the offset $\Delta V_S = -0.23$ V is added up to the nominal value V_S of the DC voltage source in oscillator 1. As may be evinced by inspecting Figure 11, the relative phase of cell 1 undergoes a sudden shift by approximately 180°, and, thereafter, the phase dynamics of the network evolve toward a new stable steady-state pattern, whereby $G(\boldsymbol{\varphi})$ is found to sit at its global minimum level, particularly -6 (see plot (b))), and the network is able to identify the chromatic number of the graph in Figure 10(a), as determined through the proposed iterative vertex coloring procedure (refer to plot (c)).

5.3. Discussion

Since a single crossover or pulse destabilisation manoeuvre may be unable to pull a more complex memristive network out of an impasse situation, or to reach the solution associated to the global minimum of the optimisation goal function, in case the dynamical system receives enough energy to move out of the solution associated to a local minimum of the optimisation goal function[25], a good approach to address this issue would be to reapply either of the two proposed strategies periodically, interchanging the connections of two distinct appropriate oscillators or applying a suitable destabilising pulse to a different ad-hoc oscillator at regular time intervals[26].

Let us provide a proof of principle of the proposed approach, focusing on the pulse destabilisation control strategy. Similar results were obtained through the periodic application of the crossover paradigm. With reference to Figure 13, plot (a) shows the time evolution of the phase shifts of the oscillators with labels running from 1 to 24 with respect to the reference cell 0 in a capacitively-coupled network of NbO_x memristor oscillators implementing a $N = 25$-node undirected graph known as queen5_5 [50]—see also the inset in plot (b)—in case compensation for the mismatch between the capacitive loads of the oscillators and for the memristor device-to-device variability is set in place, and for the case where, periodically, on the basis of the two-step strategy introduced in Section 5.2, a distinct oscillator is perturbed by means of a T_P-long pulse of appropriate height ΔV_S. As shown in plot (b) the optimisation goal function evolves progressively through various local minima before attaining the global minimum value, which is associated to the identification of the chromatic number of the queen5_5 graph, i.e., $n = 5$, as demonstrated in plot (c). After a 38 ms-long transient time interval, the network exhibits a robust oscillatory solution, given that the subsequent application of destabilisation pulse stimuli to the oscillators does no longer affect the phase dynamics.

25 In some cases, after overcoming the impasse situation, the dynamical system could approach a new oscillatory solution associated to another local minimum of $G(\boldsymbol{\varphi})$.

26 The time separation T_{int} between consecutive applications of the crossover or pulse destabilisation strategy is set to 2 ms in the simulations discussed in this section. Furthermore, in this work the estimation of the common period T of the oscillations appearing in a N-cell network, and the associated group ordering of the phase shifts of the cells $1, \ldots, N-1$ relative to the null phase of the reference cell 0 are carried out every cycle throughout the duration of any simulation. Moreover, in the first (latter) control strategy, the application of a pulse to the same oscillator (a crossover involving either oscillator from the same pair) is not allowed until at least 5 iterations of the control strategy have elapsed first. As a result, each pulse destabilisation (crossover) manoeuvre targets a different oscillator (involves a different pair of oscillators).

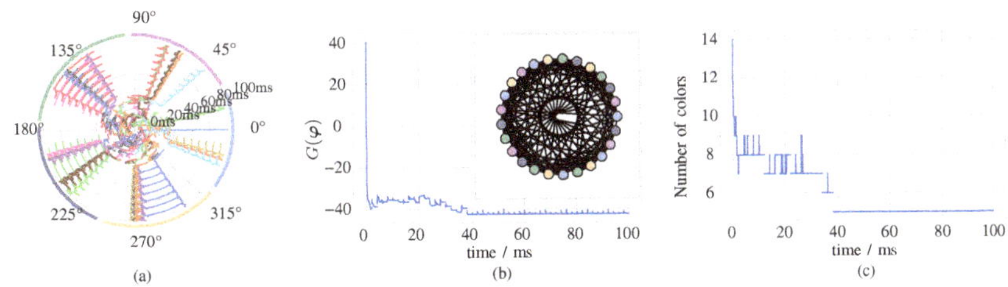

Figure 13. Evidence for the capability of a 25-cell network, preliminarily compensated for the unbalance in the number of connections per oscillator, and for the inter-device variability inherent to memristors, to converge toward the optimal solution of the vertex coloring problem for the graph queen5_5 [50]. The cyclic application of a pulse stimulus of fixed length and appropriate amplitude to an ad-hoc oscillator of the network guides the phase dynamics toward the global minimum solution. (**a**) Phase diagram visualising the time evolution of the phase of each oscillator $j \in \{1, \ldots, 24\}$ relative to the phase of the reference oscillator 0. (**b**) Time waveform of the optimisation goal function $G(\boldsymbol{\varphi})$. (**c**) Progression of the outcome of the iterative vertex coloring procedure of Section 4 over time. Throughout the second half of the simulation the minimum number of color groups, assigned to the vertices of the graph queen5_5, visualised in the inset of plot (**b**), is fixed to the chromatic number $n = 5$ of the graph itself, despite the network is subject to further pulse-based perturbations.

Table 3 shows a comparison between the solutions of the node coloring task for a number of graphs [50] pertaining to the 2nd algorithm implementation challenge for NP-hard problems in Discrete Mathematics and Theoretical Computer Science (DIMACS) [51], derived from the application of various techniques, namely an algorithmic approach known as Brélaz heuristic [52], methods based upon the analysis of the phase dynamics of capacitively-coupled arrays of locally-active memristor oscillators without a control strategy for bypassing local minima solutions, as respectively presented in [42], and in Section 4, and, finally, paradigms including either a reconfigurability of the oscillators' couplings, as discussed in Section 5.1, or a perturbation of the memristive array, as presented in Section 5.2, to enable the dynamical system to exit an impasse state, and to resume the calculations of the problem solution, thereafter.

The results obtained through the analysis of the phase dynamics in memristive oscillatory networks—refer to the approach employed in [42] and to our iterative strategy from Section 4, where no technique to bypass local minima solutions is set in place, are comparable to the corresponding ones of the Brélaz heuristic algorithm. As may be evinced from Table 3, the graph coloring paradigm implemented through capacitively-coupled memristive oscillators in [42] classifies the vertices of all investigated graphs, with the exception of the one called queen6_6 [50], into a number of color groups equal to or lower than the number of colors assigned to the nodes of the corresponding graphs through the proposed iterative strategy from Section 4. However, it should be pointed out that, while the nominal parameter setting in cell and coupling circuits is unaltered in the numerical investigations of the networks of all the graphs in Table 3, it is unclear whether the same values were assigned to the physical attributes of the components of the array for the simulations of the corresponding systems in [42]. Furthermore, while the mathematical characterisation of our NbO_x resistance switching memory is rooted on strong physics foundations, and is endowed with device-to-device variability control, the simplistic, model adopted to characterise the locally active VO_2 memristor in [42], has no physics basis, assuming that the two-terminal element may feature at any given time one of two conductance values, denoting the metallic and insulating state, respectively, depending upon the voltage falling across it, and does not account for the inherent spread in the device static and dynamic properties from sample to sample. The application of our iterative graph coloring strategy to the vertex orderings derived from the simulations of the networks from Table 3, for the case where the tendency of the oscillators' phase shifts to approach local minima solutions is counterbalanced through the implementation of either the crossover or the pulse destabilisation control paradigms, leads to an evident performance improvement. With reference to each of the graphs—namely mycie15, queen5_5, queen6_6, queen7_7, and queen8_8—whereby, according to the iterative strategy from Section 4, our memristive network is unable to identify the chromatic number n on its own, the periodic application of either of the two control paradigms from Sections 5.1 and 5.2 allows the lowest number of colors assigned to the N vertices to decrease, and, in most cases, the final phase pattern of the memristive oscillatory array allows to determine the global minimum solution. As an example, which also reveals how further studies are necessary to improve the performance of our memristive networks in coloring the vertices of complex graphs, Figure 14(a) shows the phase dynamics of a balanced network implementing the queen6_6 graph [50], for the case where the mismatch in the number of couplings per oscillator and the memristor device-to-device variability are respectively compensated via the methodologies described in Sections 3.2 and 3.3, and a periodic application of the crossover control paradigm of Section 5.1, involving a distinct pair of oscillators from cycle to cycle, is set in place. In this case, the network keeps in a transient phase throughout the simulation. As may be evinced by inspecting the time evolution of the optimisation goal function $G(\boldsymbol{\varphi})$, shown in plot (b), and the evolution of the outcome of our iterative vertex coloring strategy over time, illustrated in plot (c), the dynamical system does not exhibit a monotonic decrease toward the global minimum solution, escaping the best solution, computed around 50 ms, to approach higher local minima thereafter. With regard to our intention to enhance the local minima bypass paradigms further, the analysis of the potentially-beneficial impact of the memristor thermal noise source on the capability of the dynamical system to descend monotonically toward the global minimum solution is one of the future research activities in our agenda.

Remark 4. *The first priority in our research agenda is to realize a hardware prototype able to solve various graph coloring problems of small/medium size on the basis of the oscillators' phase dynamics. In order to allow a hardware implementation of the proposed memristive computing engine to solve a number of different graph coloring problems, the connections between the processing units need to be adjustable on a case by case basis, which calls for the use of a coupling arrangement typical of a Hopfield neural network, with the introduction of a transistor switch, controllable via some ad hoc multiplexer, in series with each capacitor* C_C. *While hardware architecture considerations for large networks are still quite premature, we believe that scaling up the size of the computing engine would require the use of an array-like structure, as typically used for memories, to implement the programmable coupling circuitry. Except for the memristors, which could be arranged in crossbar configuration across the metal layers, the rest of the circuitry, necessary to implement the oscillatory cells, would be laid out on the CMOS substrate. In later generations of the proposed hardware, also back-end of line (BEOL) transistors can be envisioned so as to further increase the area efficiency of the computing platform. From a problem-solving perspective, scaling up the network size to tackle problems of bigger dimension, envisaging, in general, a larger number of connections between the vertices of the associated graphs, shall result in an inevitable increase in the number of local minima for the optimization goal function, which complicates the operation of the control circuitry, as it tries to guide the oscillators' phases toward the optimal grouping at steady state. This is a general problem for all state-of-the-art software algorithms and hardware platforms, which aim to minimize non-convex optimization goal functions. In order to solve graph coloring problems of higher complexity, some fine tuning of the control strategies, proposed in this manuscript, as inspired by the most efficient NP-hard optimization problem solvers, available today, is expected to be necessary.*

Table 3. Comparison between the solutions of the vertex coloring problem for various graphs [50] for the 2nd algorithm implementation challenge for NP-hard problems in DIMACS [51], obtained through the application of a specific algorithm, known as Brélaz heuristic [52], by means of methods exploiting the phase dynamics of capacitively-coupled memristive networks without a control strategy for bypassing local minima solutions, namely the technique proposed in [42], and the iterative node coloring procedure, presented in Section 4, and via the iterative node coloring procedure augmented with strategies, based upon crossover and pulse destabilisation, presented in Sections 5.1 and 5.2, respectively, and aimed to overcome local minima solutions [31]. The results tabulated in the last three columns were computed through the analysis of 100 ms long numerical simulations.

			Minimum Number of Color Groups for the Classification of the Vertices of the Associated Group				
graph	vertices	n	Brélaz algorithm	[42]	iterative strategy	iterative strategy and crossover control	iterative strategy and pulse destabilisation control
myciel3	11	4	4	4	4	4	4
myciel4	20	5	5	5	5	5	5
myciel5	47	6	6	6	7	6	6
queen5_5	25	5	7	6	7	5	5
queen6_6	36	7	10	12	11	8	8
queen7_7	49	7	12	12	14	10	10
queen8_8	64	9	15	14	15	13	13

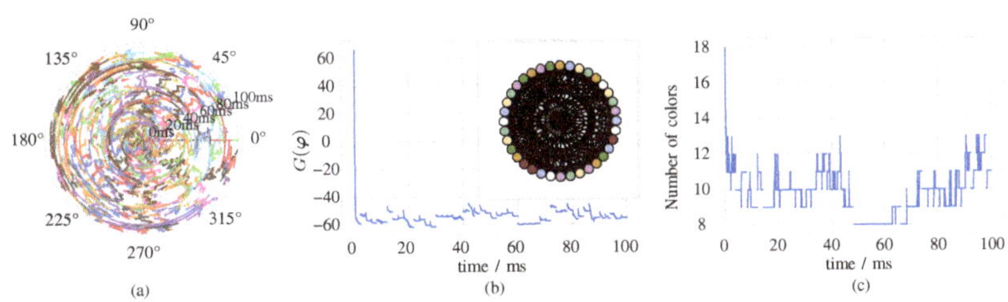

Figure 14. (**a**) Phase dynamics of the network associated to the graph queen6_6 [50], after its preliminary compensation for the unbalance in the number of connections per oscillator, and for the inter-device variability inherent to memristors, upon the periodic interchange between the couplings of two specific oscillators. In this case the phase dynamics of the network keep in a transient state throughout the 100 ms-long simulation. (**b**) Evolution of the optimisation goal function $G(\varphi)$ over time. (**c**) Minimum number of colors, assigned to the nodes of the graph queen6_6 through the iterative vertex coloring procedure of Section 4, applied once every cycle, versus time. Half way through the simulation the nodes of the graph, illustrated in the inset of plot (**b**), are classified into 8 color groups, one more than the correct number (refer to Table 3), but this solution proves to be unstable, when the network, thereafter, is subject to further crossover-based perturbations.

6. Conclusions

The local activity [26] of NbO_x memristors ([27,28]) allows the emulation of neuronal dynamics ([9,11]), the implementation of bio-inspired signal processing paradigms [53], and the reproduction of complex phenomena [30] emerging in systems from cellular biology [44]. This manuscript serves as a pedagogical tutorial to the operating principles of a cellular nonlinear network of oscillators, coupled through linear capacitors, and employing one locally-active memristor [43] each, recently introduced in [31] to solve a non-deterministic polynomial (NP)-hard combinatorial optimization problem, known as vertex coloring. While, due to page limitation, only a compact description of the signal processing paradigm, implemented by the proposed Memristor Oscillatory Network, was reported in [31], this tutorial reports all the details of the mechanisms underlying its *modus operandi*. Importantly, control methods [33] to compensate for the inherent variability of memristor devices, to counteract the imbalance between the load capacitances of the oscillators, as well as, most importantly, to prevent the bio-inspired network to attain a sub-optimal steady state, are developed and implemented in circuit form. The Memristor Oscillatory Network, endowed with the proposed control circuitry, is found to outperform state-of-the-art software and hardware competitor alternatives, identifying, for each graph from a wide selection, the lowest number of color groups for the respective vertices. As a more general conclusion, the potential of all locally-active devices, including niobium ([28,43,54]) or vanadium dioxide [55] threshold switches, and ovonic threshold switches [56,57], is expected to be subject to a thorough exploration, in the years to come, for a possible exploitation of their small-signal amplification capability for electronics applications, e.g. to build nano-oscillators with tuneable frequency ([58]), to solve NP-hard combinatorial optimization problems, as discussed in this manuscript, for reproducing complex biological phenomena [44], for exploring new forms of computing via pattern formation dynamics [59], or for designing bio-plausible neuromorphic circuits [10].

Author Contributions: All authors have read and agreed to the published version of the manuscript.

Funding: This research received no external funding.

Data Availability Statement: Not applicable.

Conflicts of Interest: The authors declare no conflict of interest.

References

1. Chua, L.O. Memristor: The missing circuit element. *IEEE Trans. Circuit Theory* **1971**, *18*, 507–519. [CrossRef]
2. Chua, L.O.; Kang, S. Memristive devices and systems. *Proc. IEEE* **1976**, *64*, 209–223. [CrossRef]
3. Strukov, D.B.; Snider, G.S.; Stewart, D.R.; Williams, R.S. The missing memristor found. *Nature* **2008**, *453*, 80–83. [CrossRef] [PubMed]
4. Prodomakis, T.; Toumazou, C.; Chua, L.O. Two Centuries of Memristors. *Nat. Mater.* **2012**, *11*, 478–481. [CrossRef] [PubMed]
5. Chua, L.O. If It's Pinched, It's a Memristor. *Semicond. Sci. Technol.* **2014**, *29*, 104001. [CrossRef]
6. Ielmini, D.; Waser, R. *Resistive Switching: From Fundamentals of Nanoionic Redox Processes to Memristive Device Applications*, 1st ed.; Wiley-VCH: Weinheim, Germany, 2016; ISBN-13: 978-3527334179.
7. Mikolajick, T.; Salinga, M.; Kund, M.; Kever, T. Nonvolatile Memory Concepts Based on Resistive Switching in Inorganic Materials. *Adv. Eng. Mater.* **2009**, *11*, 235–240. [CrossRef]
8. Indiveri, G.; Linares-Barranco, B.; Legenstein, R.; Deligeorgis, G.; Prodromakis, T. Integration of nanoscale memristor synapses in neuromorphic computing architectures. *Nanotechnology* **2013**, *24*, 384010. [CrossRef]
9. Pickett, M.D.; Medeiros-Ribeiro, G.; Williams, R.S. A scalable neuristor built with Mott memristors. *Nat. Mater.* **2013**, *12*, 114–117. [CrossRef]
10. Yi, W.; Tsang, K.K.; Lam, S.K.; Bai, X.; Crowell, J.A.; Flores, E.A. Biological plausibility and stochasticity in scalable VO_2 active memristor neurons. *Nat. Commun.* **2018**, *9*, 1–10. [CrossRef]
11. Kang, S.M.; Choi, D.; Eshraghian, J.K.; Zhou, P.; Kim, J.; Kong, B.S.; Zhu, X.; Demirkol, A.S.; Ascoli, A.; Tetzlaff, R.; Lu, W.D.; Chua, L.O. How to Build a Memristive Integrate-and-Fire Model for Spiking Neuronal Signal Generation. *IEEE Trans. Circuits Syst. I Regul. Pap.* **2021**, *68*, 4837–4850. [CrossRef]
12. Tetzlaff, R.; Ascoli, A.; Messaris, I.; Chua, L.O. Theoretical Foundations of Memristor Cellular Nonlinear Networks: Memcomputing with Bistable-like Memristors. *IEEE Trans. Circuits Syst. I Regul. Pap.* **2020**, *67*, 502–515. [CrossRef]
13. Ascoli, A.; Messaris, I.; Tetzlaff, R.; Chua, L.O. Theoretical Foundations of Memristor Cellular Nonlinear Networks: Stability Analysis with Dynamic Memristors. *IEEE Trans. Circuits Syst. I Regul. Pap.* **2020**, *67*, 1389–1401. [CrossRef]
14. Ascoli, A.; Tetzlaff, R.; Kang, S.M.; Chua, L.O. Theoretical Foundations of Memristor Cellular Nonlinear Networks: A DRM_2-based Method to Design Memcomputers with Dynamic Memristors. *IEEE Trans. Circuits Syst. I Regul. Pap.* **2020**, *67*, 2753–2766. [CrossRef]
15. Chua, L.O. (Ed.) *CNN: A Paradigm for Complexity*; World Scientific Series on Nonlinear Science: Singapore, 1998; ISBN 978-9810234836.
16. Xia, Q.; Yang, J.J. Memristive crossbar arrays for brain-inspired computing. *Nat. Mater.* **2019**, *18*, 309–323. [CrossRef] [PubMed]
17. Ventra, M.D.; Traversa, F.L. Perspective: Memcomputing: Leveraging memory and physics to compute efficiently. *J. Appl. Phys.* **2018**, *123*, 180901. [CrossRef]
18. Talati, N.; Gupta, S.; Mane, P.; Kvatinsky, S. Logic Design within Memristive Memories Using Memristor Aided loGIC (MAGIC). *IEEE Trans. Nanotechnol.* **2016**, *15*, 635–650. [CrossRef]
19. Ali, A.H.; Hur, R.B.; Wald, N.; Ronen, R.; Kvatinsky, S. Not in Name Alone: A Memristive Memory Processing Unit for Real In-Memory Processing. *IEEE Micro* **2018**, *38*, 13–21.
20. Ielmini, D.; Wong, H.-S.P. In-memory computing with resistive switching devices. *Nat. Electron.* **2018**, *1*, 333–343. [CrossRef]
21. Tzouvadaki, I.; Jolly, P.; Lu, X.; Ingebrandt, S.; de Micheli, G.; Estrela, P.; Carrara, S. Label-Free Ultrasensitive Memristive Aptasensor. *Nanoletters* **2016**, *16*, 4472–4476. [CrossRef]
22. Ibarlucea, B.; Akbar, T.F.; Kim, K.; Rim, T.; Baek, C.-K.; Ascoli, A.; Tetzlaff, R.; Baraban, L.; Cuniberti, G. Ultrasensitive Detection of Ebola Matrix Protein in a memristor mode. *NanoResearch* **2018**, *11*, 1057–1068. [CrossRef]

23. Wang, Z.; Joshi, S.; Savel'ev, S.; Song, W.; Midya, R.; Li, Y.; Rao, M.; Yan, P.; Asapu, S.; Zhuo, Y.; et al. Fully memristive neural networks for pattern classification with unsupervised learning. *Nat. Electron.* **2018**, *1*, 137–145. [CrossRef]
24. Sebastian, A.; Tuma, T.; Papandreou, N.; Gallo, M.L.; Kull, L.; Parnell, T.; Eleftheriou, E. Temporal correlation detection using computational phase-change memory. *Nat. Commun.* **2017**, *8*, 1115. [CrossRef] [PubMed]
25. Sheng, X.; Graves, C.E.; Kumar, S.; Li, X.; Buchanan, B.; Zheng, L.; Lam, S.; Li, C.; Strachan, J.P. Low-Conductance and Multilevel CMOS-Integrated Nanoscale Oxide Memristors. *Adv. Electron. Mater.* **2019**, *5*, 1800876. [CrossRef]
26. Chua, L.O. Local activity is the origin of complexity. *Int. J. Bifurc. Chaos* **2005**, *15*, 3435–3456. [CrossRef]
27. Pickett, M.D.; Williams, R.S. Sub-100 fJ and sub-nanosecond thermally driven threshold switching in niobium oxide crosspoint nanodevices. *Nanotechnology* **2012**, *23*, 215202. [CrossRef]
28. Ascoli, A.; Slesazeck, S.; Mähne, H.; Tetzlaff, R.; Mikolajick, T. Nonlinear dynamics of a locally-active memristor. *IEEE Trans. Circuits Syst. I (TCAS–I) Regul. Pap.* **2015**, *62*, 1165–1174. [CrossRef]
29. Cai, F.; Kumar, S.; Vaerenbergh, T.V.; Sheng, X.; Liu, R.; Li, C.; Liu, Z.; Foltin, M.; Yu, S.; Xia, Q.; Yang, J.J.; Beausoleil, R.; Lu, W.D.; Strachan, J.P. Power-efficient combinatorial optimization using intrinsic noise in memristor Hopfield neural networks. *Nat. Electron.* **2020**, *3*, 409–418. [CrossRef]
30. Weiher, M.; Herzig, M.; Tetzlaff, R.; Ascoli, A.; Mikolajick, T.; Slesazeck, S. Pattern formation with local active S-type NbO_x memristors. *IEEE Trans. Circuits Syst. I Regul. Pap.* **2019**, *66*, 2627–2638. [CrossRef]
31. Weiher, M.; Herzig, M.; Tetzlaff, R.; Ascoli, A.; Slesazeck, S.; Mikolajick, T. Improved Vertex Coloring With NbO_x Memristor-Based Oscillatory Networks. *IEEE Trans. Circuits Syst. I* **2021**, *68*, 2082–2095. [CrossRef]
32. Vázquez, A.R.; Fernández-Berni, J.; Leñero-Bardallo, J.A.; Vornicu, I.; Carmona-Galán, R. CMOS Vision Sensors: Embedding Computer Vision at Imaging Front-Ends. *IEEE Circuits Syst. Mag.* **2018**, *18*, 90–107. [CrossRef]
33. Ascoli, A.; Weiher, M.; Herzig, M.; Tetzlaff, R.; Slesazeck, S.; Mikolajick, T. Control Strategies to Optimize Graph Coloring via M-CNNs with Locally-Active NbO_x Memristors. In Proceedings of the International Conference on Modern Circuits and Systems Technologies (MOCAST) on Electronics and Communications, Thessaloniki, Greece, 5–7 July 2021.
34. Slesazeck, S.; Mähne, H.; Wylezich, H.; Wachowiak, A.; Radhakrishnan, J.; Ascoli, A.; Tetzlaff, R.; Mikolajick, T. Physical model of threshold switching in NbO_2 based memristors. *J. R. Soc. Chem.* **2015**, *5*, 102318–102322. [CrossRef]
35. Slesazeck, S.; Herzig, M.; Mikolajick, T.; Ascoli, A.; Weiher, M.; Tetzlaff, R. Analysis of V_{th} variability in NbO_x-based threshold switches. In Proceedings of the IEEE Nonvolatile Memory Technology Symposium (NVMTS), Pittsburgh, PA, USA, 17–19 October 2016; Carnegie Mellon University: Pittsburgh, PA, USA, 2016. [CrossRef]
36. Gibson, G.A.; Musunuru, S.; Zhang, J.; Vandenberghe, K.; Lee, J.; Hsieh, C.-C.; Jackson, W.; Jeon, Y.; Henze, D.; Li, Z.; Williams, R.S. An accurate locally active memristor model for S-type negative differential resistance in NbO_x. *Appl. Phys. Lett.* **2016**, *108*, 023505. [CrossRef]
37. Herzig, M.; Weiher, M.; Ascoli, A.; Tetzlaff, R.; Mikolajick, T.; Slesazeck, S. Multiple slopes in the negative differential resistance region of NbO_x-based threshold switches. *J. Phys. D Appl. Phys.* **2019**, *52*, 325104. [CrossRef]
38. Herzig, M.; Weiher, M.; Ascoli, A.; Tetzlaff, R.; Mikolajick, T.; Slesazeck, S. Improvement of NbO_x-based threshold switching devices by implementing multilayer stacks. *Semicond. Sci. Technol.* **2019**, *34*, 075005. [CrossRef]
39. Pickett, M.D.; Strukov, D.B.; Borghetti, J.L.; Yang, J.J.; Snider, G.S.; Stewart, D.R.; Williams, R.S. Switching dynamics in titanium dioxide memristive devices. *J. Appl. Phys.* **2009**, *106*, 074508. [CrossRef]
40. Chua, L.O. Five Non-Volatile Memristor Enigmas Solved. *Appl. Phys. A* **2018**, *124*, 563. [CrossRef]
41. Wu, C.W. Graph Coloring via Synchronization of Coupled Oscillators. *IEEE Trans. Circuits Syst. I Fundam. Theory Appl.* **1998**, *45*, 974–978.
42. Parihar, A.; Shukla, N.; Jerry, M.; Datta, S.; Raychowdhury, A. Vertex coloring of graphs via phase dynamics of coupled oscillatory networks. *Sci. Rep.* **2017**, *7*, 1–11. [CrossRef]
43. Ascoli, A.; Demirkol, A.S.; Tetzlaff, R.; Slesazeck, S.; Mikolajick, T.; Chua, L.O. On Local Activity and Edge of Chaos in a NaMLab Memristor. *Front. Neurosci.* **2021**, *15*. [CrossRef]
44. Ascoli, A.; Demirkol, A.S.; Tetzlaff, R.; Chua, L.O. Edge of Chaos Theory Resolves Smale Paradox. *IEEE Trans. Circuits Syst. I Reg. Pap.* **2022**, *69*, 252–1265. [CrossRef]
45. Mitchell, M. *An Introduction to Genetic Algorithms*; MIT Press: Cambridge, MA, USA, 1998; ISBN 978-0262631853.
46. Chibante, R. *Simulated Annealing: Theory with Applications*; Sciyo: Rijeka, Croatia, 2010; ISBN 978-953-307-134-3.
47. Ascoli, A.; Baumann, D.; Tetzlaff, R.; Chua, L.O.; Hild, M. Memristor-enhanced humanoid robot control system–Part I: Theory behind the novel memcomputing paradigm. *Int. J. Circuit Theory Appl. IJCTA* **2018**, *46*, 155–183. [CrossRef]
48. Baumann, D.; Ascoli, A.; Tetzlaff, R.; Chua, L.O.; Hild, M. Memristor-enhanced humanoid robot control system–Part II: Circuit theoretic model and performance analysis. *Int. J. Circuit Theory Appl. IJCTA* **2018**, *46*, 184–220. [CrossRef]
49. Sharma, A.A.; Bain, J.A.; Weldon, J.A. Phase coupling and control of oxide-based oscillators for neuromorphic computing. *IEEE J. Explor. Solid-State Comput. Devices Circuits* **2015**, *1*, 58–66. [CrossRef]
50. Graph Coloring Instances. Available online: https://mat.tepper.cmu.edu/COLOR/instances.html (accessed on 25 March 2022).
51. Johnson, D.S.; Trick, M.A. *Cliques, Coloring, and Satisfiability: Second DIMACS Implementation Challenge*; Based upon the Proceedings of the DIMACS Workshop, 11–13 October 1993; Series in Discrete Mathematics and Theoretical Computer Science; American Mathematical Society: Providence, RI, USA, 1996; Volume 26.
52. Brélaz, D. New Methods to Color the Vertices of a Graph. *Commun. Assoc. Comput. Mach. ACM* **1979**, *22*, 251–256. [CrossRef]

53. Pickett, M.D.; Williams, R.S. Phase transitions enable computational universality in neuristor-based cellular automata. *Nanotechnology* **2013**, *24*, 384002. [CrossRef]
54. Messaris, I.; Brown, T.D.; Demirkol, A.S.; Ascoli, A.; Chawa, M.M.A.; Williams, R.S.; Tetzlaff, R.; Chua, L.O. NbO_2-Mott Memristor: A Circuit-Theoretic Investigation. *IEEE Trans. Circuits Syst. I Regul. Pap.* **2021**, *68*, 4979–4992. [CrossRef]
55. Liu, X.; Zhang, P.; Nath, S.K.; Li, S.; Nandi, S.K.; Elliman, R.G. Understanding composite negative differential resistance in niobium oxide memristors. *J. Phys. D Appl. Phys.* **2022**, *55*, 105106. [CrossRef]
56. Callarotti, R.C.; Schmidt, P.E. Theoretical and experimental study of the operation of ovonic switches in the relaxation oscillation mode. I. The charging characteristic during the off state. *J. Appl. Phys.* **1984**, *55*, 3144. [CrossRef]
57. Callarotti, R.C.; Schmidt, P.E. Theoretical and experimental study of the operation of ovonic switches in the relaxation oscillation mode. II. The discharging characteristics and the equivalent circuits. *J. Appl. Phys.* **1984**, *55*, 3148. [CrossRef]
58. Kim, S.J.; Cho, S.W.; Lee, H.; Lee, J.; Seong, T.Y.; Kim, I.; Park, J.-K.; Kwak, J.Y.; Kim, J.; Park, J.; et al. Frequency-tunable nano-oscillator based on Ovonic Threshold Switch (OTS). arXiv:2009.13703.
59. Demirkol, A.S.; Ascoli, A.; Messaris, I.; Tetzlaff, R. Pattern formation dynamics in an MCNN structure with a numerically stable VO_2 memristor model. *Jpn. J. Appl. Phys. under review.*

MDPI
St. Alban-Anlage 66
4052 Basel
Switzerland
Tel. +41 61 683 77 34
Fax +41 61 302 89 18
www.mdpi.com

Journal of Low Power Electronics and Applications Editorial Office
E-mail: jlpea@mdpi.com
www.mdpi.com/journal/jlpea

www.ingramcontent.com/pod-product-compliance
Lightning Source LLC
Chambersburg PA
CBHW040810250926
51850CB00007BA/1134
* 9 7 8 3 0 3 6 5 6 1 8 5 1 *